T0074135

**Texts and Monographs in Symbolic Computation**

**A Series of the Research Institute for Symbolic Computation, Johannes-Kepler-University, Linz, Austria**

**Edited by B. Buchberger and G. E. Collins**

F. Winkler

# Polynomial Algorithms in Computer Algebra

SpringerWienNewYork

Dipl.-Ing. Dr. Franz Winkler
Research Institute for Symbolic Computation
Johannes-Kepler-University Linz, Linz, Austria

This work is subject to copyright.
All rights are reserved, whether the whole or part of the material is concerned, specifically those of translation, reprinting, re-use of illustrations, broadcasting, reproduction by photo-copying machines or similar means, and storage in data banks.
© 1996 Springer-Verlag/Wien

Data conversion by H.-D. Ecker, Büro für Textverarbeitung, Bonn
Printed in Austria by Novographic, Ing. Wolfgang Schmid, A-1230 Wien
Printed on acid-free and chlorine-free bleached paper

With 13 Figures

**Library of Congress Cataloging-in-Publication Data**

Winkler, Franz.
Polynomial algorithms in computer algebra / Franz Winkler.
p. cm. — (Texts and monographs in symbolic computation, ISSN 0943-853X)
Includes bibliographical references and index.
ISBN 3-211-82759-5 (alk. paper).
1. Algebra—Data processing. 2. Computer algorithms. I. Title.
II. Series.
QA155.7.E4W56 1996
512.9'42—dc20 96-7170
CIP

ISSN 0943-853X
ISBN 3-211-82759-5 Springer-Verlag Wien New York

# Preface

For several years now I have been teaching courses in computer algebra at the Universität Linz, the University of Delaware, and the Universidad de Alcalá de Henares. In the summers of 1990 and 1992 I have organized and taught summer schools in computer algebra at the Universität Linz. Gradually a set of course notes has emerged from these activities. People have asked me for copies of the course notes, and different versions of them have been circulating for a few years. Finally I decided that I should really take the time to write the material up in a coherent way and make a book out of it. Here, now, is the result of this work.

Over the years many students have been helpful in improving the quality of the notes, and also several colleagues at Linz and elsewhere have contributed to it. I want to thank them all for their effort, in particular I want to thank B. Buchberger, who taught me the theory of Gröbner bases nearly two decades ago, B. F. Caviness and B. D. Saunders, who first stimulated my interest in various problems in computer algebra, G. E. Collins, who showed me how to compute in algebraic domains, and J. R. Sendra, with whom I started to apply computer algebra methods to problems in algebraic geometry. Several colleagues have suggested improvements in earlier versions of this book. However, I want to make it clear that I am responsible for all remaining mistakes. Research of the author was partially supported by Österreichischer Fonds zur Förderung der wissenschaftlichen Forschung, project nos. P6763 (ASAG) and P8573 (SGC).

Let me give a brief overview of the contents of this book. In Chap. 1 a motivation for studying computer algebra is given, and several prerequisites for the area, such as algebraic preliminaries, representation of algebraic structures, and complexity measurement are introduced. Some of the more important basic domains of computer algebra are investigated in Chap. 2. Of course, this list is by no means exhaustive. So, for instance, power series and matrices have not been included in the list. The criterion for including a particular basic domain was its importance for the more advanced topics in the subsequent chapters. Computation by homomorphic images is presented in Chap. 3. Such homomorphic images will be of great importance in gcd computation and factorization of polynomials. These topics are dealt with in Chaps. 4 and 5. Chapter 6 contains algorithms for decomposition of polynomials. Linear systems appear often as subproblems in different areas of computer algebra. They are investigated in Chap. 7. Problems like computation of resultants, gcds, or factorizations of polynomials can be reduced to certain linear systems, so-called Hankel systems. In Chap. 8 an introduction to the theory of Gröbner bases for polynomial ideals is given, and Gröbner bases are applied to some important problems in polynomial ideal theory and solution of systems of polynomial equations. In the last three chapters, polynomial algorithms are applied to some higher level problems in computer algebra. Problems in real algebraic geometry can be decided by deciding problems in the elementary theory of real closed fields, i.e., by polynomial algorithms. Such a decision algorithm is presented in Chap. 9. Chapter 10 gives a description of Gosper's algorithm for solving summation problems. Finally, in Chap. 11, gcd computation, factorization, and solution of systems of algebraic equations are applied for deriving an algorithm for deciding whether an algebraic curve can be parametrized by rational functions, and if so for computing such a parametrization.

Clearly there are important topics in computer algebra missing from the contents of this book, such as simplification of expressions, integration of elementary functions, computer algebra solutions to differential equation problems, or algebraic computations in finite group theory. Including all these other topics would increase the size of the book beyond any reasonable bound. For this reason I limit myself to discussing that part of computer algebra, which deals with polynomials.

In recent years several books on computer algebra have been published. They all approach the field from their own particular angle. The emphasis in this book is on introducing polynomial algorithms in computer algebra from the bottom up, starting from very basic problems in computation over the integers, and finally leading to, e.g., advanced topics in factorization, solution of polynomial equations and constructive algebraic geometry. Along the way, the complexity of many of the algorithms is investigated.

I hope that this book might serve as the basis for exciting new developments in computer algebra.

Franz Winkler

# Contents

# 1 Introduction

## 1.1 What is computer algebra?

In the recent decades it has been more and more realized that computers are of enormous importance for numerical computations. However, these powerful general purpose machines can also be used for transforming, combining and computing symbolic algebraic expressions. In other words, computers can not only deal with numbers, but also with abstract symbols representing mathematical formulas. This fact has been realized much later and is only now gaining acceptance among mathematicians and engineers.

Mathematicians in the old period, say before 1850 A.D., solved the majority of mathematical problems by extensive calculations. A typical example of this type of mathematical problem solver is Euler. Even Gauss in 1801 temporarily abandoned his research in arithmetic and number theory in order to calculate the orbit of the newly discovered planetoid Ceres. It was this calculation much more than his masterpiece *Disquisitiones Arithmeticae* which became the basis for his fame as the most important mathematician of his time.

So it is not astonishing that in the 18th and beginning 19th centuries many mathematicians were real wizzards of computation. However, during the 19th century the style of mathematical research changed from quantitative to qualitative aspects. A number of reasons were responsible for this change, among them the importance of providing a sound basis for the vast theory of analysis. But the fact that computations gradually became more and more complicated certainly also played its role. This impediment has been removed by the advent of modern digital computers in general and by the development of program systems in computer algebra, in particular. By the aid of computer algebra the capacity for mathematical problem solving has been decisively improved.

Even in our days many mathematicians think that there is a natural division of labor between man and computer: a person applies the appropriate algebraic transformations to the problem at hand and finally arrives at a program which then can be left to a "number crunching" computer. But already in 1844, Lady Augusta Ada Byron, countess Lovelace, recognized that this division of labor is not inherent in mathematical problem solving and may be even detrimental. In describing the possible applications of the *Analytical Engine* developed by Charles Babbage she writes:

"Many persons who are not conversant with mathematical studies imagine that because the business of [Babbage's Analytical Engine] is to give its results in numerical notation, the nature of its processes must consequently be arithmetical and numerical rather than algebraic and analytical. This is an error. The engine can arrange and combine its numerical quantities exactly as if they were letters or any other general symbols; and in fact it might bring out its results in algebraic notation were provisions made accordingly."

And indeed a modern digital computer is a "universal" machine capable of carrying out an arbitrary algorithm, i.e., an exactly specified procedure, algebraic algorithms being no exceptions.

## An attempt at a definiton

Now what exactly is symbolic algebraic computation or, in other words, *computer algebra*? In his introduction to Buchberger et al. (1983), R. Loos made the following attempt at a definition:

> "Computer algebra is that part of computer science which designs, analyzes, implements, and applies algebraic algorithms."

While it is arguable whether computer algebra is part of computer science or mathematics, we certainly agree with the rest of the statement. In fact, in our view computer algebra is a special form of scientific computation, and it comprises a wide range of basic goals, methods, and applications. In contrast to numerical computation the emphasis is on computing with symbols representing mathematical concepts. Of course that does not mean that computer algebra is devoid of computations with numbers. Decimal or other positional representations of integers, rational numbers and the like appear in any symbolic computation. But integers or real numbers are not the sole objects. In addition to these basic numerical entities, computer algebra deals with polynomials, rational functions, trigonometric functions, algebraic numbers, etc. That does not mean that we will not need numerical algorithms any more. Both forms of scientific computation have their merits and they should be combined in a computational environment. For instance, in order to compute an approximate solution to a differential equation it might be reasonable to determine the first $n$ terms of a power series solution by exact methods from computer algebra before handing these terms over to a numerical package for evaluating the power series.

Summarizing, we might list the following characteristics of computer algebra:

1. Computer algebra is concerned with computing in algebraic structures. This might mean in basic algebraic number domains, in algebraic extensions of such domains, in polynomial rings or function fields, in differential or difference fields, in the abstract setting of group theory, or the like. Often it is more economical in terms of computation time to simplify an expression algebraically before evaluating it numerically. In this way the expression becomes simpler and less prone to numerical errors.

2. The results of computer algebra algorithms are exact and not subject to approximation errors. So, typically, when we solve a system of algebraic

equations like

$$x^4 + 2x^2y^2 + 3x^2y + y^4 - y^3 = 0$$
$$x^2 + y^2 - 1 = 0$$

we are interested in an exact representation $(\sqrt{3}/2, -1/2)$ instead of an approximative one $(0.86602\ldots, -0.5)$.

3. In general the inputs to algorithms are expressions or formulas and one also expects expressions or formulas as the result. Computer algebra algorithms are capable of giving results in algebraic form rather than numerical values for specific evaluation points. From such an algebraic expression one can deduce how changes in the parameters affect the result of the computation. So a typical result of computer algebra is

$$\int \frac{x}{x^2 - a}\,\mathrm{d}x = \frac{\ln|x^2 - a|}{2}$$

instead of

$$\int_0^{1/2} \frac{x}{x^2 - 1}\,\mathrm{d}x = 0.1438\ldots$$

As a consequence one can build decision algorithms on computer algebra, e.g., for the factorizability of polynomials, the equivalence of algebraic expressions, the solvability of integration problems in a specific class of expressions, the solvability of certain differential equation problems, the solvability of systems of algebraic equations, the validity of geometric statements, the parametrizability of algebraic curves.

Some application areas of computer algebra

1. The "piano movers problem" in robotics: The problem is to "find a path that will allow to move a given body $B$ from an initial position to a desired final position. Along this path the body $B$ should not hit any obstacles."

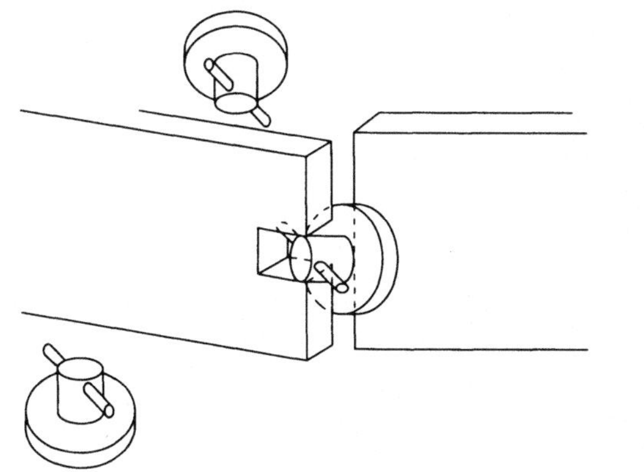

**Fig. 1**

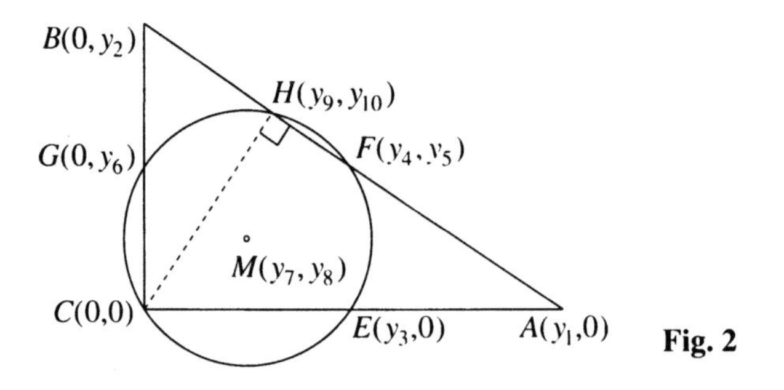

**Fig. 2**

A possible approach, described for instance in Schwarz and Sharir (1983), consists in representing the legal positions of $B$ as a semi-algebraic set $L$ in $\mathbb{R}^m$, i.e., as a union, intersection, or difference of sets

$$\{(x_1, \ldots, x_m) \mid p(x_1, \ldots, x_m) \sim 0\} ,$$

where $p$ is a polynomial with integral coefficients and $\sim \in \{=, <, >\}$. Thus the problem is reduced to the question: "Can two points $P_1$, $P_2$ in a semi-algebraic set $L$ be connected by a path, i.e., are they in the same connected component of $L$?" Collins's cad algorithm for quantifier elimination over real closed fields can answer this question.

2. Geometric theorem proving: There are several computer algebra approaches to proving theorems in Euclidean geometry which can be stated as polynomial equations. An example is: "The altitude pedal of the hypothenuse of a right-angled triangle and the midpoints of the three sides of the triangle are cocircular" (Fig. 2). The hypotheses of this geometric statement, describing a correct drawing of the corresponding figure, are polynomial equations in the coordinates of the points in the figure. The same holds for the conclusion.

Hypotheses:

$h_1 \equiv 2y_3 - y_1 = 0$ ($E$ is the midpoint of $\overline{AC}$),

$h_2 \equiv (y_7 - y_3)^2 + y_8^2 - (y_7 - y_4)^2 - (y_8 - y_5)^2 = 0$

($\overline{EM}$ and $\overline{FM}$ are equally long),

$\vdots$

$h_m$.

Conclusion:

$c \equiv (y_7 - y_3)^2 + y_8^2 - (y_7 - y_9)^2 - (y_8 - y_{10})^2 = 0$

($\overline{EM}$ and $\overline{HM}$ are equally long).

So the geometric problem is reduced to showing that the conclusion polynomial $c$ vanishes on all the common roots of the hypothesis polynomials, i.e., $c$ is contained in the radical ideal generated by $h_1, \ldots, h_m$. This question can be determined by a Gröbner basis computation.

3. Analysis of algebraic varieties: We consider the tacnode curve (Fig. 3), a

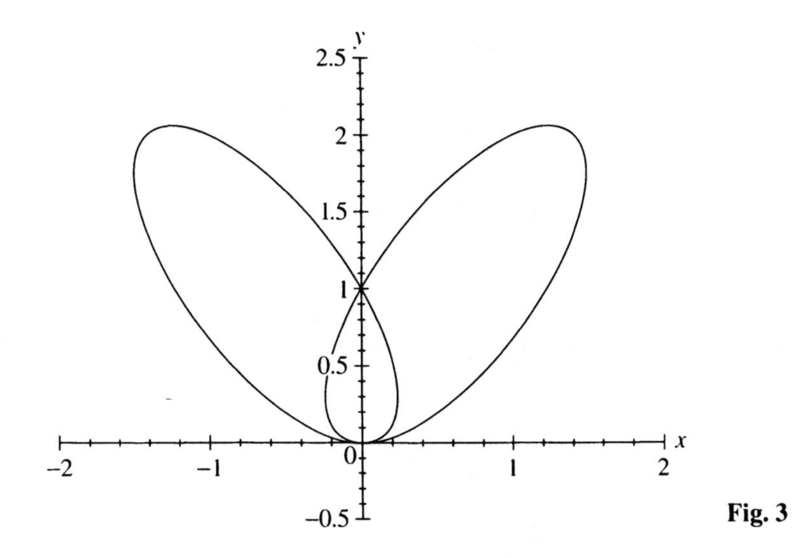


Fig. 3

plane algebraic curve defined by the equation

$$f(x, y) = 2x^4 - 3x^2y + y^4 - 2y^3 + y^2 = 0 .$$

The tacnode is an irreducible curve, which can be checked by trying to factor $f(x, y)$ over $\mathbb{C}$ by the use of a computer algebra system. The tacnode has two singular points, where branches intersect. The coordinates of these singular points are the solutions of the system of algebraic equations

$$\begin{aligned} f(x, y) &= 0 , \\ \frac{\partial f}{\partial x}(x, y) &= 8x^3 - 6xy = 0 , \\ \frac{\partial f}{\partial y}(x, y) &= 4y^3 - 3x^2 - 6y^2 + 2y = 0 . \end{aligned}$$

By a Gröbner basis computation this system is transformed into the equivalent system

$$\begin{aligned} 3x^2 + 2y^2 - 2y &= 0 , \\ xy &= 0 , \\ x^3 &= 0 \end{aligned}$$

from which the singular points $(0, 0)$ and $(0, 1)$ can immediately be read off. We get the tangents at a singular point $(a, b)$ by moving it to the origin with the transformation $T(x, y) = (x+a, y+b)$, factoring the form of lowest degree, and applying the inverse transformation $T^{-1}(x, y) = (x-a, y-b)$. So the tangents at $(0, 1)$ are $y = 1 \pm \sqrt{3}x$ and there is one tangent $y = 0$ of multiplicity 2 at $(0, 0)$.

A global rational parametrization of the tacnode

$$x(t) = \frac{t^3 - 6t^2 + 9t - 2}{2t^4 - 16t^3 + 40t^2 - 32t + 9} ,$$
$$y(t) = \frac{t^2 - 4t + 4}{2t^4 - 16t^3 + 40t^2 - 32t + 9}$$

can be computed.

Or we might just be interested in power series approximations of the branches through the singular point $(0, 0)$. They are

$$x(t) = t, \quad y(t) = t^2 - 2t^4 + \mathcal{O}(t^5) \quad \text{and}$$
$$x(t) = t, \quad y(t) = 2t^2 + 16t^4 + \mathcal{O}(t^5) .$$

4. Modelling in science and technology: Very often problems in these fields are posed as integration problems or differential equation problems, e.g.,

$$-6\frac{\partial q}{\partial x}(x) + \frac{\partial^2 p}{\partial x^2}(x) - 6\sin(x) = 0 ,$$
$$6\frac{\partial^2 q}{\partial x^2}(x) + a^2\frac{\partial p}{\partial x}(x) - 6\cos(x) = 0$$

with initial values $p(0) = 0$, $q(0) = 1$, $p'(0) = 0$, $q'(0) = 1$.

An application of computer algebra algorithms will yield the formal solution

$$p(x) = -\frac{12\sin(ax)}{a(a^2 - 1)} - \frac{6\cos(ax)}{a^2} + \frac{12\sin(x)}{a^2 - 1} + \frac{6}{a^2} ,$$
$$q(x) = \frac{\sin(ax)}{a} - \frac{2\cos(ax)}{a^2 - 1} + \frac{(a^2 + 1)\cos(x)}{a^2 - 1}$$

for $a \notin \{-1, 0, 1\}$.

Usually scientific theories rest on certain mathematical models. In order to test such theories it is necessary to compute in the mathematical model and thus derive theoretic predictions which can then be compared with actual experiments. Computer algebra offers a tool for the scientist to carry out the often extensive algebraic computations in mathematical models of scientific theories. So computer algebra contributes to shifting the frontiers of scientific intractability.

## Limitations of computer algebra

So if computer algebra can do all the above, why hasn't it completely superseded numerical computation? The reason is that computer algebra – just like any other theory or collection of methods – has its limitations. There are problems for which the computation of an exact symbolic solution is prohibitively expensive; there are other problems for which no exact symbolic solution is known; and then there are problems for which one can rigorously prove that no exact symbolic solution exists. Let us look at examples for each of these situations.

1. Elimination theory: The problem consists in "finding" the solutions to a system of algebraic equations

$$
\begin{aligned}
f_1(x_1, \ldots, x_n) &= 0 , \\
&\vdots \\
f_m(x_1, \ldots, x_n) &= 0 ,
\end{aligned}
$$

where the $f_i$'s are polynomials with, say, integral coefficients. If there are only finitely many solutions of this system, then a symbolic method consists of first triangularizing the system, i.e., finding a univariate polynomial $g_{1,1}(x_1)$, whose roots $\alpha$ are the $x_1$-coordinates of the solutions of the system, bivariate polynomials $g_{2,1}(x_1, x_2), \ldots, g_{2,i_2}(x_1, x_2)$, such that the $x_2$-coordinates of the solutions of the system are the roots of $g_{2,1}(\alpha, x_2), \ldots, g_{2,i_2}(\alpha, x_2)$, etc., and then lifting solutions of the simplified problems to solutions of the whole system.

There are general symbolic methods for solving the elimination problem, but their complexity is at least exponential in the number of the variables $n$. So, consequently, they are applicable only to systems in relatively few variables.

Nevertheless, there are approximative numerical approaches to "solving" systems in high numbers of variables, such as homotopy methods.

2. Differential equations: There are simple types of differential equations, such as integration problems, or homogeneous linear differential equations, for which the existence of Liouvillian solutions can be decided and such solutions can be computed if they exist (see Kovacic 1986, Singer 1981). However, not much is known in terms of symbolic algorithms for other types of differential equations, in particular partial differential equations.

3. Canonical simplification: Often symbolic expressions need to be simplified in order to avoid an enormous swell of intermediate expressions, or also for making decisions about equality of expressions.

The class of radical expressions *ER* is built from variables $x_1, \ldots, x_n$, rational constants, the arithmetic function symbols $+, -, \cdot, /$, and the radical sign $\sqrt[q]{\ }$, or, equivalently, rational powers ("radicals") $s^r$ for $r \in \mathbb{Q}$. We call two radical expressions equivalent iff they describe the same meromorphic functions. So, for instance,

$$\frac{\sqrt{2}}{\sqrt{x+1} \cdot \sqrt[4]{24x+24}} \sim \frac{\sqrt[4]{6^3} \cdot \sqrt[4]{x+1}}{6x+6} .$$

The equivalence of unnested radical expressions, i.e., radicals do not contain other radicals, can be decided by an algorithm due to B. F. Caviness and R. J. Fateman (Caviness 1970, Fateman 1972, Caviness and Fateman 1976).

Now let us consider the class of transcendental expressions *ET*, built from one variable $x$, rational constants, the transcendental constant $\pi$, and the function symbols $+, \cdot, \sin(.), | \ . \ |$ (absolute value). Two expressions are equivalent iff they describe the same functions on $\mathbb{R}$. Based on work by D. Richardson and J. Matijasevic on the undecidability of Hilbert's 10th problem, B. F. Caviness (1970) proved that the equivalence of expressions in *ET* is undecidable.

## 1.2 Program systems in computer algebra

The first beginnings of the development of program systems for computer algebra date back to the 1950s. In 1953 H. G. Kahrimanian wrote a master's thesis on analytic differentiation at Temple University in Philadelphia. He also wrote corresponding assembler programs for the UNIVAC I. At the end of the 1950s and the beginning of the 1960s a lot of effort at the Massachusetts Institute of Technology was directed towards research that paved the way for computer algebra systems as we know them today. An example of this is J. McCarthy's work on the programming language LISP. Other people implemented list processing packages in existing languages. In the early 1960s G. E. Collins created the system PM, which later developed into the computer algebra system Aldes/SAC-II, and more recently into the library SACLIB written in C.

Currently there exist a large number of computer algebra systems. Most of them are written for narrowly specified fields of applications, e.g., for high energy physics, celestial mechanics, general relativity, and algebraic geometry. Instead of listing a great number of rather specialized systems, we concentrate on the few ones which offer most of the existing computer algebra algorithms and which are of interest to a general user.

SAC: Starting in the late 1960s, the SAC computer algebra system was developed mainly at the University of Wisconsin at Madison under the direction of G. E. Collins. Currently the center of development is at RISC-Linz. The system has gone through various stages, SAC-I, SAC-II, and now SACLIB, which is written in C. Being a research system, SAC does not offer an elaborate user interface. The emphasis is on the implementation and experimentation with the newest and fastest algorithms for computing with polynomials and algebraic numbers.

Macsyma: Starting in the late 1960s, Macsyma was developed at the Massachusetts Institute of Technology (MIT) under the direction of J. Moses. Macsyma is one of the truly general computer algebra systems. The system contains one of the biggest libraries of algebraic algorithms available in any computer algebra system. Currently there are various versions of Macsyma in existence.

Reduce: Also in the late 1960s, the development of Reduce was started at the University of Utah under the direction of A. Hearn. Currently the center of development is at the Rand corporation. Reduce started out as a specialized system for physics, with many of the special functions needed in this area. In the meantime it has changed into a general computer algebra system.

Magma: In the 1970s, J. Cannon at Sidney started the development of the Cayley system, which ultimately led to the present system Magma. Its main emphasis is on group theoretic computations and finite geometries.

Derive: This is the only general purpose computer algebra system which has been written specifically for the limited resources available on PCs and other small machines. D. Stoutemeyer has been developing the system (and its predecessor muMath) at the University of Hawaii.

Maple: The system has been developed at the University of Waterloo by a group directed by K. O. Geddes and G. H. Gonnet, starting around 1980. It is designed to have a relatively small kernel, so that many users can be supported at the same time. Additional packages have to be loaded as needed. Maple is currently one of the most widely used computer algebra systems.

Mathematica: This is a relatively young computer algebra system; the first versions were available just a few years ago. It has been developed by S. Wolfram Research Inc. Notable are its links to numerical computation and graphical output.

Axiom: At the IBM research center at Yorktown Heights a group directed by R. D. Jenks has for a long time been developing the Scratchpad system. Recently Scratchpad has been renamed Axiom and its distribution is now organized by The Numerical Algorithms Group (NAG) in Oxford. Axiom features a very modern approach to computer algebra systems in several ways, providing generic algorithms and a natural mathematical setting in which to implement algorithms.

## A sample session of Maple

```
> maple
    |\^/|     Maple V Release 3 (University of Linz)
._|\|   |/|_. Copyright (c) 1981-1994 by Waterloo Maple Software and the
 \  MAPLE  /  University of Waterloo. All rights reserved. Maple and Maple
 <____ ____>  V are registered trademarks of Waterloo Maple Software.
      |       Type ? for help.
>
> # as any other computer algebra system, Maple computes with long
> # integers
>
> bigint:= sum(i^2, i=1..99999999);
                      bigint := 333333328333333350000000
> ifactor(bigint);
                7        8
           (2)  (3) (5)  (11) (73) (89) (101) (137) (1447) (1553)
>
> # now let's see some examples of computations with polynomials
>
> pol:=expand( (x-2)^2 * (x+1) * (x^3-23*x+1)^2 );
     pol :=
      9       7        6        5         4        3      8         2
     x  - 46 x  + 144 x  + 523 x  - 1817 x  + 147 x  - 3 x  + 2113 x
      - 184 x + 4
> po2:= expand( (x-2) *(x^5+31*x^3+2*x) );
                     6      4      2      5       3
          po2 := x  + 31 x  + 2 x  - 2 x  - 62 x  - 4 x
> gcd(pol,po2);
                                 x - 2
> gcdex(pol,po2,x,'s','t');
                                 x - 2
```

```
> s;
              388828112807569325  2     582350036308715       12543465117426833  4
- 1/2 - ------------------ x  - ----------------- x - ----------------- x
              48979363028959228        24489681514479614     48979363028959228

           9412368064896     3
      - ----------------- x
         12244840757239807
> t;
 7767142427983062623  3   7240172238838992023  2   12715368265072921137
 ------------------- x  - ------------------- x  - -------------------- x
  48979363028959228        48979363028959228         48979363028959228

     1113115808836205007   301071640470705033  5   150074005808315194  4
   + ------------------- - ----------------- x  + ----------------- x
      48979363028959228     24489681514479614       12244840757239807

     12543465117426833  7   12505815645167249  6
   + ----------------- x  - ----------------- x
     48979363028959228      48979363028959228
> simplify(po1*s+po2*t);
                                   x - 2
> # a square free factorization of po1 is computed
>
> sqrfree(po1);
                                4      3       2
             [1, [[x + 1, 1], [x  - 2 x  - 23 x  + 47 x - 2, 2]]]
>
> # now let's factor po1 completely over the integers modulo 3 and
> # over the rational numbers, and over an algebraic extension of the
> # rational numbers
>
> Factor(po1) mod 3;
                           2          2        3        2
                         (x  + x + 2)  (x + 1)  (x + 2)
> factor(po1);
                                 2            3            2
                        (x - 2)  (x + 1) (x  - 23 x + 1)
> minpo:= subs(x=y,op(3,"));
                                        3            2
                            minpo := (y  - 23 y + 1)
> alias(alpha = RootOf(minpo)):
> factor(po1,alpha);
                        2                          2 2              2        2
            (x + 1) (x  + alpha x - 23 + alpha )  (x - alpha)  (x - 2)
> # the following polynomial is irreducible over the rationals, but
> # factors over the complex numbers
> po3:= x^2+y^2;
                                          2    2
                                  po3 := x  + y
> factor(po3);
                                     2    2
                                    x  + y
> evala(AFactor(po3));
                             2                         2
          (x - RootOf(_Z  + 1) y) (x + RootOf(_Z  + 1) y)
>
> # for doing linear algebra we load the package "linalg"
>
> with(linalg):
Warning: new definition for   norm
Warning: new definition for   trace
>
> A:=matrix([[1,-3,3],[3,-5,3],[6,-6,4]]);
                                     [ 1  -3  3 ]
                                     [          ]
                                A := [ 3  -5  3 ]
                                     [          ]
                                     [ 6  -6  4 ]
```

```
> det(A);
                                        16

> charpoly(A,x);
                                   3
                                  x  - 12 x - 16

> eigenvals(A);
                                    4, -2, -2

> eigenvects(A);
        [4, 1, {[ 1, 1, 2 ]}], [-2, 2, {[ 1, 1, 0 ], [ -1, 0, 1 ]}]
> ffgausselim(A,'r','d');      # fraction free Gaussian elimination
                                  [ 1  -3   3 ]
                                  [           ]
                                  [ 0   4  -6 ]
                                  [           ]
                                  [ 0   0  16 ]
> r;         # the rank of A
                                        3

> d;         # the determinant of A
                                        16

> B:=matrix(3,3,[1,2,3,1,2,3,1,5,6]);
                                      [ 1  2  3 ]
                                      [         ]
                                 B := [ 1  2  3 ]
                                      [         ]
                                      [ 1  5  6 ]
> linsolve(B,[0,0,0]);
                           [ - _t[1], - _t[1], _t[1] ]

>
> # maple can do indefinite summation
>
> sum(i^2, i);        # 1^2 + ... + i^2
                                  3         2
                             1/3 i  - 1/2 i  + 1/6 i

> sum(i*a^i, i);       # 1*a^1 + ... + i*a^i
                                  i
                                 a  (i a - i - a)
                                 ----------------
                                              2
                                      (a - 1)

>
> # differentiation, integration
>
> expr1:= x^(x^x);
                                                x
                                              (x )
                                  expr1 := x

> diff(expr1,x);
                           x
                         (x ) /  x                    x \
                        x     | x  (ln(x) + 1) ln(x) + -- |
                              \                        x  /
> integrate(",x);
                                          x
                                        (x )
                                       x

> expr2:=1/(x^3+a*x^2+x);
                                            1
                              expr2 := ------------
                                        3      2
                                       x  + a x  + x

> integrate(expr2,x);
                                                           2 x + a
                                          a arctanh(-------------)
                                                             2 1/2
                                                     (- 4 + a )
        ln(x) - 1/2 ln(x^2 + a x + 1) + --------------------------
                                                        2 1/2
                                                 (- 4 + a )
```

```
> diff(",x);
                        2 x + a                          a
          1/x - 1/2 ------------- + 2 -------------------------------
                     2                                  /             2\
                    x  + a x + 1             2          |    (2 x + a) |
                                        (- 4 + a ) |1 - -----------|
                                                        |          2   |
                                                        \    - 4 + a   /

> simplify(");
                                          1
                                  ----------------
                                    2
                                  (x  + a x + 1) x

> expr3:=1/(sqrt(a^2-x^2));
                                            1
                             expr3 := -------------
                                        2    2 1/2
                                      (a  - x )

> int(expr3,x=0..a);
                                      1/2 Pi

>
> # some differential equations can be solved symbolically
>
> de1:=-6*diff(q(x),x)+diff(p(x),x,x)-6*sin(x);
                                        /  2      \
                           /  d      \  | d       |
               de1 := - 6 |---- q(x)| + |----- p(x)| - 6 sin(x)
                           \ dx      /  |   2     |
                                        \ dx      /

> de2:=6*diff(q(x),x,x)+a^2*diff(p(x),x)-6*cos(x);
                       /  2      \
                       | d       |     2 /  d      \
              de2 := 6 |----- q(x)| + a  |---- p(x)| - 6 cos(x)
                       |   2     |       \ dx      /
                       \ dx      /

> desolu:=dsolve({de1,de2,p(0)=0,q(0)=1,D(p)(0)=0,D(q)(0)=1},{p(x),q(x)});
desolu := {
q(x) =
                                2                          2
1/2 (cos(a x) cos(- x + a x) a  - cos(a x) cos(x + a x) a  - 2 cos(x) a

                                                             5
 - cos(a x) cos(x + a x) + cos(a x) cos(- x + a x) + 2 cos(x) a

                             2                          2
 + sin(a x) sin(- x + a x) a  - sin(a x) sin(x + a x) a

                             3
 + sin(a x) sin(- x + a x) a  + sin(a x) sin(- x + a x) a

                             3
 + cos(a x) cos(- x + a x) a  + cos(a x) cos(- x + a x) a

                           3
 + sin(a x) sin(x + a x) a  + sin(a x) sin(x + a x) a

                           3
 + cos(a x) cos(x + a x) a  + cos(a x) cos(x + a x) a

                                                              5
                                                             a  cos(a x)
 + sin(a x) sin(- x + a x) - sin(a x) sin(x + a x) - 4 -----------
                                                              2
                                                             a  - 1

      3
     a  cos(a x)      4              2              /   3   2
 + 4 ----------- + 2 a  sin(a x) - 2 a  sin(a x))  /  (a  (a  - 1)),
        2
       a  - 1

p(x) =
             3
(12 sin(x) a  - 12 sin(x) a + 6 sin(a x) cos(x + a x)
 - 6 sin(a x) cos(x + a x) a + 6 sin(a x) cos(- x + a x) a
 + 6 sin(a x) cos(- x + a x) - 6 cos(a x) sin(- x + a x) a
 - 6 cos(a x) sin(- x + a x) - 6 cos(a x) sin(x + a x)

                                                   4                   2
                                        3         a  sin(a x)         a  sin(a x)
 + 6 cos(a x) sin(x + a x) a + 6 a  - 6 a - 12 ----------- + 12 -----------
                                                   2                   2
                                                  a  - 1              a  - 1

     3                             /    3   2
 - 6 a  cos(a x) + 6 cos(a x) a)  /   (a  (a  - 1))

}
```

```
>
> # computation of Groebner basis, determining singularities of an
> # algebraic curve by solving a system of algebraic equations
>
> with(grobner);
    [finduni, finite, gbasis, gsolve, leadmon, normalf, solvable, spoly]
> curve:=2*x^4-3*x^2*y+y^2-2*y^3+y^4;
                                 4      2       2      3    4
                  curve := 2 x  - 3 x  y + y  - 2 y  + y
> gbasis({curve,diff(curve,x),diff(curve,y)},[x,y],plex);
                          2      2                2    3
                     [3 x  + 2 y  - 2 y, x y, - y  + y ]
>
> # plotting
>
> interface(plotdevice=postscript,plotoutput=figure4);
> funct4:=b^2+27/(2*b)-45/2;
                                      2    27
                         funct4 := b  + --- - 45/2
                                        2 b
> plot(funct4,b=-8..8,-30..30);
> interface(plotoutput=figure5);
> funct5:=x^2+3*BesselJ(0,y^2)*exp(1-x^2-y^2);
                        2                 2            2    2
             funct5 := x  + 3 BesselJ(0, y ) exp(1 - x  - y )
> plot3d(funct5,x=-2..2,y=-2..2,axes=FRAME);
>
>  # writing programs in Maple, e.g. for extended Euclidean algorithm
>
> myee:=proc(f,g,x)
>          local h,s,t;
>          h:=gcdex(f,g,x,'s','t');
>          RETURN([h,eval(s),eval(t)])
> end:
>
> f1:=expand((x+1)*(x+2));
                                         2
                                 f1 := x  + 3 x + 2
> f2:=expand((x+1)*(x-1));
                                           2
                                   f2 := x  - 1
> myee(f1,f2,x);
                                [x + 1, 1/3, -1/3]
> simplify("[2]*f1 + "[3]*f2);
                                      x + 1
> quit;
```

## 1.3 Algebraic preliminaries

For a thorough introduction to algebra we refer the reader to any of a great number of classical textbooks, e.g., Lang (1984), MacLane and Birkhoff (1979), van der Waerden (1970), or Zariski and Samuel (1958). Here we simply introduce some definitions and basic facts that will be useful in subsequent chapters.

Throughout this book we will denote the set of natural numbers by $\mathbb{N}$, the natural numbers with 0 by $\mathbb{N}_0$, the integers by $\mathbb{Z}$, the rational numbers by $\mathbb{Q}$, the real numbers by $\mathbb{R}$, and the complex numbers by $\mathbb{C}$. Furthermore, we will denote the integers without 0 by $\mathbb{Z}^*$, and similarly for other domains.

A *semigroup* $(S, \circ)$ is a set $S$ together with an associative binary operation $\circ$

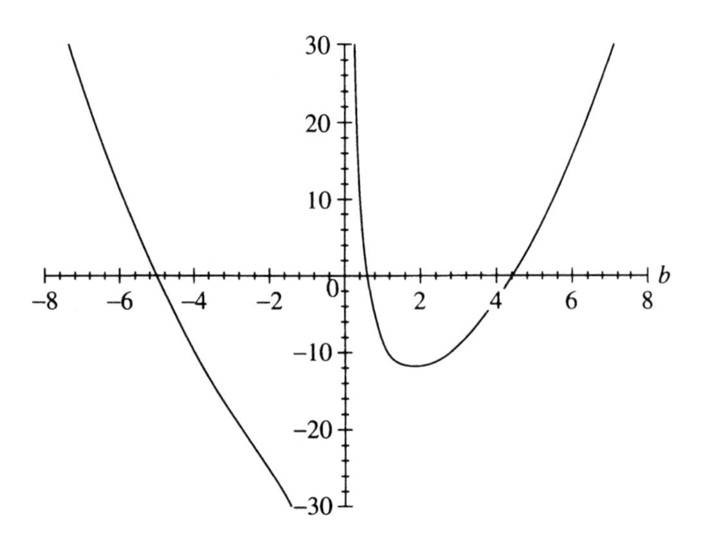


**Fig. 4.** Maple plot output, figure 4

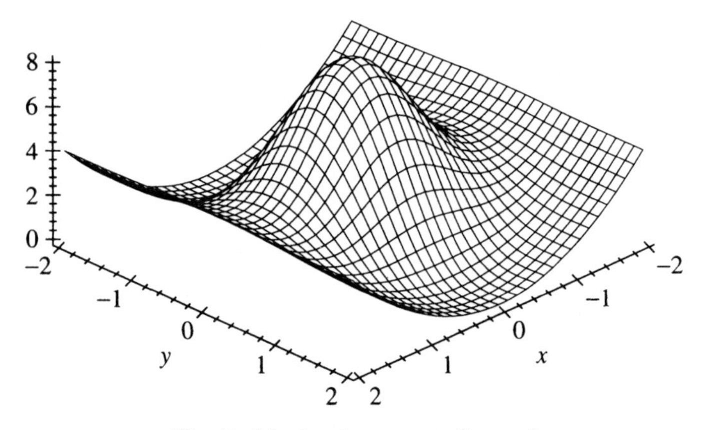


**Fig. 5.** Maple plot output, figure 5

on $S$. A *monoid* $(S, \circ, e)$ is a semigroup with an *identity element* $e$; that is, $e \circ x = x \circ e = x$ for all $x \in S$. Alternatively we could define a monoid as a semigroup with a nullary operation which yields the identity element. A semigroup or a monoid is *commutative* iff the operation $\circ$ is commutative. If the operations are understood from the context, then we often speak of the semigroup or the monoid $S$ without directly mentioning the operations.

A *group* $(G, \circ, \diamond, e)$ is a monoid $(G, \circ, e)$ together with a unary *inverse* operation $\diamond$, i.e., $x \circ (\diamond x) = e = (\diamond x) \circ x$ for all $x \in G$. $G$ is a *commutative* or *abelian* group if $\circ$ is commutative.

A *ring* $(R, +, -, \cdot, 0)$ is an abelian group $(R, +, -, 0)$ and a semigroup $(R, \cdot)$ satisfying the laws of distributivity $x \cdot (y+z) = x \cdot y + x \cdot z$ and $(x+y) \cdot z = x \cdot z + y \cdot z$. A *commutative* ring is one in which the operation $\cdot$ is commutative. A *ring with identity* is a ring $R$ together with an element $1$ $(\neq 0)$, such that

$(R, \cdot, 1)$ is a monoid. Unless stated otherwise, we will always use the symbols $+, -, \cdot, 0, 1$ for the operations of a ring. We call these operations *addition*, *minus*, *multiplication*, *zero*, and *one*. The *subtraction* operation (also written as $-$) is defined as $x - y := x + (-y)$ for $x, y \in R$. Multiplication is usually denoted simply by concatenation. The *characteristic* of a commutative ring with identity $R$, $\mathrm{char}(R)$, is the least positive integer $m$ such that

$$\underbrace{1 + \cdots + 1}_{m \text{ times}} = 0 . \tag{1.3.1}$$

$\mathrm{char}(R) = 0$ if no such $m$ exists.

Let $(R, +, -, \cdot, 0)$ and $(R', +', -', \cdot', 0')$ be rings. A *homomorphism of rings* $h$ is a function from $R$ to $R'$ satisfying the conditions

$$h(0) = 0', \quad h(r + s) = h(r) +' h(s), \quad h(r \cdot s) = h(r) \cdot' h(s) . \tag{1.3.2}$$

If $R$ and $R'$ are rings with identities 1 and $1'$, respectively, then $h$ also has to satisfy

$$h(1) = 1' . \tag{1.3.3}$$

A homomorphism $h$ is an *isomorphism* from $R$ to $R'$ iff $h$ is one-to-one and onto. In this case we say that $R$ and $R'$ are *isomorphic*, $R \cong R'$.

Let $R$ be a commutative ring with identity. A subset $I$ of $R$ is an *ideal* in $R$ if $a + b \in I$ and $ac \in I$ for all $a, b \in I$ and $c \in R$. $I$ is a *proper* ideal if $\{0\} \neq I \neq R$. $I$ is a *maximal* ideal if it is not contained in a bigger proper ideal. $I$ is a *prime* ideal if $ab \in I$ implies $a \in I$ or $b \in I$. $I$ is a *primary* ideal if $ab \in I$ implies $a \in I$ or $b^n \in I$ for some $n \in \mathbb{N}$. $I$ is a *radical* ideal if $a^n \in I$ for some $n \in \mathbb{N}$ implies $a \in I$. The *radical* of the ideal $I$, radical$(I)$ or $\sqrt{I}$, is the ideal $\{a \mid a^n \in I \text{ for some } n \in \mathbb{N}\}$. A set $B \subseteq R$ *generates* the ideal $I$ or $B$ is a *generating set* or *basis* for $I$ if

$$I = \left\{ \sum_{i=1}^{n} r_i b_i \mid n \in \mathbb{N}_0,\ r_1, \ldots, r_n \in R,\ b_1, \ldots, b_n \in B \right\} . \tag{1.3.4}$$

In this case we say that $I$ is the *ideal generated by* $B$, $I = \mathrm{ideal}(B) = \langle B \rangle$. $I$ is *finitely generated* if it has a finite generating set. $I$ is a *principal* ideal if it has a generating set of cardinality 1.

An ideal $I$ in $R$ generates a congruence relation $\equiv_I$ on $R$ by $a \equiv_I b$ or $a \equiv b \bmod I$ iff $a - b \in I$ ($a$ is *congruent to* $b$ *modulo* $I$). The factor ring $R_{/I}$ (consisting of the congruence classes w.r.t. $\equiv_I$) inherits the operations of $R$ in a natural way. If $R$ is a commutative ring with 1 and $I$ is prime, then $R_{/I}$ is an integral domain. If $I$ is maximal, then $R_{/I}$ is a field.

In the following considerations let us take non-zero elements of a commutative ring $R$ with identity 1. Invertible elements of $R$ are called *units*. If $a = b \cdot u$ for a unit $u$, then $a$ and $b$ are called *associates*. $b$ *divides* $a$ iff $a = b \cdot c$ for some $c \in R$. If $c$ divides $a - b$ we say that $a$ is *congruent* to $b$ *modulo* $c$, $a \equiv b \bmod c$.

For every $c$ the congruence modulo $c$ is an equivalence relation. An element $a$ of $R$ is *irreducible* iff every $b$ dividing $a$ is either a unit or an associate of $a$. An element $a$ of $R$ is *prime* iff $a$ is not a unit, and whenever $a$ divides a product $b \cdot c$, then $a$ divides either $b$ or $c$. In general prime and irreducible elements can be different, e.g., 6 has two different factorizations into irreducibles in $\mathbb{Z}[\sqrt{-5}]$ and none of these factors is prime (compare Exercise 2).

A *zero divisor* in a commutative ring $R$ is a non-zero element $a \in R$ such that for some non-zero $b \in R$ we have $ab = 0$. An *integral domain* or simply *domain* $D$ is a commutative ring with identity having no zero divisors. An integral domain $D$ satisfies the cancellation law:

$$ab = ac \quad \text{and} \quad a \neq 0 \Longrightarrow b = c\ . \tag{1.3.5}$$

If $R$ is an integral domain, then also the ring of polynomials $R[x]$ over $R$ is an integral domain. A *principal ideal domain* is a domain in which every ideal is principal.

An integral domain $D$ is a *unique factorization domain* (ufd) iff every non-unit of $D$ is a finite product of irreducible factors and every such factorization is unique up to reordering and unit factors. In a unique factorization domain prime and irreducible elements are the same. Moreover, any pair of elements $a$, $b$ (not both elements being equal to 0) has a *greatest common divisor* (gcd) $d$ satisfying (i) $d$ divides both $a$ and $b$, and (ii) if $c$ is a common divisor of $a$ and $b$, then $c$ divides $d$. The gcd of $a$ and $b$ is determined up to associates. If $\gcd(a, b) = 1$ we say that $a$ and $b$ are *relatively prime*. We list some important properties of gcds:

(GCD 1) $\gcd(\gcd(a, b), c) = \gcd(a, \gcd(b, c))$.
(GCD 2) $\gcd(a \cdot c, b \cdot c) = c \cdot \gcd(a, b)$.
(GCD 3) $\gcd(a + b \cdot c, c) = \gcd(a, c)$.
(GCD 4) If $c = \gcd(a, b)$ then $\gcd(a/c, b/c) = 1$.
(GCD 5) If $\gcd(a, b) = 1$ then $\gcd(a, b \cdot c) = \gcd(a, c)$.

In $\mathbb{Z}$ we have the well-known Euclidean algorithm for computing a gcd. In general, an integral domain $D$ in which we can execute the Euclidean algorithm, i.e., we have division with quotient and remainder such that the remainder is less than the divisor, is called a *Euclidean domain*.

A *field* $(K, +, -, \cdot, ^{-1}, 0, 1)$ is a commutative ring with identity $(K, +, -, \cdot, 0, 1)$ and simultaneously a group $(K \setminus \{0\}, \cdot, ^{-1}, 1)$. If all the operations on $K$ are computable, we call $K$ a *computable field*. If $D$ is an integral domain, the *quotient field* $Q(D)$ of $D$ is defined as

$$Q(D) = \left\{ \frac{a}{b} \mid a, b \in D,\ b \neq 0 \right\} / \sim$$
$$\text{where} \quad \frac{a}{b} \sim \frac{a'}{b'} \Longleftrightarrow ab' = a'b\ . \tag{1.3.6}$$

The operations $+, -, \cdot, ^{-1}$ can be defined on representatives of the elements of $Q(D)$ in the following way:

$$\begin{array}{ll} \frac{a}{b} + \frac{c}{d} = \frac{ad+bc}{bd} & \frac{a}{b} \cdot \frac{c}{d} = \frac{ac}{bd} \\ -\frac{a}{b} = \frac{-a}{b} & \left(\frac{a}{b}\right)^{-1} = \frac{b}{a} \end{array} \tag{1.3.7}$$

The equivalence classes of $0/1$ and $1/1$ are the zero and one in $Q(D)$, respectively. $Q(D)$ is the smallest field containing $D$.

Let $R$ be a ring. A *(univariate) polynomial over* $R$ is a mapping $p\colon \mathbb{N}_0 \to R$, $n \mapsto p_n$, such that $p_n = 0$ nearly everywhere, i.e., for all but finitely many values of $n$. If $n_1 < n_2 < \ldots < n_r$ are the nonnegative integers for which $p$ yields a non-zero result, then we usually write $p = p(x) = \sum_{i=1}^{r} p_{n_i} x^{n_i}$. $p_j$ is the *coefficient* of $x^j$ in the polynomial $p$. We write coeff$(p, j)$ for $p_j$. The set of all polynomials over $R$ together with the usual addition and multiplication of polynomials form a ring. We denote this ring of polynomials over $R$ as $R[x]$. (In fact, as can be seen from the formal definition, the polynomial ring does not really depend on $x$, which just acts as a convenient symbol for denoting polynomials.) Many properties of a ring $R$ are inherited by the polynomial ring $R[x]$. Examples of such inherited properties are commutativity, having a multiplicative identity, being an integral domain, or being a ufd. Let $p$ be a non-zero element of $R[x]$. The *degree* of $p$, deg$(p)$, is the maximal $n \in \mathbb{N}_0$ such that $p_n \neq 0$. The *leading term* of $p$, lt$(p)$, is $x^{\deg(p)}$. The *leading coefficient* of $p$, lc$(p)$, is the coefficient of lt$(p)$ in $p$. The polynomial $p(x)$ is *monic* if lc$(p) = 1$.

An *n-variate polynomial over* the ring $R$ is a mapping $p\colon \mathbb{N}_0^n \to R$, $(i_1, \ldots, i_n) \mapsto p_{i_1,\ldots,i_n}$, such that $p_{i_1,\ldots,i_n} = 0$ nearly everywhere. $p$ is usually written as $\sum p_{i_1,\ldots,i_n} x_1^{i_1} \ldots x_n^{i_n}$, where the formal summation ranges over all tuples $(i_1, \ldots, i_n)$ on which $p$ does not vanish. The set of all $n$-variate polynomials over $R$ form a ring, $R[x_1, \ldots, x_n]$. This $n$-variate polynomial ring can be viewed as built up successively from $R$ by adjoining one polynomial variable at a time. In fact, $R[x_1, \ldots, x_n]$ is isomorphic to $(R[x_1, \ldots, x_{n-1}])[x_n]$. The *(total) degree* of an $n$-variate polynomial $p \in R[x_1, \ldots, x_n]^*$ is defined as deg$(p)$ $:= \max\{\sum_{j=1}^{n} i_j \mid p_{i_1,\ldots,i_n} \neq 0\}$. We write coeff$(p, x_n, j)$ for the coefficient of $x_n^j$ in $p$, where $p$ is considered in $(R[x_1, \ldots, x_{n-1}])[x_n]$. The *degree* of $p = \sum_{i=0}^{m} p_i(x_1, \ldots, x_{n-1}) x_n^i \in (R[x_1, \ldots, x_{n-1}])[x_n]^*$ *in the variable n*, $\deg_{x_n}(p)$, is $m$, if $p_m \neq 0$. By reordering the set of variables we get $\deg_{x_i}(p)$ for all $1 \le i \le n$. In a similar way we get $\mathrm{lt}_{x_i}(p)$ and $\mathrm{lc}_{x_i}(p)$.

Let $p(x) = p_n x^n + \ldots + p_0$ be a polynomial of degree $n$ over $\mathbb{R}$. For measuring the size of $p$ we will use various *norms*, e.g.,

$$\|p(x)\|_1 = \sum_{i=0}^{n} |p_i|, \quad \|p(x)\|_2 = \sqrt{p_n^2 + \ldots + p_0^2}\,,$$

$$\|p(x)\|_\infty = \max\{|p_n|, \ldots, |p_0|\}\,.$$

The *resultant* $\mathrm{res}_x(f, g)$ of two univariate polynomials $f(x), g(x)$ over an integral domain $D$ is the determinant of the *Sylvester matrix* of $f$ and $g$, consisting of shifted lines of coefficients of $f$ and $g$. $\mathrm{res}_x(f, g)$ is a constant in $D$. For $m = \deg(f), n = \deg(g)$, we have $\mathrm{res}_x(f, g) = (-1)^{mn}\mathrm{res}_x(g, f)$, i.e., the resultant is symmetric up to sign. If $a_1, \ldots, a_m$ are the roots of $f$ and $b_1, \ldots, b_n$ are the roots of $g$ in their common splitting field, then

$$\mathrm{res}_x(f, g) = \mathrm{lc}(f)^n \mathrm{lc}(g)^m \prod_{i=1}^{m} \prod_{j=1}^{n} (a_i - b_j) .$$

The resultant has the important property that, for non-zero polynomials $f$ and $g$, $\mathrm{res}_x(f, g) = 0$ if and only if $f$ and $g$ have a common root, and in fact, if $D$ is a ufd, $f$ and $g$ have a common divisor of positive degree in $D[x]$. If $f$ and $g$ have positive degrees, then there exist polynomials $a(x), b(x)$ over $I$ such that $af + bg = \mathrm{res}_x(f, g)$. The *discriminant* of $f(x)$ is

$$\mathrm{discr}_x(f) = (-1)^{m(m-1)/2}\mathrm{lc}(f)^{2(m-1)} \prod_{i \neq j} (a_i - a_j) .$$

We have the relation $\mathrm{res}_x(f, f') = (-1)^{m(m-1)/2}\mathrm{lc}(f)\mathrm{discr}_x(f)$, where $f'$ is the derivative of $f$.

Let $K$ be a field. A *power series* $A(x)$ over $K$ is a mapping $A\colon \mathbb{N}_0 \to K$. We usually write the power series $A$ as $A(x) = \sum_{i=0}^{\infty} a_i x^i$, where $a_i$ is the image of $i$ under the mapping $A$. The power series over $K$ form a commutative ring with 1 and we denote this ring by $K[[x]]$. The *order* of the power series $A$ is the smallest $i$ such that $a_i \neq 0$.

Let $K, L$ be fields such that $K \subset L$. Let $\alpha \in L \setminus K$ such that $f(\alpha) = 0$ for some irreducible $f \in K[x]$. Then $\alpha$ is called *algebraic* over $K$ of *degree* $\deg(f)$. $f$ is determined up to a constant and is called the *minimal polynomial* for $\alpha$ over $K$. By $K(\alpha)$ we denote the smallest field containing $K$ and $\alpha$. $K(\alpha)$ is an *algebraic extension field* of $K$.

### Exercises

1. Prove: If $D$ is an integral domain, then also the polynomial ring $D[x]$ is an integral domain.
2. Let $R$ be the ring $\mathbb{Z}[\sqrt{-5}]$, i.e., the ring of complex numbers of the form $a + b\sqrt{-5}$, where $a, b \in \mathbb{Z}$. Show
   a. $R$ is an integral domain,
   b. $R$ is not a unique factorization domain (e.g., 6 and 9 do not have unique factorizations).

## 1.4 Representation of algebraic structures

Before we can hope to compute with algebraic expressions, we have to devise a representation of these algebraic expressions suitable for the operations we want to perform. Whereas in numerical computation floating point or extended precision numbers and arrays are the representation of choice, these data structures prove to be totally inadequate for the purpose of symbolic algebraic computation. Let us demonstrate this fact by a few typical examples.

As a first example we take the computation of the greatest common divisor (gcd) of two polynomials $f(x, y), g(x, y)$ with integral coefficients. For two primitive polynomials their gcd can be computed by constructing a polynomial remainder sequence $r_0, r_1, \ldots, r_k$ (basically by pseudodivision) starting with $f$ and $g$ (see Chap. 4). The primitive part of the last non-zero element $r_k$ in this sequence is the gcd of $f$ and $g$. For the two relatively small polynomials

$$\begin{aligned}
f(x, y) &= y^6 + xy^5 + x^3y - xy + x^4 - x^2 , \\
g(x, y) &= xy^5 - 2y^5 + x^2y^4 - 2xy^4 + xy^2 + x^2y
\end{aligned}$$

we get the polynomial remainder sequence

$$\begin{aligned}
r_0 &= f , \\
r_1 &= g, \\
r_2 &= (2x - x^2)y^3 + (2x^2 - x^3)y^2 + (x^5 - 4x^4 + 3x^3 + 4x^2 - 4x)y \\
&\quad + x^6 - 4x^5 + 3x^4 + 4x^3 - 4x^2 , \\
r_3 &= (-x^7 + 6x^6 - 12x^5 + 8x^4)y^2 + (-x^{13} + 12x^{12} - 58x^{11} + 136x^{10} \\
&\quad - 121x^9 - 117x^8 + 362x^7 - 236x^6 - 104x^5 + 192x^4 - 64x^3)y \\
&\quad - x^{14} + 12x^{13} - 58x^{12} + 136x^{11} - 121x^{10} - 116x^9 + 356x^8 \\
&\quad - 224x^7 - 112x^6 + 192x^5 - 64x^4 , \\
r_4 &= (-x^{28} + 26x^{27} - 308x^{26} + 2184x^{25} - 10198x^{24} + 32188x^{23} \\
&\quad - 65932x^{22} + 68536x^{21} + 42431x^{20} - 274533x^{19} + 411512x^{18} \\
&\quad - 149025x^{17} - 431200x^{16} + 729296x^{15} - 337472x^{14} - 318304x^{13} \\
&\quad + 523264x^{12} - 225280x^{11} - 78848x^{10} + 126720x^9 - 53248x^8 \\
&\quad + 8192x^7)y - x^{29} + 26x^{28} - 308x^{27} + 2184x^{26} - 10198x^{25} \\
&\quad + 32188x^{24} - 65932x^{23} + 68536x^{22} + 42431x^{21} - 274533x^{20} \\
&\quad + 411512x^{19} - 149025x^{18} - 431200x^{17} + 729296x^{16} - 337472x^{15} \\
&\quad - 318304x^{14} + 523264x^{13} - 225280x^{12} - 78848x^{11} + 126720x^{10} \\
&\quad - 53248x^9 + 8192x^8 .
\end{aligned}$$

The gcd of $f$ and $g$ is the primitive part (with respect to $y$) of $r_4$, which is $y + x$. So we see that although the two input polynomials are of moderate size, the intermediate expressions get bigger and bigger. The final result, however, is again a small polynomial. Actually the biggest polynomial in this computation occurs in the pseudo-division of $r_3$ by $r_4$. The intermediate polynomial has degree 70 in $x$!

Very similar phenomena occur in many algorithms of computer algebra, e.g., in integration algorithms or Gröbner basis computations. Therefore we need dynamic storage allocation in computer algebra. The data structures used for algebraic computation must be able to reflect the expansion and shrinking of the objects during computation. A data structure that has all these properties is the list structure.

*Definition 1.4.1.* Let $A$ be a set. The set of *lists over* $A$, list($A$), is defined as the smallest set containing the *empty list* [ ] (different from any element of $A$) and the list $[a_1, \ldots, a_n]$ for all $a_1, \ldots, a_n \in A \cup \text{list}(A)$.

list($A$) is equipped with the following (partial) operations:

- EMPTY: $\text{list}(A) \to \{T, F\}$ maps [ ] to T and all other lists to F.
- FIRST: $\text{list}(A) \to \text{list}(A) \cup A$ maps $[a_1, \ldots, a_n]$ to $a_1$ and is undefined for [ ].
- REST: $\text{list}(A) \to \text{list}(A)$ maps $[a_1, a_2, \ldots, a_n]$ to $[a_2, \ldots, a_n]$ and is undefined for [ ].
- CONS: $A \cup \text{list}(A) \times \text{list}(A) \to \text{list}(A)$ maps $(a, [a_1, \ldots, a_n])$ to $[a, a_1, \ldots, a_n]$.
- APPEND: $\text{list}(A) \times \text{list}(A) \to \text{list}(A)$ maps $([a_1, \ldots, a_n], [b_1, \ldots, b_m])$ to $[a_1, \ldots, a_n, b_1, \ldots, b_m]$.
- LENGTH: $\text{list}(A) \to \mathbb{N}_0$ maps [ ] to 0 and $[a_1, \ldots, a_n]$ to $n$.
- INV: $\text{list}(A) \to \text{list}(A)$ maps $[a_1, \ldots, a_n]$ to $[a_n, \ldots, a_1]$.
- INIT: $\mathbb{N}_0 \times \text{list}(A) \to \text{list}(A)$ is defined as $\text{INIT}(0, L) = [\,]$, $\text{INIT}(m, L) = L$ for $m > \text{LENGTH}(L)$, and $\text{INIT}(m, [a_1, \ldots, a_n]) = [a_1, \ldots, a_m]$ for $m \leq \text{LENGTH}(L)$.
- DEL: $\mathbb{N}_0 \times \text{list}(A) \to \text{list}(A)$ is defined as $\text{DEL}(m, L) = [\,]$ for $m \geq \text{LENGTH}(L)$ and $\text{DEL}(m, [a_1, \ldots, a_n]) = [a_{m+1}, \ldots, a_n]$ for $m < n$.
- SHIFT: $\mathbb{N}_0 \times A \times \text{list}(A) \to \text{list}(A)$ is defined as $\text{SHIFT}(0, a, L) = L$ and $\text{SHIFT}(n, a, L) = \text{CONS}(a, \text{SHIFT}(n-1, a, L))$ for $n > 0$.

Observe that all lists over a set $A$ can be constructed by successive application of CONS, starting with the list [ ].

Lists are suited for representing objects of variable length because they are not fixed in size. Whenever we have an object represented by a list, e.g., a polynomial $x^5 + 2x^3 - x^2 + 1$ represented by [[5, 1], [3, 2], [2, −1], [0, 1]], the object can be enlarged by adding a new first element, e.g., by adding the term $3x^6$ to the polynomial, yielding the representation [[6, 3], [5, 1], ..., [0, 1]].

The final goal of computer algebra is to implement algebraic structures and algorithms on a computer. Of course, we cannot describe here the implementation on a specific make of computer. Instead, we describe the implementation on a hypothetical machine. The model we use is a *random access machine* (RAM) as specified in Aho et al. (1974). We suppose that the machine has set aside a part of its memory for list processing, and that this part of the memory is initially organized in a so-called *available-space-list* (ASL). Each pair of adjacent words in the ASL is combined into a *list cell*. The first word in such a cell, called the *information field*, is used for storing information, the second word, called the *address* or *pointer field*, holds the address of the next list cell. The

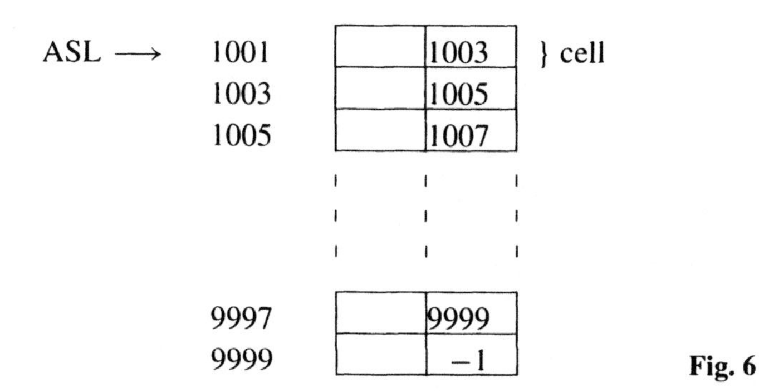


**Fig. 6**

last cell in the ASL holds an invalid address, e.g., −1, indicating that there is no successor to this cell. The ASL is then identified with the address of its first cell. See Fig. 6.

The specific addresses in the address fields of the ASL are not really of interest to us. What matters only is that the successive cells of the ASL are linked together by *pointers* and ASL points to the first cell in this collection. So graphically we represent the ASL as a sequence of cells connected by pointers:

In order to store the list $L = [a_1, \ldots, a_k]$, $k$ cells are removed from the ASL and linked together. The address of the first of these cells is stored in $L$. These removed cells need not be consecutive elements of ASL:

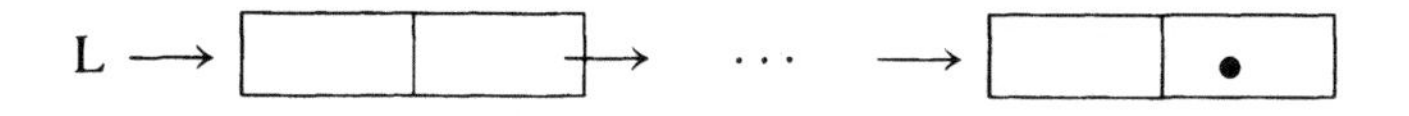


Now let us describe the general situation. The empty list is represented by the empty pointer • (an invalid address, e.g., −1). If $L$ is represented by

L ⟶ [ | ]→ ... ⟶ [ | • ]

then the list $L' = [a, L]$ is represented by

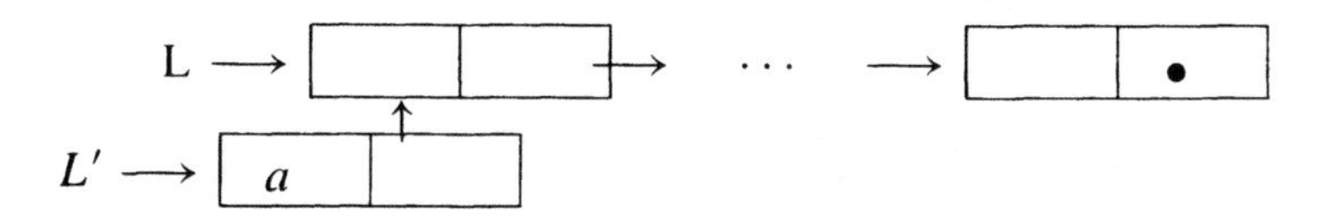


i.e., as a pointer to a cell containing $a$ in the information field and a pointer to $L$ in the address field. If $a$ itself is a list, then a pointer to this list is stored in the information field of the first cell of $L'$.

Of course the problem arises how to distinguish between actual information, i.e., an element of $A$ if we are considering lists over $A$, and addresses to sublists in the information field of a list. This can be achieved by sacrificing 1 bit of the information field. From now on we tacitly assume that we can always make this distinction.

As an example let us consider the list

$$L = [\ [10, []],\ 3,\ [[5], 2],\ 7]$$

over $\mathbb{N}$. Its machine representation is

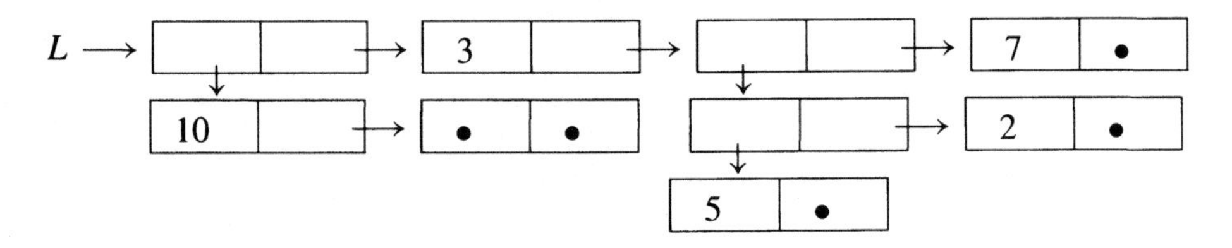


All the operations defined on lists can be carried out rather efficiently on this representation. For computing EMPTY($L$) we just have to check whether the pointer stored in $L$ is a valid one. FIRST($L$) has to extract the contents of the information field of the first cell of $L$. REST($L$) has to extract the contents of the address field of the first cell of $L$. CONS($a$, $L$) is computed by retrieving a new cell $C$ from the ASL, storing $a$ in the information field and $L$ in the address field of $C$ and returning the address of $C$. All these operations are clearly independent of the length of $L$, since only the pointer $L$ or the first cell of the list have to be inspected or changed. Their computation time is a constant. For computing APPEND($L_1$, $L_2$), the list $L_1$ is copied into $L_1'$ and the contents of the address field of the last cell of $L_1'$ is changed to $L_2$. This takes time proportional to the length of $L_1$. Also for computing LENGTH($L$), the successive cells of $L$ have to be counted off, which clearly takes time proportional to the length of $L$.

The basic concepts of lists and operations on them have been introduced in Newell et al. (1957). The language LISP by J. McCarthy is based on list processing (McCarthy et al. 1962). A thorough treatment of list processing can be found in Knuth (1973) or Horowitz and Sahni (1976).

There are, of course, also other possible representations of algebraic objects. The basic concept and use of straight-line programs in algebraic computation are, for instance, described in Strassen (1972), Freeman et al. (1986), von zur Gathen (1987), and Kaltofen (1988). The computer algebra system Maple uses dynamic arrays instead of linked lists.

## 1.5 Measuring the complexity of algorithms

The complexity analysis of algorithms is an important research area in its own right. The reader is referred to Aho et al. (1974), Book (1986), Kaltofen (1990). For our purposes the following simple approach is sufficient. Our model of computation is a RAM equipped with the storage management for lists described in the previous section. Nevertheless, we will not use the language of a RAM for describing algorithms, but we will employ a certain pseudocode language, the so-called *algorithmic language*. We do not give a formal specification of the algorithmic language but just say that it is a PASCAL like language, with assignments, "if_then_else," "for" and "while" loops, and recursive calls. It will always be clear how an algorithm $\mathcal{A}$ described in the algorithmic language can be translated into a program $\mathcal{A}'$ on a RAM. We measure the complexity of an algorithm $\mathcal{A}$ by measuring the complexity of its translation $\mathcal{A}'$ on a RAM. By $t_{\mathcal{A}}(x)$ we denote the time, i.e., the number of basic steps, needed for executing the algorithm $\mathcal{A}$ on the input $x$. The following definition can be found in Collins (1973).

*Definition 1.5.1.* Let $X$ be the set of inputs of an algorithm $\mathcal{A}$, and let $\mathcal{P} = \{X_j\}_{j\in J}$ be a partition of $X$ into finite sets, such that $X_i \cap X_j = \emptyset$ for $i \neq j$ and $X = \bigcup_{j\in J} X_j$.

By $t^+_{\mathcal{A}}(j) := \max\{t_{\mathcal{A}}(x) \mid x \in X_j\}$ we denote the *maximum computing time function* or the *maximum time complexity function* of $\mathcal{A}$ (with respect to the partition $\mathcal{P}$).

By $t^-_{\mathcal{A}}(j) := \min\{t_{\mathcal{A}}(x) \mid x \in X_j\}$ we denote the *minimum computing time function* or the *minimum time complexity function* of $\mathcal{A}$ (with respect to the partition $\mathcal{P}$).

By $t^*_{\mathcal{A}}(j) := \sum_{x\in X_j} t_{\mathcal{A}}(x)/|X_j|$ we denote the *average computing time function* or the *average time complexity function* of $\mathcal{A}$ (with respect to the partition $\mathcal{P}$).

For a given algorithm $\mathcal{A}$ and a partition $\mathcal{P} = \{X_j\}_{j\in J}$ of the input set $X$ of $\mathcal{A}$, the complexity functions $t^+_{\mathcal{A}}$ and $t^-_{\mathcal{A}}$ are functions from $J$ to $\mathbb{N}$ and $t^*_{\mathcal{A}}$ is a function from $J$ to $\mathbb{Q}^+$, the positive rational numbers.

*Definition 1.5.2.* Let $f$ and $g$ be functions from a set $S$ to $\mathbb{R}^+$, the positive real numbers. $f$ *is dominated by* $g$ or $g$ *dominates* $f$ or $f$ *is of order* $g$, in symbols $f \preceq g$ or $f = \mathcal{O}(g)$, iff there is a positive real number $c$ such that $f(x) \leq c \cdot g(x)$ for all $x \in S$. $f$ and $g$ are *codominant* or *proportional*, in symbols $f \sim g$, iff $f \preceq g$ and $g \preceq f$. If $f \preceq g$ and not $g \preceq f$ then we say that $g$ *strictly dominates* $f$, in symbols $f \prec g$.

$\preceq$ is a partial ordering on $(\mathbb{R}^+)^S$ and $\sim$ is an equivalence relation on $(\mathbb{R}^+)^S$. We will often use the following properties, which can be easily verified.

**Lemma 1.5.1.** Let $f, g$ be functions from $S$ to $\mathbb{R}^+$.
a. If $f(x) \leq c \cdot g(x)$ for $c \in \mathbb{R}^+$ and for all but a finite number of $x \in S$, then $f = \mathcal{O}(g)$.
b. If $f = \mathcal{O}(g)$ then $g + f \sim g$.

**Lemma 1.5.2.**
a. If $c, d \in \mathbb{N}_0$, $c < d$, and $n \in \mathbb{N}$, then $n^c \prec n^d$.
b. If $c, d \in \mathbb{N}$, $1 < c, d$, and $x \in \mathbb{N}$, then $\log_c x \sim \log_d x$.

Only for a few algorithms will we give an analysis of the minimum or aver-

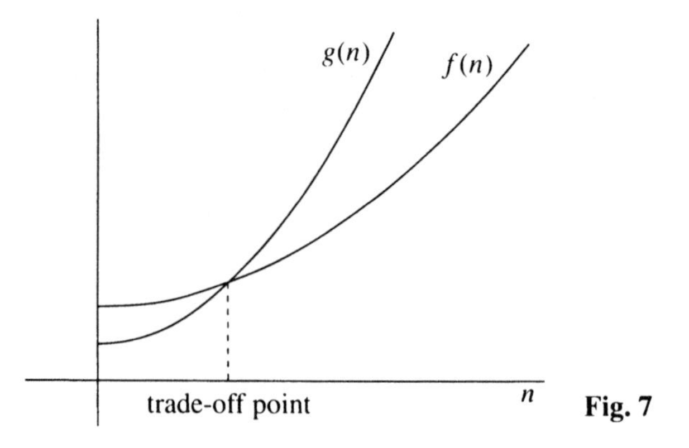


**Fig. 7**

age complexity functions. Usually it is much simpler to determine the maximum complexity function, and in many cases this is the best we can achieve. So when we talk of the complexity of an algorithm $\mathcal{A}$, we usually mean the maximum complexity.

The order of the complexity function of an algorithm tells us the asymptotic time requirement for executing the algorithm on large inputs. Let us assume that $\mathcal{A}_1, \mathcal{A}_2, \mathcal{A}_3$ are algorithms for solving the same problem, and their complexities are, respectively, $\mathcal{O}(n)$, $\mathcal{O}(n^3)$, $\mathcal{O}(2^n)$. Suppose in time $t$ (e.g., 1 second) we can solve a problem of size $s_i$ by the algorithm $\mathcal{A}_i$. Then in time $10t$ we can solve a problem of size $10s_1$ by $\mathcal{A}_1$, a problem of size $3.16s_2$ by the algorithm $\mathcal{A}_2$, and a problem of size only $s_3 + 3.3$ by the algorithm $\mathcal{A}_3$.

However, we should be careful about relying too heavily on the theoretical complexity of algorithms, if we supply only inputs of small or moderate size. So, for instance, the Karatsuba algorithm for multiplying integers has a complexity of $\mathcal{O}(n^{\log_2 3})$, whereas the Schönhage–Strassen algorithm has a complexity of $\mathcal{O}(n \cdot \log n \cdot \log\log n)$. Nevertheless, in all the major computer algebra systems only the Karatsuba algorithm is used, because the superior complexity of the Schönhage–Strassen algorithm becomes effective only for integers of astronomical size. For smaller inputs the constant of the Schönhage–Strassen algorithm determines the practical complexity. A similar phenomenon can be observed in factorization of polynomials, where in practice the Berlekamp–Hensel algorithm is preferred to the theoretically better Lenstra–Lenstra–Lovász algorithm.

In general, if $f(n)$, $g(n)$ are the complexity functions of two competing algorithms, where $f \prec g$, we need to determine the *trade-off point*, i.e., the value of $n$ for which $f(m) < g(m)$, for $m > n$. See Fig. 7.

### 1.6 Bibliographic notes

There are several general articles and books on computer algebra and related topics. Some of them are Akritas (1989), Boyle and Caviness (1990), Buchberger et al. (1983), Caviness (1986), Cohen (1993), Davenport et al. (1988), Geddes et al. (1992), Knuth (1981), Kutzler et al. (1992), Lidl and Pilz (1984), Lipson (1981), Mignotte (1992), Mishra (1993), Moses (1971a, b), Pavelle et al. (1981),

Pohst and Zassenhaus (1989), Rand and Armbruster (1987), Rolletschek (1991), Winkler (1987, 1988b, 1993), Zippel (1993). The use of computer algebra for education in mathematics is described in Karian (1992).

For further information on particular computer algebra systems we refer to Buchberger et al. (1993), Cannon (1984), Char et al. (1991a, b, 1992), Collins and Loos (1982), Fitch (1985), Gonnet and Gruntz (1993), Harper et al. (1991), Heck (1993), Hehl et al. (1992), Jenks and Sutor (1992), MacCallum and Wright (1991), Maeder (1991), Pavelle and Wang (1985), Rand (1984), Rayna (1987), Rich et al. (1988), Simon (1990), Wolfram (1991).

There were or are several series of conferences in computer algebra and related topics, e.g., SYMSAC, EUROSAM, EUROCAL, ISSAC, AAECC, DISCO, as well as single conferences and workshops organized for particular areas or applications. We list some of the most important ones: Bobrow (1968), Bronstein (1993), Buchberger (1985a), Calmet (1982, 1986), Caviness (1985), Char (1986), Chudnovsky and Jenks (1990), Cohen (1991), Cohen and van Gastel (1992), Cohen et al. (1993), Davenport (1989), Della Dora and Fitch (1989), Fateman (1977), Fitch (1984), Floyd (1966), Gianni (1989), Gonnet (1989), Grossman (1989), Huguet and Poli (1989), Janßen (1987), Jenks (1974, 1976), Kaltofen and Watt (1989), Leech (1970), Levelt (1995), Lewis (1979), Mattson and Mora (1991), Mattson et al. (1991), Miola (1993), Mora (1989), Mora and Traverso (1991), Ng (1979), Pavelle (1985), Petrick (1971), Sakata (1991), Shirkov et al. (1991), van Hulzen (1983), von zur Gathen and Giesbrecht (1994), Wang (1981, 1992), Watanabe and Nagata (1990), Watt (1991), Yun (1980).

Periodical publications specializing in computer algebra are *Journal of Symbolic Computation (JSC)*, editor B. F. Caviness (formerly B. Buchberger), published by Academic Press; *Applicable Algebra in Engineering, Communication, and Computing (AAECC)*, editor J. Calmet, published by Springer-Verlag; *SIGSAM Bulletin*, editor R. Grossman, published by ACM Press.

# 2 Arithmetic in basic domains

## 2.1 Integers

For the purposes of exact algebraic computation integers have to be represented exactly. In practice, of course, the size of the machine memory bounds the integers that can be represented. But it is certainly not acceptable to be limited by the word length of the machine, say $2^{32}$.

*Definition 2.1.1.* Let $\beta \geq 2$ be a natural number. A *$\beta$-digit* is an integer $b$ in the range $-\beta < b < \beta$. Every positive integer $a$ can be written uniquely as $a = \sum_{i=0}^{n-1} a_i \beta^i$ for some $n \in \mathbb{N}$, $a_0, \ldots, a_{n-1}$ nonnegative $\beta$-digits and $a_{n-1} > 0$. In the *positional number system* with *basis* (or *radix*) $\beta$ the number $a$ is represented by the uniquely determined list $a_{(\beta)} = [+, a_0, \ldots, a_{n-1}]$. In an analogous way a negative integer $a$ is represented by $a_{(\beta)} = [-, a_0, \ldots, a_{n-1}]$, where $a = \sum_{i=0}^{n-1}(-a_i)\beta^i$, $n \in \mathbb{N}$, $a_0, \ldots, a_{n-1}$ nonnegative $\beta$-digits and $a_{n-1} > 0$. The number 0 is represented by the empty list [ ].

If the integer $a$ is represented by the list $[\pm, a_0, \ldots, a_{n-1}]$, then $L_\beta(a) := n$ is the *length of a w.r.t. $\beta$*. $L_\beta(0) := 0$.

So, for example, in the positional number system with basis 10000, the integer 2000014720401 is represented by the list $[+, 401, 1472, 0, 2]$. By representing integers as lists, we can exactly represent integers of arbitrary size. The memory needed can be allocated dynamically, "as the need arises." For all practical purposes we will choose $\beta$ such that a $\beta$-digit can be stored in a single machine word.

In measuring the complexity of arithmetic operations on integers we will often give the complexity functions as functions of $L_\beta$, the length of the integer arguments of the operations w.r.t. to the basis $\beta$ of the number system, i.e., we decompose $\mathbb{Z}$ as

$$\mathbb{Z} = \bigcup_{i=0}^{\infty} \mathbb{Z}_{(n)} ,$$

where $\mathbb{Z}_{(n)} = \{a \in \mathbb{Z} \mid L_\beta(a) = n\}$. Similarly, for algorithms taking two integers as inputs we decompose

$$\mathbb{Z} \times \mathbb{Z} = \bigcup_{m,n \in \mathbb{N}_0} \mathbb{Z}_{(m,n)} ,$$

where $\mathbb{Z}_{(m,n)} = \mathbb{Z}_{(m)} \times \mathbb{Z}_{(n)}$.

It is crucial to note that for two different radices $\beta$ and $\gamma$ the associated length functions are proportional. So we will often speak of the length of an integer without referring to a specific radix. The proof of the following lemma is left as an exercise.

**Lemma 2.1.1.** Let $\beta$ and $\gamma$ be two radices for positional number systems in $\mathbb{Z}$.
a. $L_\beta(a) = \lfloor \log_\beta(|a|) \rfloor + 1$ for $a \neq 0$.
b. $L_\beta \sim L_\gamma$.

For future reference we note that obviously the algorithm INT_SIGN for computing the sign $\pm 1$ or 0 of an integer $a$ takes constant time. So signs do not present a problem in integer arithmetic, and we will omit their computation in subsequent algorithms.

In fact, we could also omit the sign bit in the representation of integers, and instead use non-positive digits for representing negative numbers. This, of course, means that we need a new algorithm for determining the sign of an integer. Such an algorithm is developed in the exercises and in fact one can prove that this determination of the sign, although in the worst case proportional to the length of the integer, in average is only proportional to a constant.

Addition

The "classical" addition algorithm INT_SUMC considers both inputs $a$ and $b$ as numbers of equal length (with leading zeros, if necessary) and adds corresponding digits with carry until both inputs are exhausted. So $t_{\mathrm{INT\_SUMC}}(m, n) \sim \max(m, n)$, where $m$ and $n$ are the lengths of the inputs. A closer analysis, however, reveals that after the shorter input is exhausted the carry has to be propagated only as long as the corresponding digits of the longer input are $\beta - 1$ or $-\beta + 1$, respectively. This fact is used for constructing a more efficient algorithm, whose computing time depends only linearly on the length of the shorter of the two summands. We assume that an algorithm DIGIT_SUM is available, which adds the contents of two machine words, i.e., two $\beta$-digits $a, b$, yielding a digit $c$ and a digit $d \in \{-1, 0, 1\}$, such that $(a+b)_{(\beta)} = [c, d]$. Obviously the complexity function of DIGIT_SUM is constant.

First, let us simplify our problem by assuming that the two integers to be added have the same sign.

**Algorithm INT_SUM1**(in: $a, b$; out: $c$);
[$a, b$ are integers, $\mathrm{sign}(a) = \mathrm{sign}(b) \neq 0$; $c = a + b$]
1. [the sign of $c$ is initialized, the signs of $a, b$ are removed]
   $c :=$ CONS(FIRST($a$), [ ]); $e := 0$; $a' :=$ REST($a$); $b' :=$ REST($b$);
2. [while $a'$ and $b'$ are not exhausted, add successive digits of $a'$ and $b'$]
   while $a' \neq$ [ ] and $b' \neq$ [ ] do
   {$d_1 :=$ FIRST($a'$); $a' :=$ REST($a'$); $d_2 :=$ FIRST($b'$); $b' :=$ REST($b'$);
   $(d, f) :=$ DIGIT_SUM($d_1, d_2$);
   if $f \neq 0$
   then $(d, f') :=$ DIGIT_SUM($d, e$) [$f' = 0$ in this case]

```
      else (d, f) := DIGIT_SUM(d, e);
      c := CONS(d, c); e := f};
3. [the carry is propagated, until it disappears or both numbers are exhausted]
   if a' = [ ] then g := b' else g := a';
   while e ≠ 0 and g ≠ [ ] do
      {d1 := FIRST(g); g := REST(g);
      (d, e) := DIGIT_SUM(d1, e); c := CONS(d, c)};
4. if e = 0
   then {c := INV(c); c := APPEND(c, g)}
   else {c := CONS(e, c); c := INV(c)};
   return.
```

**Lemma 2.1.2.** For $a, b \in \mathbb{Z}$ with $\text{sign}(a) = \text{sign}(b) \neq 0$, INT_SUM1 correctly computes $a + b$.

*Proof.* We only consider the case $a, b > 0$. For $a, b < 0$ the correctness can be proved analogously. Let $m = L_\beta(a)$, $n = L_\beta(b)$, w.l.o.g. $m \leq n$. After the "while" loop in step (2) has been executed $i$ times, $0 \leq i \leq m$, we have

$$a + b = \text{INV}(c) + \beta^i \cdot e + \beta^i \cdot (a' + b') \ .$$

So when step (2) is finished,

$$a + b = \text{INV}(c) + \beta^m \cdot e + \beta^m \cdot b' \ .$$

After the "while" loop in step (3) has been executed $j$ times, $0 \leq j \leq n - m$, we have

$$a + b = \text{INV}(c) + \beta^{m+j} \cdot e + \beta^{m+j} \cdot b' \ .$$

When the carry $e$ becomes 0 we only need to combine $\text{INV}(c)$ and $b'$. Otherwise $b' = [\ ]$ after $n - m$ iterations and we only need to add $e$ as the highest digit to the result. □

**Theorem 2.1.3.** The maximum, minimum, and average complexity functions of INT_SUM1 are proportional to $\max(m, n)$, $\min(m, n)$, and $\min(m, n)$, respectively, where $m$ and $n$ are the $\beta$-lengths of the inputs.

*Proof.* The maximum and minimum complexity functions for INT_SUM1 are obvious. So let us consider the average complexity.

Let $a, b$ be the inputs of INT_SUM1, and let $m = L(a)$, $n = L(b)$. Obviously the complexity of steps (1), (2) is proportional to $\min(m, n)$. We will show that the average complexity function of step (3) is constant. This will imply that the length of $c$ in step (4) will be proportional to $\min(m, n)$, and therefore also the average complexity of this step is proportional to $\min(m, n)$.

W.l.o.g. assume that the inputs are positive and that $m < n$. Let $k = n - m$. If $\beta$ is the radix of the number system, then there are $(\beta - 1)\beta^{k-1}$ possible assignments for the $k$ highest digits of $b$. The carry has to be propagated exactly up to position $m+i$ of $b$ for $i < k$, if the digits in positions $m+1, \ldots, m+i-1$ of $b$ all are $\beta - 1$ and the digit in position $m+i$ is less than $\beta - 1$. So there are $(\beta - 1)^2 \beta^{k-i-1}$ possible assignments of the digits in positions $m+1, \ldots, m+k$ of $b$, for which exactly $i$ iterations through step (3) are required. There are $\beta - 2$ assignments for which the carry is propagated exactly up to position $m+k$, and for one assignment the propagation is up to position $m+k+1$. Summation over the total time for all these assignments yields

$$\sum_{i=1}^{k-1} i \cdot (\beta - 1)^2 \beta^{k-i-1} + k \cdot (\beta - 2) + (k+1) \cdot 1$$
$$= (\beta^k - \beta k + k - 1) + \beta k - k + 1 = \beta^k .$$

So the average complexity for step (3) is

$$\frac{\beta^k}{(\beta - 1)\beta^{k-1}} = \frac{\beta}{\beta - 1} \leq 2 ,$$

i.e., it is constant. □

By similar considerations one can develop an algorithm INT_SUM2 for adding two nonzero integers with opposite signs in maximum, minimum, and average time proportional to the maximum, minimum, and minimum of the lengths of the inputs, respectively (see Exercises). The combination of these two algorithms leads to an addition algorithm INT_SUM for adding two arbitrary integers. This proves the following theorem.

**Theorem 2.1.4.** There is an addition algorithm INT_SUM for integers with maximum, minimum, and average complexity functions proportional to $\max(m, n)$, $\min(m, n)$, and $\min(m, n)$, respectively, where $m$ and $n$ are the $\beta$-lengths of the inputs.

The algorithm INT_NEG for computing the additive inverse $-a$ of an integer $a$ is obviously of constant complexity. The difference $a - b$ of two integers $a$ and $b$ can be computed as INT_SUM($a$, INT_NEG($b$)). So the algorithm INT_DIFF for computing the difference of two integers has the same complexity behaviour as INT_SUM. The algorithm INT_ABS for computing the absolute value of an integer is either of constant complexity or proportional to the length of the input, depending on which representation of integers we use.

### Multiplication

Now we approach the question of how fast we can multiply two integers. Here we can give only a first answer. We will come back to this question later in Sect. 3.3. The "classical" multiplication algorithm INT_MULTC proceeds by multiplying every digit of the first input by every digit of the second input and adding the results after appropriate shifts. The complexity of INT_MULTC is proportional to the product of the lengths of the two inputs, and if the inputs are of the same length $n$, then the complexity of INT_MULTC is proportional to $n^2$.

A faster multiplication algorithm has been discovered by A. Karatsuba and Yu. Ofman (1962). The basic idea in the Karatsuba algorithm is to cut the two inputs $x, y$ of length $\leq n$ into pieces of length $\leq n/2$ such that

$$x = a \cdot \beta^{n/2} + b, \qquad y = c \cdot \beta^{n/2} + d \ . \tag{2.1.1}$$

A usual divide-and-conquer approach would reduce the product of two integers of length $n$ to four products of integers of length $n/2$. The complexity of this algorithm would still be $\mathcal{O}(n^2)$. Karatsuba and Ofman, however, noticed that one of the four multiplications can be dispensed with.

$$\begin{aligned} x \cdot y &= ac\beta^n + (ad + bc)\beta^{n/2} + bd \\ &= ac\beta^n + ((a+b)(c+d) - ac - bd)\,\beta^{n/2} + bd \ . \end{aligned} \tag{2.1.2}$$

So three multiplications of integers of length $n/2$ and a few shifts and additions are sufficient for computing the product $x \cdot y$.

In the Karatsuba algorithm INT_MULTK we neglect the signs of the integers. Their handling is rather obvious and only obscures the statement of the algorithm.

**Algorithm INT_MULTK**(in: $x, y$; out: $z$);
[$x, y$ integers; $z = x \cdot y$]
$n :=$ max(LENGTH($x$), LENGTH($y$));
if $n = 1$ then {$z :=$ INT_MULTC($x, y$); return};
if $n$ is odd then $n := n + 1$;
$(a, b) :=$ (DEL($n/2, x$), INIT($n/2, x$));
$(c, d) :=$ (DEL($n/2, y$), INIT($n/2, y$));
$u :=$ INT_MULTK($a + b, c + d$);
$v :=$ INT_MULTK($a, c$);
$w :=$ INT_MULTK($b, d$);
$z := v\beta^n + (u - v - w)\beta^{n/2} + w$;
return.

**Theorem 2.1.5.** The complexity of the Karatsuba algorithm INT_MULTK is $\mathcal{O}(n^{\log_2 3})$, where $n$ is the length of the inputs.

*Proof.* Initially we assume that $n$ is a power of 2. Let $x$ and $y$ be integers of length not exceeding $n$, and let $a, b, c, d$ be the parts of $x, y$ as in (2.1.1). During the execution of the Karatsuba algorithm we have to compute the products $(a + b)(c + d)$, $ac$, $bd$. All the other operations are additions and shifts, which take time proportional to $n$. The factors in $ac$ and $bd$ are of length not exceeding

$n/2$, whereas the factors in $(a+b)(c+d)$ might be of length $n/2+1$. We write the factors as

$$a+b = a_1\beta^{n/2} + b_1, \quad c+d = c_1\beta^{n/2} + d_1 , \tag{2.1.3}$$

where $a_1$ and $c_1$ are the leading digits of $a+b$ and $c+d$, respectively. Now

$$(a+b)(c+d) = a_1c_1\beta^n + (a_1d_1 + b_1c_1)\beta^{n/2} + b_1d_1 . \tag{2.1.4}$$

In the product $b_1d_1$ the factors are of length not exceeding $n/2$. All the other operations are multiplications by a single digit or shifts, and together their complexity is proportional to $n$.

So if we denote the time for multiplying two integers of length $n$ by $M(n)$, we get the recursion equation

$$M(n) = \begin{cases} k & \text{for } n = 1, \\ 3M(n/2) + kn & \text{for } n > 1. \end{cases} \tag{2.1.5}$$

Here we have taken the constant $k$ to be a bound for the complexity of multiplication of digits as well as for the constant factor in the linear complexity functions of the addition and shift operations. The solution to (2.1.5) is

$$M(n) = 3kn^{\log_2 3} - 2kn , \tag{2.1.6}$$

which can easily be verified by induction. This proves the assertion for all $n$ which are powers of 2.

Finally let us consider the general case, where $n$ is an arbitrary positive integer. In this case we could, theoretically, increase the length of the inputs to the next higher power of 2 by adding leading zeros. The length of the multiplicands is at most doubled in this process. In the asymptotic complexity, however, the factor 2 is negligible, since $(2n)^{\log_2 3}$ is $\mathcal{O}(n^{\log_2 3})$. □

The Karatsuba algorithm is practically used in computer algebra systems. In fact, the idea of Karatsuba and Ofman can be generalized to yield a multiplication algorithm of complexity $n^{1+\epsilon}$ for any positive real $\epsilon$. We do not go into details here, but rather refer to the excellent exposition in Knuth (1981: sect. 4.3.3). We will, however, describe a theoretically even faster method, based on the fast Fourier transform (see Sect. 3.3). For these faster methods, however, the overhead is so enormous, that a practical importance seems unlikely.

That the complexity of multiplication depends mainly on the smaller of the two inputs is explained by the following theorem.

**Theorem 2.1.6.** Let IM be a multiplication algorithm for integers with complexity $t^+_{\mathrm{IM}}(m)$ for multiplying two integers of lengths not greater than $m$, such that $m \preceq t^+_{\mathrm{IM}}(m)$. Then there exists a multiplication algorithm IM′ with

$$t^+_{\mathrm{IM}'}(m,n) \preceq \begin{cases} (m/n)\cdot t^+_{\mathrm{IM}}(n) & \text{for } m \geq n, \\ (n/m)\cdot t^+_{\mathrm{IM}}(m) & \text{for } m < n \end{cases}$$

for inputs of lengths $m$ and $n$, respectively.

*Proof.* Let $a, b$ be the integers to be multiplied, and $m = L(a), n = L(b)$. W.l.o.g. assume that $m \geq n$. IM′ decomposes $a$ into pieces $a_0, \ldots, a_{l-1}$ of length $\leq n$, such that $a = \sum_{i=0}^{l-1} a_i \cdot \beta^{ni}$. The number of pieces can be chosen as $l = \lceil m/n \rceil \leq (m/n) + 1$. Now each piece $a_i$ is multiplied by $b$ by algorithm IM and finally these partial results are shifted and added. Thus for some positive constant $c$

$$t^+_{\mathrm{IM}'}(m,n) \leq \left(\frac{m}{n}+1\right)\cdot t^+_{\mathrm{IM}}(n) + \left(\frac{m}{n}+1\right)\cdot cn \preceq (m/n)\cdot t^+_{\mathrm{IM}}(n)\ ,$$

which completes the proof. □

Division

The problem of integer division consists of computing the uniquely determined integral quotient, $q = \mathrm{quot}(a,b)$, and remainder, $r = \mathrm{rem}(a,b)$, for $a, b \in \mathbb{Z}$, $b \neq 0$, such that

$$a = q\cdot b + r \quad \text{and} \quad \begin{cases} 0 \leq r < |b| & \text{for } a \geq 0, \\ -|b| < r \leq 0 & \text{for } a < 0. \end{cases}$$

If $|a| < \beta^j \cdot |b|$ for $j \in \mathbb{N}$, then $q$ has at most $j$ digits. $j$ will be approximately $L(a) - L(b) + 1$. For determining the highest digit in the quotient one certainly does not need more than linear time in $L(b)$, even if all the possible digits are tried. So we get

**Theorem 2.1.7.** There is an algorithm INT_DIV for computing the quotient and remainder of two integers $a, b$ of lengths $m, n$, respectively, $m \geq n$, whose complexity $t^+_{\mathrm{INT\_DIV}}(m,n)$ is $\mathcal{O}(n\cdot(m-n+1))$.

In fact we need not really try all the possible digits of the quotient, but there is a very efficient algorithmic way of "guessing" the highest digit. Such a method has been described in Pope and Stein (1960), where the following theorem is proved. See also Collins et al. (1983).

**Theorem 2.1.8.** Let $a_{(\beta)} = [+, a_0, a_1, \ldots, a_{m-1}]$, $b_{(\beta)} = [+, b_0, \ldots, b_{n-1}]$, $\beta^j b \leq a < \beta^{j+1} b$ for $j \in \mathbb{N}$, $m \geq n$ and $b_{n-1} \geq \lfloor \beta/2 \rfloor$. If $\bar{q}$ is maximal in $\mathbb{Z}$ with $\bar{q}\beta^j b \leq a$ and $q^* = \lfloor (a_{n+j}\beta + a_{n+j-1})/b_{n-1} \rfloor$ (we set $a_i = 0$ for $i \geq L(a)$), then $\bar{q} \leq q^* \leq \bar{q} + 2$.

By a successive application of Theorem 2.1.8 the digits in the quotient $q$ of $a$ and $b$ can be computed. Let $m = L(a)$, $n = L(b)$, $0 \leq a$, $0 < b$, and $b_{n-1} \geq \lfloor \beta/2 \rfloor$. Then $a < \beta^{m-n+1}b$, so $q$ has at most $m - n + 1$ digits. First the highest digit $q_{m-n}$ is determined from the guess $q^*$. We need at most 2 correction steps of subtracting 1 from the initial guess. Collins and Musser (1977) have shown that the probabilities of $q^*$ being $q_{m-n} + i$ for $i = 0, 1, 2$ are 0.67, 0.32, and 0.01, respectively. Now $a - \beta^{m-n}q_{m-n}b < \beta^{m-n}b$ and the process can be continued to yield $q_{m-n-1}$ and so on.

The condition $b_{n-1} \geq \lfloor \beta/2 \rfloor$ can be satisfied by replacing $a$ and $b$ by $a' = a \cdot d$, $b' = b \cdot d$, respectively, where $d = \lfloor \beta/(b_{n-1} + 1) \rfloor$. This does not change the quotient $q$ and $\mathrm{rem}(a, b) = (a' - q \cdot b')/d$.

These considerations lead to a better division algorithm INT_DIV, the *Pope–Stein algorithm*. The theoretical complexity function of the Pope–Stein algorithm, however, is still $n(m - n + 1)$, as in Theorem 2.1.7.

In Aho et al. (1974), the relation of the complexity of integer multiplication, division, and some other operations is investigated. It is shown that the complexity functions for multiplication of integers of length $\leq n$ and division of integers of length $\leq 2n$ by integers of length $\leq n$ are proportional.

### Conversion

We assume that we have the arithmetic operations for integers in $\beta$-representation available. There are two types of conversions that we need to investigate: (1) conversion of an integer $a$ from $\gamma$-representation into $\beta$-representation, and (2) conversion of $a$ from $\beta$-representation into $\gamma$-representation.

It is quite obvious how we can do arithmetic with radix $\beta^j$, if we can do arithmetic with radix $\beta$. So in conversion problem 1 we may assume that $\gamma < \beta$, i.e., $\gamma$ is a $\beta$-digit. If $a_{(\gamma)} = [\pm, a_0, \ldots, a_{n-1}]$, then we get $a_{(\beta)}$ by Horner's rule

$$a = (\ldots((a_{n-1}\gamma + a_{n-2})\gamma + a_{n-3})\gamma + \ldots + a_1)\gamma + a_0 .$$

Every multiplication by $\gamma$ takes time linear in the length of the multiplicand, and every addition of a $\gamma$-digit $a_i$ takes constant time. So the maximum complexity of conversion of type 1 is proportional to

$$\sum_{i=1}^{n-1} i \sim n^2 = L_\gamma(a)^2 .$$

Conversion problem 2 can be solved by successive division by $\gamma = \gamma_{(\beta)}$. Every such division step reduces the length of the input by a constant, and takes time proportional to the length of the intermediate result, i.e., the maximum complexity of conversion of type 2 is proportional to $L_\beta(a)^2$.

Computation of greatest common divisors

$\mathbb{Z}$ is a unique factorization domain. So for any two integers $x, y$ which are not both equal to 0, there is a greatest common divisor (gcd) $g$ of $x$ and $y$. $g$ is determined up to multiplication by units, i.e., up to sign. Usually we mean the positive greatest common divisor when we speak of "the greatest common divisor." For the sake of completeness let us define $\gcd(0, 0) := 0$.

But in addition to mere existence of gcds in $\mathbb{Z}$, there is also a very efficient algorithm due to Euclid ($\approx$ 330–275 B.C.) for computing the gcd. This is probably the oldest full fledged non-trivial algorithm in the history of mathematics. In later chapters we will provide an extension of the scope of Euclid's algorithm to its proper algebraic setting. But for the time being, we are just concerned with integers.

Suppose we want to compute $\gcd(x, y)$ for $x, y \in \mathbb{N}$. We divide $x$ by $y$, i.e., we determine the quotient $q$ and the remainder $r$ of $x$ divided by $y$, such that

$$x = q \cdot y + r, \quad \text{with } r < y .$$

Now $\gcd(x, y) = \gcd(y, r)$, i.e., the size of the problem has been reduced. This process is repeated as long as $r \neq 0$. Thus we get the so-called *Euclidean remainder sequence*

$$r_1, r_2, \ldots, r_n, r_{n+1} ,$$

with $r_1 = x$, $r_2 = y$, $r_i = \text{rem}(r_{i-2}, r_{i-1})$ for $3 \leq i \leq n+1$ and $r_{n+1} = 0$. Clearly $\gcd(x, y) = r_n$. Associated with this remainder sequence we get a sequence of quotients

$$q_1, \ldots, q_{n-1} ,$$

such that

$$r_i = q_i \cdot r_{i+1} + r_{i+2} \quad \text{for } 1 \leq i \leq n-1 .$$

Thus in $\mathbb{Z}$ greatest common divisors can be computed by the *Euclidean algorithm* INT_GCDE.

**Algorithm INT_GCDE**(in: $x, y$; out: $g$);
[$x, y$ are integers; $g = \gcd(x, y)$, $g \geq 0$]
1. $r' :=$ INT_ABS($x$); $r'' :=$ INT_ABS($y$);
2. while $r'' \neq 0$ do
   {($q, r$) := INT_DIV($r', r''$);
   $r' := r''$; $r'' := r$};
3. $g := r'$;
   return.

The computation of gcds of integers is an extremely frequent operation in any computation in computer algebra. So we must carefully analyze its complexity. G. Lamé proved already in the 19th century that for positive inputs bounded by $n$ the number of division steps in the Euclidean algorithm is at most $\lceil \log_\phi(\sqrt{5}n) \rceil - 2$, where $\phi = \frac{1}{2}(1 + \sqrt{5})$. See Knuth (1981: sect. 4.5.3).

**Theorem 2.1.9.** Let $l_1, l_2$ be the lengths of the inputs $x, y$ of INT_GCDE, and let $k$ be the length of the output. Then $t^+_{\mathrm{INT\_GCDE}}(l_1, l_2, k)$ is $\mathcal{O}(\min(l_1, l_2) \cdot (\max(l_1, l_2) - k + 1))$.

*Proof.* Steps (1) and (3) take constant time. So it remains to investigate the complexity behaviour of step (2), $t_2^+(l_1, l_2, k)$.

Let $r_1, r_2, \ldots, r_{n+1}$ be the remainder sequence and $q_1, \ldots, q_{n-1}$ the quotient sequence computed by INT_GCDE for the inputs $x, y$. If $|x| < |y|$ then the first iteration through the loop in (2) results in a reversal of the input pair. In this case the first iteration through the loop takes time proportional to $\min(l_1, l_2)$. So in the sequel we assume that $|x| \geq |y| > 0$. By Theorem 2.1.7

$$t_2^+(l_1, l_2, k) \preceq \sum_{i=1}^{n-1} L(q_i) L(r_{i+1}) \leq L(r_2) \cdot \left(\sum_{i=1}^{n-2} L(q_i + 1) + L(q_{n-1})\right) . \quad (2.1.7)$$

$q_i \geq 1$ for $1 \leq i \leq n-2$ and $q_{n-1} \geq 2$. By Exercise 5

$$\sum_{i=1}^{n-2} L(q_i + 1) + L(q_{n-1}) \sim L\left(q_{n-1} \cdot \prod_{i=1}^{n-2} (q_i + 1)\right) . \quad (2.1.8)$$

For $1 \leq i \leq n-2$ we have $r_{i+2}(q_i + 1) < r_{i+1} q_i + r_{i+2} = r_i$, and therefore $q_i + 1 < r_i / r_{i+2}$. Furthermore $q_{n-1} = r_{n-1}/r_n$. Thus

$$q_{n-1} \cdot \prod_{i=1}^{n-2} (q_i + 1) < \frac{r_{n-1} \cdot r_1 \cdot r_2}{r_n \cdot r_{n-1} \cdot r_n} \leq \left(\frac{r_1}{r_n}\right)^2 . \quad (2.1.9)$$

Combining (2.1.7), (2.1.8), and (2.1.9) we finally arrive at

$$t_2^+(l_1, l_2, k) \preceq \min(l_1, l_2) \cdot L\left(\left(\frac{r_1}{r_n}\right)^2\right) \sim \min(l_1, l_2) \cdot (\max(l_1, l_2) - k + 1) .$$

So $t^+_{\mathrm{INT\_GCDE}}(l_1, l_2, k) \preceq \min(l_1, l_2) \cdot (\max(l_1, l_2) - k + 1)$. □

The greatest common divisor $g$ of $x$ and $y$ generates an ideal $\langle x, y \rangle$ in $\mathbb{Z}$, and $g \in \langle x, y \rangle$. So in particular $g$ can be written as a linear combination of $x$ and $y$,

$$g = u \cdot x + v \cdot y .$$

These linear coefficients can be computed by a straightforward extension of INT_GCDE, the *extended Euclidean algorithm* INT_GCDEE. Throughout the algorithm INT_GCDEE the invariant

$$r' = u' \cdot x + v' \cdot y \quad \text{and} \quad r'' = u'' \cdot x + v'' \cdot y$$

is preserved.

**Algorithm INT_GCDEE**(in: $x, y$; out: $g, u, v$);
[$x, y$ are integers; $g = \gcd(x, y) = u \cdot x + v \cdot y$, $g \geq 0$]
1. $(r', u', v') :=$ (INT_ABS($x$), INT_SIGN($x$), 0);
   $(r'', u'', v'') :=$ (INT_ABS($y$), 0, INT_SIGN($y$));
2. while $r'' \neq 0$ do
   {$q :=$ INT_QUOT($r', r''$);
   $(r, u, v) := (r', u', v') - q \cdot (r'', u'', v'')$;
   $(r', u', v') := (r'', u'', v'')$;
   $(r'', u'', v'') := (r, u, v)$; }
3. $(g, u, v) := (r', u', v')$;
   return.

### Exercises

1. Prove Lemma 2.1.1.
2. Assume that in the representation of integers as lists we omit the information about the sign, and we use nonnegative digits for positive integers and nonpositive digits for negative numbers. Design an algorithm which computes the sign of an integer in this representation in constant average time.
3. Develop an algorithm INT_SUM2 for adding two nonzero integers of different sign with average complexity proportional to the minimum of the lengths of the inputs.
4. Implement the multiplication algorithm of Karatsuba and Ofman and determine the trade-off point between the "classical" algorithm INT_MULTC and INT_MULTK.
5. Let $a_1, a_2, \ldots$ be an infinite sequence of integers with $a_i \geq 2$ for all $i$. Define the functions $f, g$ from $\mathbb{N}$ to $\mathbb{R}$ as $f(n) := \sum_{i=1}^{n} L(a_i)$, $g(n) := L(\prod_{i=1}^{n} a_i)$. Prove that $f(n) \sim g(n)$ by uniform constants independent of the $a_i$'s, i.e., for some $c, C \in \mathbb{R}^+$, $c \cdot f(n) \leq g(n) \leq C \cdot f(n)$ for all $n \in \mathbb{N}$.
6. Design a division algorithm for positive integers based on Theorem 2.1.8 for guessing the digits in the quotient. Presuppose that there is an algorithm for determining $\lfloor (a\beta + b)/c \rfloor$ for $\beta$-digits $a, b, c$, i.e., an algorithm for dividing an integer of length 2 by a digit. Many computers provide an instruction for dividing a double-precision integer by a single precision integer.
7. Prove that the outputs of INT_GCDEE are no larger than the two inputs.
8. Let $a, b, a', b', a'', b''$ be positive integers such that $a'/b' \leq a/b \leq a''/b''$, and let $q = \text{quot}(a, b)$, $q' = \text{quot}(a', b')$, $q'' = \text{quot}(a'', b'')$, $r' = \text{rem}(a', b')$. Prove: If $q' = q''$ and $r' > 0$, then

$$\frac{b'}{a' - q'b'} \geq \frac{b}{a - qb} \geq \frac{b''}{a'' - q''b''} .$$

9. Based on Exercise 8 devise an algorithm which computes the remainder sequence of two arbitrarily long integers $x, y$ using only divisions by single digits. (Such an algorithm is originally due to D. H. Lehmer (1938).)

## 2.2 Polynomials

We consider polynomials over a commutative ring $R$ in finitely many variables $x_1, \ldots, x_n$, i.e., our domain of computation is $R[x_1, \ldots, x_n]$. Before we can design algorithms on polynomials, we need to introduce some notation and suitable representations.

### Representations

A representation of polynomials can be either recursive or distributive, and it can be either dense or sparse. Thus, there are four basically different representations of multivariate polynomials.

In a *recursive representation* a nonzero polynomial $p(x_1, \ldots, x_n)$ is viewed as an element of $(R[x_1, \ldots, x_{n-1}])[x_n]$, i.e., as a univariate polynomial in the main variable $x_n$,

$$p(x_1, \ldots, x_n) = \sum_{i=0}^{m} p_i(x_1, \ldots, x_{n-1})x_n^i, \qquad \text{with } p_m \neq 0 .$$

In the *dense recursive representation* $p$ is represented as the list

$$p_{(dr)} = [(p_m)_{(dr)}, \ldots, (p_0)_{(dr)}] ,$$

where $(p_i)_{(dr)}$ is in turn the dense recursive representation of the coefficient $p_i(x_1, \ldots, x_{n-1})$. If $n = 1$ then the coefficients $p_i$ are elements of the ground ring $R$ and are represented as such. The dense representation makes sense if many coefficients are different from zero. On the other hand, if the set of support of a polynomial is sparse, then a *sparse recursive representation* is better suited, i.e.,

$$p(x_1, \ldots, x_n) = \sum_{i=0}^{k} p_i(x_1, \ldots, x_{n-1})x_n^{e_i} ,$$

$$e_0 > \cdots > e_k \text{ and } p_i \neq 0 \text{ for } 0 \leq i \leq k ,$$

is represented as the list

$$p_{(sr)} = [e_0, (p_0)_{(sr)}, \ldots, e_k, (p_k)_{(sr)}] .$$

Again, in the base case $n = 1$ the coefficients $p_i$ are elements of $R$ and are represented as such.

In a *distributive representation* a nonzero polynomial $p(x_1, \ldots, x_n)$ is viewed as an element of $R[x_1, \ldots, x_n]$, i.e., a function from the set of power products in $x_1, \ldots, x_n$ into $R$. An exponent vector $(j_1, \ldots, j_n)$ of a power product $x_1^{j_1} \cdots x_n^{j_n}$ is mapped to a coefficient in $R$. In a *dense distributive representation* we need a bijection $e\colon \mathbb{N} \longrightarrow \mathbb{N}_0^n$. A polynomial

$$p(x_1, \ldots, x_n) = \sum_{i=0}^{r} a_i x^{e(i)} \quad \text{with } a_r \neq 0$$

is represented as the list

$$p_{(dd)} = [a_r, \ldots, a_0] \ .$$

A *sparse distributive representation* of

$$p(x_1, \ldots, x_n) = \sum_{i=0}^{s} a_i x^{e(k_i)} \quad \text{with } a_i \neq 0 \text{ for } 0 \leq i \leq s$$

is the list

$$p_{(sd)} = [e(k_0), a_0, \ldots, e(k_s), a_s] \ .$$

Which representation is actually employed depends of course on the algorithms that are to be applied. In later chapters we will see examples for algorithms that depend crucially on a recursive representation and also for algorithms that need a distributive representation. However, only very rarely will there be a need for dense representations in computer algebra. If the set of support of multivariate polynomials is dense, then the number of terms even in polynomials of modest degree is so big, that in all likelihood no computations are possible any more.

In the sequel we will mainly analyze the complexity of operations on polynomials in recursive representation. So if not explicitly stated otherwise, the representation of polynomials is assumed to be recursive.

### Addition and subtraction

The algorithms for addition and subtraction of polynomials are obvious: the coefficients of like powers have to be added or subtracted, respectively. If $p$ and $q$ are $n$-variate polynomials in dense representation, with $\max(\deg_{x_i}(p), \deg_{x_i}(q)) \leq d$ for $1 \leq i \leq n$, then the complexity of adding $p$ and $q$ is $\mathcal{O}(A(p,q) \cdot (d+1)^n)$, where $A(p,q)$ is the maximal time needed for adding two coefficients of $p$ and $q$ in the ground ring $R$. If $p$ and $q$ are in sparse representation, and $t$ is a bound for the number of terms $x_i^m$ with nonzero coefficient in $p$ and $q$, for $1 \leq i \leq n$, then the complexity of adding $p$ and $q$ is $\mathcal{O}(A(p,q) \cdot t^n)$.

### Multiplication

In the classical method for multiplying univariate polynomials $p(x) = \sum_{i=0}^{m} p_i x^i$ and $q(x) = \sum_{j=0}^{n} q_j x^j$ the formula

$$p(x) \cdot q(x) = \sum_{l=0}^{m+n} \Big( \sum_{i+j=l} p_i \cdot q_j \Big) x^l \tag{2.2.1}$$

is employed. If $p$ and $q$ are $n$-variate polynomials in dense representation with $d$ as above, then the complexity of multiplying $p$ and $q$ is $\mathcal{O}(M(p,q) \cdot (d+1)^{2n})$, where $M(p,q)$ is the maximal time needed for multiplying two coefficients of $p$ and $q$ in the ground ring $R$. Observe that $(d+1)^n$ is a good measure of the size of the polynomials, when the size of the coefficients is neglected.

As for integer multiplication one can apply the Karatsuba method. That is, the multiplicands $p$ and $q$ are decomposed as

$$p(x) = p_1(x) \cdot x^{\lceil d/2 \rceil} + p_0(x), \qquad q(x) = q_1(x) \cdot x^{\lceil d/2 \rceil} + q_0(x) ,$$

and the product is computed as

$$\begin{aligned} p(x) \cdot q(x) = {} & p_1 \cdot q_1 \cdot x^{2\lceil d/2 \rceil} \\ & + \big((p_1 + p_0) \cdot (q_1 + q_0) - p_1 \cdot q_1 - p_0 \cdot q_0\big) \cdot x^{\lceil d/2 \rceil} + p_0 \cdot q_0 . \end{aligned}$$

Neglecting the complexity of operations on elements of the ground ring $R$, we get that the complexity of multiplying $p$ and $q$ is $\mathcal{O}((d+1)^{n \log_2 3})$.

For multiplying the sparsely represented polynomials $p(x) = \sum_{i=0}^{t} p_i x^{e_i}$ and $q(x) = \sum_{j=0}^{t} q_j x^{f_j}$, one basically has to (1) compute $p \cdot q_j x^{f_j}$ for $j = 0, \ldots, t$, and (2) add this to the already computed partial result, which has roughly $(j-1)t$ terms, if $t \ll \deg(p), \deg(q)$. So the overall time complexity of multiplying polynomials in sparse representation is

$$\sum_{i=1}^{t} \big( \underbrace{M(p,q) \cdot t}_{(1)} + \underbrace{(i-1)t}_{(2)} \big) \preceq M(p,q) \cdot t^3 .$$

### Division

First let us assume that we are dealing with univariate polynomials over a field $K$. If $b(x)$ is a non-zero polynomial in $K[x]$, then every other polynomial $a(x) \in K[x]$ can be divided by $b(x)$ in the sense that one can compute a *quotient* $q(x) = \mathrm{quot}(a,b)$ and a *remainder* $r(x) = \mathrm{rem}(a,b)$ such that

$$a(x) = q(x) \cdot b(x) + r(x) \quad \text{and} \quad (r(x) = 0 \text{ or } \deg(r) < \deg(b)) . \tag{2.2.2}$$

The quotient $q$ and remainder $r$ in (2.2.2) are unique. The algorithm POL_DIVK computes the quotient and remainder for densely represented polynomials. It can easily be modified for sparsely represented polynomials.

**Algorithm POL_DIVK**(in: $a, b$; out: $q, r$);
$[a, b \in K[x]$, $b \neq 0$; $q = \mathrm{quot}(a, b)$, $r = \mathrm{rem}(a, b)$. $a$ and $b$ are assumed to be in dense representation, the results $q$ and $r$ are likewise in dense representation]
1. $q := [\ ]$; $a' := a$; $c := \mathrm{lc}(b)$; $m := \deg(a')$; $n := \deg(b)$;
2. while $m \geq n$ do
   $\{d := \mathrm{lc}(a')/c$; $q := \mathrm{CONS}(d, q)$; $a' := a' - d \cdot x^{m-n} \cdot b$;
   for $i = 1$ to $\min\{m - \deg(a') - 1, m - n\}$ do $q := \mathrm{CONS}(0, q)$;
   $m := \deg(a')\}$;
3. $q := \mathrm{INV}(q)$; $r := a'$; return.

**Theorem 2.2.1.** Let $a(x), b(x) \in K[x]$, $b \neq 0$, $m = \deg(a)$, $n = \deg(b)$, $m \geq n$. The number of field operations in executing POL_DIVK on the inputs $a$ and $b$ is $\mathcal{O}((n+1)(m-n+1))$.

*Proof.* The "while"-loop is executed $m - n + 1$ times. The number of field operations in one pass through the loop is $\mathcal{O}(n+1)$. □

The algorithm POL_DIVK is not applicable any more, if the underlying domain of coefficients is not a field. In this case, the leading coefficient of $a$ may not be divisible by the leading coefficient of $b$. Important examples of such polynomial rings are $\mathbb{Z}[x]$ or multivariate polynomial rings. In fact, there are no quotient and remainder satisfying Eq. (2.2.2). However, it is possible to satisfy (2.2.2) if we allow to normalize the polynomial $a$ by a certain power of the leading coefficient of $b$.

**Theorem 2.2.2.** Let $R$ be an integral domain, $a(x), b(x) \in R[x]$, $b \neq 0$, and $m = \deg(a) \geq n = \deg(b)$. There are uniquely defined polynomials $q(x), r(x) \in R[x]$ such that

$$\begin{aligned} &\mathrm{lc}(b)^{m-n+1} \cdot a(x) = q(x) \cdot b(x) + r(x) \quad \text{and} \\ &(r(x) = 0 \ \text{ or } \ \deg(r) < \deg(b)) \ . \end{aligned} \tag{2.2.3}$$

*Proof.* $R$ being an integral domain guarantees that multiplication of a polynomial by a non-zero constant does not change the degree.

For proving the existence of $q$ and $r$ we proceed by induction on $m - n$. For $m - n = 0$ the polynomials $q(x) = \mathrm{lc}(a)$, $r(x) = \mathrm{lc}(b) \cdot a - \mathrm{lc}(a) \cdot b$ obviously satisfy (2.2.3).

Now let $m - n > 0$. Let

$$c(x) := \mathrm{lc}(b) \cdot a(x) - x^{m-n} \cdot \mathrm{lc}(a) \cdot b(x) \quad \text{and} \quad m' := \deg(c) \ .$$

Then $m' < m$. For $m' < n$ we can set $q' := 0$, $r := \mathrm{lc}(b)^{m-n} \cdot c$ and we get $\mathrm{lc}(b)^{m-n} \cdot c(x) = q'(x) \cdot b(x) + r(x)$. For $m' \geq n$ we can use the induction hypothesis on $c$ and $b$, yielding $q_1, r_1$ such that

$$\mathrm{lc}(b)^{m'-n+1} \cdot c = q_1 \cdot b + r_1 \quad \text{and} \quad (r_1 = 0 \text{ or } \deg(r_1) < \deg(b)) .$$

Now we can multiply both sides by $\mathrm{lc}(b)^{m-m'-1}$ and we get

$$\mathrm{lc}(b)^{m-n} \cdot c(x) = q'(x) \cdot b(x) + r(x), \quad \text{where } r = 0 \text{ or } \deg(r) < \deg(b) .$$

Back substitution for $c$ yields (2.2.3).

For establishing the uniqueness of $q$ and $r$, we assume to the contrary that both $q_1, r_1$ and $q_2, r_2$ satisfy (2.2.3). Then $q_1 \cdot b + r_1 = q_2 \cdot b + r_2$, and $(q_1 - q_2) \cdot b = r_2 - r_1$. For $q_1 \neq q_2$ we would have $\deg((q_1 - q_2) \cdot b) \geq \deg(b) > \deg(r_1 - r_2)$, which is impossible. Therefore $q_1 = q_2$ and consequently also $r_1 = r_2$. □

*Definition 2.2.1.* Let $R, a(x), b(x), m, n$ be as in Theorem 2.2.2. Then the uniquely defined polynomials $q(x)$ and $r(x)$ satisfying (2.2.3) are called the *pseudoquotient* and the *pseudoremainder*, respectively, of $a$ and $b$. We write $q = \mathrm{pquot}(a, b)$, $r = \mathrm{prem}(a, b)$.

**Algorithm POL_DIVP**(in: $a, b$; out: $q, r$);
[$a, b \in R[x], b \neq 0; q = \mathrm{pquot}(a, b), r = \mathrm{prem}(a, b)$. $a$ and $b$ are assumed to be in dense representation, the results $q$ and $r$ are likewise in dense representation]
1. $q := [\ ]$; $a' := a$; $c := \mathrm{lc}(b)$; $m := \deg(a')$; $n := \deg(b)$;
2. $c^{(1)} := c$; for $i = 2$ to $m - n$ do $c^{(i)} := c \cdot c^{(i-1)}$;
3. while $m \geq n$ do
   {$d := \mathrm{lc}(a') \cdot c^{(m-n)}$; $q := \mathrm{CONS}(d, q)$; $a' := c \cdot a' - \mathrm{lc}(a') \cdot x^{m-n} \cdot b$;
   for $i = 1$ to $\min\{m - \deg(a') - 1, m - n\}$ do
     {$q := \mathrm{CONS}(0, q)$; $a' := c \cdot a'$};
   $m := \deg(a')$};
4. $q := \mathrm{INV}(q)$; $r := a'$; return.

The algorithm POL_DIVP computes the pseudoquotient and pseudoremainder of two polynomials over an integral domain $R$.

**Theorem 2.2.3.** Let $a(x), b(x) \in \mathbb{Z}[x]$, $m = \deg(a) \geq n = \deg(b)$, $l = L(\max\{\|a\|_\infty, \|b\|_\infty\})$. Then the complexity of POL_DIVP executed on $a$ and $b$ is $\mathcal{O}(m \cdot (m - n + 1)^2 \cdot l^2)$.

*Proof.* The precomputation of powers of $c$ in step (2) takes time $\mathcal{O}((m-n)^2 l^2)$.

The "while"-loop is executed $m - n + 1$ times, if we assume that the drop in degree of $a'$ is always 1, which is clearly the worst case. At the beginning of the $i$-th iteration through the loop, $L(\|a'\|_\infty)$ is $\mathcal{O}(il)$. So the computation of $d$

takes time $\mathcal{O}(i(m-n+1-i)l^2)$. The computation of $a'$ takes time $\mathcal{O}(i \cdot m \cdot l^2)$. The overall complexity is of the order

$$(m-n)^2 l^2 + \sum_{i=1}^{m-n+1} i \cdot m \cdot l^2 \sim (m-n+1)^2 \cdot m \cdot l^2 . \qquad \square$$

As we will see later, pseudoremainders can be used in a generalization of Euclid's algorithm. The following is an important technical requirement for this generalization.

**Lemma 2.2.4.** Let $R, a(x), b(x), m, n$ be as in Theorem 2.2.2. Let $\alpha, \beta \in R$. Then $\text{pquot}(\alpha \cdot a, \beta \cdot b) = \beta^{m-n} \cdot \alpha \cdot \text{pquot}(a, b)$ and $\text{prem}(\alpha \cdot a, \beta \cdot b) = \beta^{m-n+1} \cdot \alpha \cdot \text{prem}(a, b)$.

Evaluation

Finally we consider the problem of evaluating polynomials. Let $p(x) = p_n x^n + \ldots + p_0 \in R[x]$ for a commutative ring $R$ and $a \in R$. We want to compute $p(a)$.

Successive computation and addition of $p_0, p_1 x, \ldots, p_n x^n$ requires $2n-1$ multiplications and $n$ additions in $R$. A considerable improvement is obtained by *Horner's rule*, which evaluates $p$ at $a$ according to the scheme

$$p(a) = (\ldots(p_n \cdot a + p_{n-1}) \cdot a + \ldots) \cdot a + p_0 ,$$

requiring $n$ multiplications and $n$ additions in $R$. One get's Horner's rule from the computation of $\text{rem}(p, x-a)$, by using the relation $p(a) = \text{rem}(p, x-a)$. In fact, $p(a) = \text{rem}(p, f)(a)$ for every polynomial $f$ with $f(a) = 0$. In particular, for $f(x) = x^2 - a^2$ one gets the *2nd order Horner's rule*, which evaluates the polynomial

$$p(x) = \underbrace{\sum_{j=0}^{\lfloor n/2 \rfloor} p_{2j} x^{2j}}_{p^{(\text{even})}} + \underbrace{\sum_{j=0}^{\lceil n/2 \rceil -1} p_{2j+1} x^{2j+1}}_{p^{(\text{odd})}}$$

at $a$ as

$$p^{(\text{even})} = (\ldots(p_{2\lfloor n/2 \rfloor} \cdot a^2 + p_{2(\lfloor n/2 \rfloor -1)}) \cdot a^2 + \ldots) \cdot a^2 + p_0 ,$$
$$p^{(\text{odd})} = ((\ldots(p_{2\lceil n/2 \rceil -1} \cdot a^2 + p_{2\lceil n/2 \rceil -3}) \cdot a^2 + \ldots) \cdot a^2 + p_1) \cdot a .$$

The second order Horner's rule requires $n+1$ multiplications and $n$ additions in $R$, which is no improvement over the 1st order Horner's rule. However, if both $p(a)$ and $p(-a)$ are needed, then the second evaluation can be computed by just one more addition.

### Exercises

1. Analyze the complexity of adding $n$-variate polynomials $p, q \in \mathbb{Z}[x_1, \ldots, x_n]$ in dense representation, where $\deg_{x_i}(p), \deg_{x_i}(q) \leq d$ for $1 \leq i \leq n$ and $L(\max\{\|p\|_\infty, \|q\|_\infty\}) = l$.
2. With the notation of Exercise 1, analyze the complexity of multiplying $p$ and $q$ by (2.2.1).
3. Let $a(x), b(x) \in \mathbb{Q}[x]$, $b \neq 0$, $m = \deg(a)$, $n = \deg(b)$, $m \geq n$, and $L(v) \leq D$ for every coefficient $v$ of the numerator or denominator of $a$ or $b$. What is the complexity of POL_DIVK executed on $a$ and $b$?
4. Prove Lemma 2.2.4.
5. What is the complexity of POL_DIVP, if we count only the number of ring operations in $R$?
6. Derive Horner's rule and the 2nd order Horner's rule from division by $x - a$ and $x^2 - a^2$.

## 2.3 Quotient fields

Let $D$ be an integral domain and $Q(D)$ its quotient field. The arithmetic operations in $Q(D)$ can be based on (1.3.7). If $D$ is actually a Euclidean domain, then we can compute normal forms of quotients by eliminating the gcd of the numerator and the denominator. We say that $r \in Q(D)$ is *in lowest terms* if numerator and denominator of $r$ are relatively prime. The rational numbers $\mathbb{Q} = Q(\mathbb{Z})$ and the rational functions $K(x) = Q(K[x])$, for a field $K$, are important examples of such quotient fields.

The efficiency of arithmetic depends on a clever choice of when exactly the gcd is eliminated in the result. In a classical approach numerators and denominators are computed according to (1.3.7) and afterwards the result is transformed into lowest terms. P. Henrici (1956) has devised the fastest known algorithms for arithmetic in such quotients fields. The so-called *Henrici algorithms* for addition and multiplication of $r_1/r_2$ and $s_1/s_2$ in $Q(D)$ rely on the following facts.

**Theorem 2.3.1.** Let $D$ be a Euclidean domain, $r_1, r_2, s_1, s_2 \in D$, $\gcd(r_1, r_2) = \gcd(s_1, s_2) = 1$.
a. If $d = \gcd(r_2, s_2)$, $r_2' = r_2/d$, $s_2' = s_2/d$, then $\gcd(r_1 s_2' + s_1 r_2', r_2 s_2') = \gcd(r_1 s_2' + s_1 r_2', d)$.
b. If $d_1 = \gcd(r_1, s_2)$, $d_2 = \gcd(s_1, r_2)$, $r_1' = r_1/d_1$, $r_2' = r_2/d_2$, $s_1' = s_1/d_2$, $s_2' = s_2/d_1$, then $\gcd(r_1' s_1', r_2' s_2') = 1$.

**Algorithm QF_SUMH**(in: $r = (r_1, r_2)$, $s = (s_1, s_2)$; out: $t = (t_1, t_2)$);
[$r, s \in Q(D)$ in lowest terms. $t$ is a representation of $r + s$ in lowest terms.]
1. if $r_1 = 0$ then $\{t := s;$ return$\}$;
   if $s_1 = 0$ then $\{t := r;$ return$\}$;
2. $d := \gcd(r_2, s_2)$;

3. if $d = 1$
   then $\{t_1 := r_1 s_2 + r_2 s_1;\ t_2 := r_2 s_2\}$
   else
     $\{r_2' := r_2/d;\ s_2' := s_2/d;\ t_1' := r_1 s_2' + s_1 r_2';\ t_2' := r_2 s_2';$
     if $t_1' = 0$
     then $\{t_1 := 0;\ t_2 := 1\}$
     else $\{e := \gcd(t_1', d);$
       if $e = 1$
       then $\{t_1 := t_1';\ t_2 := t_2'\}$
       else $\{t_1 := t_1'/e;\ t_2 := t_2'/e\}\ \}\ \}$
   return.

Since the majority of the computing time is spent in extracting the gcd from the result, the Henrici algorithms derive their advantage from replacing one gcd computation of large inputs by several gcd computations for smaller inputs.

**Algorithm QF_MULTH**(in: $r = (r_1, r_2), s = (s_1, s_2)$; out: $t = (t_1, t_2)$);
[$r, s \in Q(D)$ in lowest terms. $t$ is a representation of $r \cdot s$ in lowest terms.]
1. if $r_1 = 0$ or $s_1 = 0$ then $\{t_1 := 0;\ t_2 := 1;$ return$\}$;
2. $d_1 := \gcd(r_1, s_2);\ d_2 := \gcd(s_1, r_2);$
3. if $d_1 = 1$
   then $\{r_1' := r_1;\ s_2' := s_2\}$
   else $\{r_1' := r_1/d_1;\ s_2' := s_2/d_1\}$;
   if $d_2 = 1$
   then $\{s_1' := s_1;\ r_2' := r_2\}$
   else $\{s_1' := s_1/d_2;\ r_2' := r_2/d_2\}$;
4. $t_1 := r_1' s_1';\ t_2 := r_2' s_2';$
   return.

Let us compare the complexities of the classical algorithm QF_SUMC versus the Henrici algorithm for addition in $\mathbb{Q}$. We will only take into account the gcd computations, since they are the most expensive operations in any algorithm.

Suppose $r = r_1/r_2, s = s_1/s_2 \in \mathbb{Q}$ and the numerators and denominators are bounded by $n$ in length. In QF_SUMC we have to compute a gcd of 2 integers of length $2n$ each. In QF_SUMH we first compute a gcd of 2 integers of length $n$ each, and, if $d = \gcd(r_2, s_2) \neq 1$ and $k = L(d)$, a gcd of integers of length $2n - k$ and $k$, respectively. We will make use of the complexity function for gcd computation stated in Theorem 2.1.9, i.e., $t^+_{\mathrm{INT\_GCD}}(l_1, l_2, k)$ is $\mathcal{O}(\min(l_1, l_2) \cdot (\max(l_1, l_2) - k + 1))$.

If $d = 1$, then the computing time for the gcd in QF_SUMC is roughly $4n^2$, whereas the gcd in QF_SUMH takes time roughly $n^2$. So QF_SUMH is faster than QF_SUMC by a factor of 4.

Now let us assume that $d \neq 1$, $k = n/2$, and $e = 1$. In this case the computation time for the gcd in QF_SUMC is $2n(2n - n/2) = 3n^2$. The times for the gcd computations in QF_SUMH are $n(n - n/2) = n^2/2$ and $(n/2)(3n/2) = 3n^2/4$. So in this case QF_SUMH is faster than QF_SUMC by a factor of 12/5.

The advantage of the Henrici algorithms over the classical ones becomes even more pronounced with increasing costs of gcd computations, e.g., in multivariate function fields like $\mathbb{Q}(x_1, \ldots, x_n) = Q(\mathbb{Q}[x_1, \ldots, x_n])$.

Whether we use the classical algorithms or the Henrici algorithms for arithmetic in $Q(D)$, it is clear that arithmetic in the quotient field is considerably more expensive than the arithmetic in the underlying domain. So whenever possible, we will try to avoid working in $Q(D)$. In particular, this is possible in gcd computation and factorization of polynomials over the integers.

### Exercises

1. Prove Theorem 2.3.1.
2. Apply the classical and the Henrici algorithms for computing
   a. $1089/140 + 633/350$ and
   b. $\frac{x^3+x^2-x-1}{x^2-4} \cdot \frac{x^2+5x+6}{x^3-x^2-x+1}$.
3. Compare the classical and the Henrici algorithm for multiplication in $\mathbb{Q}$ in a similar way as we have done for addition.
4. Conduct a series of experiments to find out about the mean increase in length in addition of rational numbers.

## 2.4 Algebraic extension fields

Let $K$ be a field and $\alpha$ algebraic over $K$. Let $f(x) \in K[x]$ be the minimal polynomial of $\alpha$ and $m = \deg(f)$. For representing the elements in the algebraic extension field $K(\alpha)$ of $K$ we use the isomorphism $K(\alpha) \cong K[x]_{/\langle f(x)\rangle}$. Every polynomial $p(x)$ can be reduced modulo $f(x)$ to some $r(x)$ with $\deg(r) < m$. On the other hand, two different polynomials $r(x), s(x)$ with $\deg(r), \deg(s) < m$ cannot be congruent modulo $f(x)$, since otherwise $r - s$, a non-zero polynomial of degree less than $m$, would be a multiple of $f$. Thus, every element $a \in K(\alpha)$ has a unique representation

$$a = \underbrace{a_{m-1}x^{m-1} + \cdots + a_1 x + a_0}_{a(x)} + \langle f(x)\rangle, \qquad a_i \in K\ .$$

We call $a(x)$ the *normal representation* of $a$, and sometimes we also write $a(\alpha)$.

From this unique normal representation we can immediately deduce that $K(\alpha)$ is a vector space over $K$ of dimension $m$ and $\{1, \alpha, \alpha^2, \ldots, \alpha^{m-1}\}$ is a basis of this vector space.

Consider, for instance, the field $\mathbb{Q}$ and let $\alpha$ be a root of $x^3 - 2$. $\mathbb{Q}(\alpha) = \mathbb{Q}[x]_{/\langle x^3-2\rangle}$ is an algebraic extension field of $\mathbb{Q}$, in which $x^3 - 2$ has a root, namely $\alpha$, whose normal representation is $x$. So $\mathbb{Q}(\alpha) = \mathbb{Q}(\sqrt[3]{2})$ can be represented as $\{a_2x^2 + a_1x + a_0 \mid a_i \in \mathbb{Q}\}$. In $\mathbb{Q}(\alpha)[x]$ the polynomial $x^3 - 2$ factors into $x^3 - 2 = (x - \alpha)(x^2 + \alpha x + \alpha^2)$.

Addition and subtraction in $K(\alpha)$ can obviously be carried out by simply adding and subtracting the normal representations of the arguments. The result is again a normal representation. Multiplication of normal representations and subsequent reduction (i.e., remainder computation) by the minimal polynomial $f(x)$ yields the normal representation of the product of two elements in $K(\alpha)$.

If we assume that the complexity of field operations in $K$ is proportional to 1, then the complexity for addition and subtraction in $K(\alpha)$ is $\mathcal{O}(m)$. The complexity of multiplication is dominated by the complexity of the reduction modulo $f(x)$. A polynomial of degree $< 2m$ has to be divided by a polynomial of degree $m$, i.e., the complexity of multiplication in $K(\alpha)$ is $\mathcal{O}(m^2)$.

The inverse $a^{-1}$ of $a \in K(\alpha)$ can be computed by an application of the extended Euclidean algorithm E_EUCLID (see Sect. 3.1) to the minimal polynomial $f(x)$ and the normal representation $a(x)$ of $a$. Since $f(x)$ and $a(x)$ are relatively prime, the extended Euclidean algorithm yields the gcd 1 and linear factors $u(x), v(x) \in K[x]$ such that

$$u(x)f(x) + v(x)a(x) = 1 \quad \text{and} \quad \deg(v) < m .$$

So $v(x)$ is the normal representation of $a^{-1}$.

For example, let $\mathbb{Q}(\alpha)$ be as above, i.e., $\alpha$ a root of $x^3 - 2$. Let $a, b \in \mathbb{Q}(\alpha)$ with normal representations $a(x) = 2x^2 - x + 1$ and $b(x) = x + 2$, respectively. Then $a + b = 2x^2 + 3$, $a \cdot b = \mathrm{rem}(2x^3 + 3x^2 - x + 2, x^3 - 2) = 3x^2 - x + 6$. For computing $a^{-1}$ we apply E_EUCLID to $x^3 - 2$ and $a(x)$, getting

$$\tfrac{1}{43}(2x - 19)(x^3 - 2) + \tfrac{1}{43}(-x^2 + 9x + 5)a(x) = 1 .$$

So $a^{-1}$ has the normal representation $(-x^2 + 9x + 5)/43$.

An algebraic extension $K(\alpha)$ over $K$ with minimal polynomial $f(x)$ is separable if and only if $f(x)$ has no multiple roots or, in other words, $f'(x) \neq 0$. In characteristic 0 every algebraic extension is separable. Let $K(\alpha_1)\ldots(\alpha_n)$ be a multiple algebraic extension of $K$. So $\alpha_i$ is the root of an irreducible polynomial $f_i(x) \in K(\alpha_1)\ldots(\alpha_{i-1})[x]$. For every such multiple separable algebraic field extension there exists an algebraic element $\gamma$ over $K$ such that

$$K(\alpha_1)\ldots(\alpha_n) = K(\gamma) ,$$

i.e., every multiple separable algebraic extension can be rewritten as a simple algebraic extension. $\gamma$ is a *primitive element* of this algebraic extension. For an algorithmic determination of primitive elements we refer to Sect. 5.4.

## Exercises

1. Let $R$ be the ring $\mathbb{Q}[x]_{/\langle f(x)\rangle}$, where $f(x) = x^3 - x^2 + 2x - 2$. Decide whether $p(x)$ (or, more precisely, the equivalence class of $p(x)$) has an inverse in $R$, and if so, compute $p^{-1}$.
   a. $p(x) = x^2 + x + 1$,
   b. $p(x) = x^2 + x - 2$.
2. Let $\mathbb{Z}_5(\alpha)$ be the algebraic extension of $\mathbb{Z}_5$ by a root $\alpha$ of the irreducible

polynomial $x^5 + 4x + 1$. Compute the normal representation of $(a \cdot b)/c$, where $a = \alpha^3 + \alpha + 2$, $b = 3\alpha^4 + 2\alpha^2 + 4\alpha$, $c = 2\alpha^4 + \alpha^3 + 2\alpha + 1$.

## 2.5 Finite fields

### Modular arithmetic in residue class rings

Every integer $m$ generates an ideal $\langle m \rangle$ in $\mathbb{Z}$. Two integers $a, b$ are *congruent* modulo $\langle m \rangle$ iff $a - b \in \langle m \rangle$, or in other words, $m | a - b$. In this case we write

$$a \equiv b \bmod m \quad \text{or} \quad a \equiv_{\bmod m} b \ .$$

So obviously $a$ and $b$ are congruent modulo $m$ if and only if they have the same residue modulo $m$. Since $\equiv_{\bmod m}$ is an equivalence relation, we get a decomposition of the integers into equivalence classes, the *residue classes* modulo $m$.

Let us consider the residue classes of integers modulo any positive integer $m$. In general, this is not a field, not even an integral domain, but just a commutative ring. This commutative ring is called the *residue class ring modulo m* and it is denoted by $\mathbb{Z}_{/\langle m \rangle}$ or just $\mathbb{Z}_{/m}$. The residue class ring is a field if and only if the modulus $m$ is a prime number.

For the purpose of computation in such a residue class ring we need to choose representations of the elements. There are two natural representations for $\mathbb{Z}_{/m}$, namely as the residue classes corresponding to

$$\{0, 1, \ldots, m-1\} \qquad \text{(least non-negative representation)}$$

or the residue classes corresponding to

$$\left\{a \in \mathbb{Z} \;\middle|\; -\frac{m}{2} < a \leq \frac{m}{2}\right\} \qquad \text{(zero-centered representation).}$$

Both representations are useful in specific applications. Of course a change of representations is trivial.

The canonical homomorphism $H_m$ which maps an integer $a$ to its representation in $\mathbb{Z}_{/m}$ is simply the computation of the remainder of $a$ w.r.t. $m$. So according to Theorem 2.1.7 it takes time proportional to $L(m)(L(a)-L(m)+1)$.

Addition $+_m$ and multiplication $\cdot_m$ in $\mathbb{Z}_{/m}$ are defined as follows on the representatives

$$a +_m b = H_m(a+b), \qquad a \cdot_m b = H_m(a \cdot b) \ .$$

Using this definition and the bounds of Sect. 2.1, we see that the obvious algorithms MI_SUM and MI_MULT for $+_m$ and $\cdot_m$, respectively, have the complexities $t^+_{\text{MI\_SUM}}, t^*_{\text{MI\_SUM}} \preceq L(m)$ and $t^+_{\text{MI\_MULT}} \preceq L(m)^2 + L(m)(2L(m) - L(m) + 1) \sim L(m)^2$. So even if we use faster multiplication methods, the bound for MI_MULT does not decrease.

An element $a \in \mathbb{Z}_{/m}$ has an inverse if and only if $\gcd(m, a) = 1$, otherwise $a$ is a zero-divisor. So we can decide whether $a$ can be inverted and if so compute $a^{-1}$ by applying INT_GCDEE to $m$ and $a$. If $a$ is invertible we will get a linear combination $u \cdot m + v \cdot a = 1$, and $v$ will be the inverse of $a$. The time complexity for computing the inverse is $\mathcal{O}(L(m)^2)$.

An important property of modular arithmetic is captured in Fermat's "little" theorem. Theoretically this provides an alternative method for computing inverses modulo primes.

**Theorem 2.5.1** (Fermat's little theorem). If $p$ is a prime and $a$ is an integer not divisible by $p$, then $a^{p-1} \equiv 1 \bmod p$.

Arithmetic in finite fields

Now let us turn to finite fields. As we have seen above, $\mathbb{Z}_{/p}$ will be a finite field if and only if $p$ is a prime number. In general a finite field need not have prime cardinality. However, every finite field has cardinality $p^n$, for $p$ a prime. On the other hand, for every prime $p$ and natural number $n$ there exists a unique (up to isomorphism) finite field of order $p^n$. This field is usually denoted $GF(p^n)$, the *Galois field* of order $p^n$.

From what we have derived in previous sections, it is not difficult to construct Galois fields and to compute in them. Let $f(x)$ be an irreducible polynomial of degree $n$ in $\mathbb{Z}_p[x]$. We will see later how to check irreducibility efficiently, and in fact for every choice of $p$ and $n$ there is a corresponding $f$. Then $f$ determines an algebraic field extension of $\mathbb{Z}_p$ of degree $n$, i.e.,

$$GF(p^n) \cong \mathbb{Z}_p[x]_{/\langle f(x) \rangle} \ .$$

So the arithmetic operations can be handled as in any algebraic extension field.

For a thorough introduction to the theory of finite fields, we refer to Lidl and Niederreiter (1983) and Lidl and Pilz (1984). Here we list only some facts that will be useful in subsequent chapters.

$GF(p^n)$ is the splitting field of $x^{p^n} - x$ over $\mathbb{Z}_p$, i.e.,

$$x^{p^n} - x = \prod_{a \in GF(p^n)} (x - a) \ .$$

Every $\beta \in GF(p^n)$ is algebraic over $\mathbb{Z}_p$. If $s$ is the smallest positive integer such that $\beta^{p^s} = \beta$, then $m_\beta(x) = \prod_{i=0}^{s-1}(x - \beta^{p^i})$ is the minimal polynomial of $\beta$ over $\mathbb{Z}_p$. The multiplicative group of $GF(p^n)$ is cyclic. A generating element of this cyclic group is called a *primitive element* of the Galois field.

An important property of Galois fields is the "freshman's dream." In fact, this theorem holds for any field of characteristic $p$.

**Theorem 2.5.2.** Let $a(x), b(x) \in GF(p^n)[x]$. Then $(a(x) + b(x))^p = a(x)^p + b(x)^p$.

*Proof.* In the binomial expansion of the left-hand side

$$\sum_{i=0}^{p} \binom{p}{i} a(x)^i b(x)^{p-i}$$

all the binomial coefficients except the first and the last are divisible by $p$ and for those we have $\binom{p}{0} = 1 = \binom{p}{p}$. □

**Corollary.** Let $a(x) \in \mathbb{Z}_p[x]$. Then $a(x)^p = a(x^p)$.

*Proof.* Let $a(x) = \sum_{i=0}^{m} a_i x^i$. By Theorem 2.5.2 and Fermat's little theorem we have $a(x)^p = \sum_{i=0}^{m} (a_i x^i)^p = \sum_{i=0}^{m} a_i x^{ip} = a(x^p)$. □

*Example 2.5.1.* Let us carry out some of these constructions in $GF(2^4)$. The polynomial $f(x) = x^4 + x + 1$ is irreducible over $\mathbb{Z}_2$, so

$$GF(2^4) \cong \mathbb{Z}_2[x]_{/\langle x^4+x+1 \rangle}$$

**Table 1.** $GF(2^4)$. Elements $\beta = b_0 + b_1\alpha + b_2\alpha^2 + b_3\alpha^3$ of $\mathbb{Z}_2(\alpha)$, where $\alpha^4 + \alpha + 1 = 0$

| $\beta$ | $b_0$ | $b_1$ | $b_2$ | $b_3$ | minimal polynomial of $\beta$ over $\mathbb{Z}_2$ |
|---|---|---|---|---|---|
| $0$ | 0 | 0 | 0 | 0 | $x$ |
| $1$ | 1 | 0 | 0 | 0 | $x + 1$ |
| $\alpha$ | 0 | 1 | 0 | 0 | $x^4 + x + 1$ |
| $\alpha^2$ | 0 | 0 | 1 | 0 | $x^4 + x + 1$ |
| $\alpha^3$ | 0 | 0 | 0 | 1 | $x^4 + x^3 + x^2 + x + 1$ |
| $1 + \alpha = \alpha^4$ | 1 | 1 | 0 | 0 | $x^4 + x + 1$ |
| $\alpha + \alpha^2 = \alpha^5$ | 0 | 1 | 1 | 0 | $x^2 + x + 1$ |
| $\alpha^2 + \alpha^3 = \alpha^6$ | 0 | 0 | 1 | 1 | $x^4 + x^3 + x^2 + x + 1$ |
| $1 + \alpha + \alpha^3 = \alpha^7$ | 1 | 1 | 0 | 1 | $x^4 + x^3 + 1$ |
| $1 + \alpha^2 = \alpha^8$ | 1 | 0 | 1 | 0 | $x^4 + x + 1$ |
| $\alpha + \alpha^3 = \alpha^9$ | 0 | 1 | 0 | 1 | $x^4 + x^3 + x^2 + x + 1$ |
| $1 + \alpha + \alpha^2 = \alpha^{10}$ | 1 | 1 | 1 | 0 | $x^2 + x + 1$ |
| $\alpha + \alpha^2 + \alpha^3 = \alpha^{11}$ | 0 | 1 | 1 | 1 | $x^4 + x^3 + 1$ |
| $1 + \alpha + \alpha^2 + \alpha^3 = \alpha^{12}$ | 1 | 1 | 1 | 1 | $x^4 + x^3 + x^2 + x + 1$ |
| $1 + \alpha^2 + \alpha^3 = \alpha^{13}$ | 1 | 0 | 1 | 1 | $x^4 + x^3 + 1$ |
| $1 + \alpha^3 = \alpha^{14}$ | 1 | 0 | 0 | 1 | $x^4 + x^3 + 1$ |
| $1 = \alpha^{15}$ | 1 | 0 | 0 | 0 | $x + 1$ |

Let $\alpha$ be a root of $f$. Every $\beta \in GF(2^4)$ has a unique representation as

$$\beta = b_0 + b_1\alpha + b_2\alpha^2 + b_3\alpha^3, \qquad b_i \in \mathbb{Z}_2 .$$

For computing the minimal polynomial $m_\beta(x)$ for $\beta = \alpha^6$, we consider the powers $\beta^2 = \alpha^{12}$, $\beta^4 = \alpha^9$, $\beta^8 = \alpha^3$, $\beta^{16} = \alpha^6 = \beta$ (Table 1), so

$$m_\beta(x) = \prod_{i=0}^{3}(x - \beta^{p^i}) = x^4 + x^3 + x^2 + x + 1 .$$

Every $\beta \in GF(2^4)$ is a power of $\alpha$, so $\alpha$ is a primitive element of $GF(2^4)$. However, not every irreducible polynomial has a primitive element as a root. For instance, $g(x) = x^4 + x^3 + x^2 + x + 1$ also is irreducible over $\mathbb{Z}_2[x]$, so $GF(2^4) \cong \mathbb{Z}_2[x]_{/\langle g(x)\rangle}$. But $\beta$, a root of $g$, is not a primitive element of $GF(16)$, since $\beta^5 = 1$.

### Exercises

1. Check the "freshman's dream" by computing both sides of $(a(x) + b(x))^2 = a(x)^2 + b(x)^2$, where $a(x), b(x)$ are the polynomials $a(x) = (y^3 + y)x^3 + (y^2 + y)x^2$, $\quad b(x) = (y^2 + y + 1)x^2 + (y^3 + 1)$ over the Galois field $GF(2^4) = \mathbb{Z}_2[y]_{/\langle y^4+y+1\rangle}$.
2. What is the complexity of computing $a(x)^p$ in $\mathbb{Z}_p[x]$, where $\deg(a) = m$?
3. Consider the finite field $\mathbb{Z}_3(\alpha) = \mathbb{Z}_3[x]_{/\langle x^4+x+2\rangle}$ $(\cong GF(3^4))$. Determine the minimal polynomial for $\beta = 2\alpha^3 + \alpha$.

## 2.6 Bibliographic notes

Further material on arithmetic in basic domains can be found in Heindel (1971), Caviness and Collins (1976), Collins et al. (1983), and Jebelean (1993). Complexity of polynomial multiplication over finite fields is considered in Kaminski and Bshouty (1989), sparse polynomial division in Kaminski (1987). Handling of power series in computer algebra is described in Koepf (1992).

# 3 Computing by homomorphic images

## 3.1 The Chinese remainder problem and the modular method

The Chinese remainder method has already been investigated by Chinese mathematicians more than 2000 years ago. For a short introduction to the history we refer to Knuth (1981). The main idea consists of solving a problem over the integers by solving this problem in several homomorphic images modulo various primes, and afterwards combining the solutions of the modular problems to a solution of the problem over the integers. In fact, the method can be generalized to work over arbitrary Euclidean domains, i.e., domains in which we can compute greatest common divisors by the Euclidean algorithm. An interesting list of different statements of the Chinese remainder theorem is given in Davis and Hersh (1981).

### Euclidean domains

*Definition 3.1.1.* A *Euclidean domain (ED)* $D$ is an integral domain together with a *degree function* deg: $D^* \to \mathbb{N}_0$, such that

a. $\deg(a \cdot b) \geq \deg(a)$ for all $a, b \in D^*$,

b. (division property) for all $a, b \in D$, $b \neq 0$, there exists a *quotient* $q$ and a *remainder* $r$ in $D$ such that $a = q \cdot b + r$ and ($r = 0$ or $\deg(r) < \deg(b)$).

When we write "$r = a \bmod b$" we mean that $r$ is a remainder of $a$ and $b$ as in (a). In other words, the function mod $b$ returns a remainder of its argument modulo $b$. In the same way we will consider functions quot and rem, yielding $q$ and $r$, respectively, for inputs $a$ and $b$ as in Definition 3.1.1.

*Example 3.1.1.* a. $\mathbb{Z}$ with $\deg(a) = |a|$ is an ED. If $a = q \cdot b + r$ and $0 < r < |b|$, then also $q + 1$ and $r - b$ are a possible pair of quotient and remainder for $a$ and $b$. So in an ED quotients and remainders are not uniquely defined.

b. Every field $K$ with $\deg(a) = 1$ for all $a \in K^*$ is an ED.

c. For every field $K$ the univariate polynomial ring $K[x]$, where the degree function deg returns the usual degree of a polynomial (canonical degree function), is an ED. In fact, quotient and remainder can be computed by the algorithm POL_DIVK.

d. If the coefficient ring of $R[x]$ is not a field, then $R[x]$ with the canonical degree function is not an ED. Consider $ax = q \cdot (bx) + r$, where $a, b \in R$. For $q$ and $r$ to be a quotient and remainder of $ax$ and $bx$, $q$ would have to be an element of $R$ satisfying $a = q \cdot b$. This equation, however, is not solvable for arbitrary $a, b \in R^*$. So, for instance, polynomials over the integers and multivariate polynomials over a field together with the canonical degree function do not form Euclidean domains.

In an ED $D$ we have $\deg(1) \leq \deg(a)$ for all non-zero $a$, and $\deg(1) = \deg(a)$ if and only if $a$ is a unit. If $c$ is not a unit and non-zero and $a = b \cdot c$, then $\deg(b) < \deg(a)$.

**Theorem 3.1.1.** Any two non-zero elements $a, b$ of an ED $D$ have a greatest common divisor $g$ which can be written as a linear combination $g = s \cdot a + t \cdot b$ for some $s, t \in D$.

*Proof.* Let $I = \langle a, b \rangle$, the ideal generated by $a, b$ in $D$. Let $g$ be a non-zero element of $I$ with minimal degree, i.e., for all $c \in I^*$, $\deg(g) \leq \deg(c)$. So $g = s \cdot a + t \cdot b$ for some $s, t \in D$. Obviously $\langle g \rangle \subseteq I$. On the other hand, let $c \in I$. There are a quotient $q$ and a remainder $r$ such that $c = q \cdot g + r$ and $r = 0$ or $\deg(r) < \deg(g)$. But $r \in I$, so we have $r = 0$. Thus, $\langle g \rangle = I$. So $g$ is a common divisor of $a$ and $b$. Now let $c$ be any common divisor of $a$ and $b$. Then $c$ divides $s \cdot a + t \cdot b = g$. □

*Definition 3.1.2.* For $a, b \in D^*$, $D$ an ED, and $g = \gcd(a, b)$, the equation $g = s \cdot a + t \cdot b$ is called the *Bezout equality* and $s, t$ are *Bezout cofactors.*

The Bezout equality can obviously be generalized to arbitrary elements of $D$, if we set $\gcd(0, 0) = 0$.

Using Theorem 3.1.1 it is rather straightforward to show that any ED is a ufd, cf. Exercise 3. So in an ED any non-zero $a$ can be written as $a = p_1 \cdot \ldots \cdot p_s$, where $p_1, \ldots, p_s$ are prime. This representation is unique up to units and reordering of the factors.

An ED in which quotient and remainder are computable by algorithms quot and rem admits an algorithm for computing the greatest common divisor $g$ of any two elements $a, b$. This algorithm has originally been stated by Euclid for the domain of the integers. In fact, it can be easily extended to compute not only the gcd but also the coefficients $s, t$ in the linear combination $g = s \cdot a + t \cdot b$, i.e., the Bezout cofactors.

**Algorithm E_EUCLID**(in: $a, b$; out: $g, s, t$);
[$a, b$ are elements of the Euclidean domain $D$; $g$ is the greatest common divisor of $a, b$ and $g = s \cdot a + t \cdot b$]

1. $(r_0, r_1, s_0, s_1, t_0, t_1) := (a, b, 1, 0, 0, 1)$;
   $i := 1$;
2. while $r_i \neq 0$ do
   $\{q_i := \text{quot}(r_{i-1}, r_i)$;
   $(r_{i+1}, s_{i+1}, t_{i+1}) := (r_{i-1}, s_{i-1}, t_{i-1}) - q_i \cdot (r_i, s_i, t_i)$;
   $i := i + 1\}$;
3. $(g, s, t) := (r_{i-1}, s_{i-1}, t_{i-1})$; return.

The extended Euclidean algorithm, E_EUCLID, terminates, because $\deg(r_i)$ decreases in every iteration. Throughout the algorithm the relation $r_i = s_i \cdot a + t_i \cdot b$ is preserved, so at termination we have $g = s \cdot a + t \cdot b$. Also, throughout the algorithm $\gcd(r_i, r_{i+1}) = \gcd(r_{i+1}, r_{i+2})$. So, when finally a remainder $r_i = 0$ is reached, then the previous $r_{i-1}$ must be the desired greatest common divisor.

**Theorem 3.1.2.** Let $K$ be a field, $a, b \in K[x]$, $\deg(a) \geq \deg(b) > 0$, $a$ and $b$ not associates. Let $g, s, t$ be the result of applying E_EUCLID to $a$ and $b$. Then $\deg(s) < \deg(b) - \deg(g)$ and $\deg(t) < \deg(a) - \deg(g)$.

*Proof.* Let $r_0, r_1, \ldots, r_{k-1}, r_k = 0$ be the sequence of remainders computed by E_EUCLID, and similarly $q_1, \ldots, q_{k-1}$ the sequence of quotients and $s_0, s_1, \ldots, s_k, t_0, t_1, \ldots, t_k$ the sequences of linear coefficients. Obviously $\deg(q_i) = \deg(r_{i-1}) - \deg(r_i)$ for $1 \leq i \leq k-1$.

For $k = 2$ the statement obviously holds. If $k > 2$, then for $2 \leq i \leq k-1$ we have $\deg(r_i) = \deg(r_1) - \sum_{l=2}^{i} \deg(q_l) < \deg(r_1) - \sum_{l=2}^{i-1} \deg(q_l)$, $\deg(s_i) \leq \sum_{l=2}^{i-1} \deg(q_l)$ and $\deg(t_i) \leq \sum_{l=1}^{i-1} \deg(q_l)$. So $\deg(r_i) + \deg(s_i) < \deg(r_1)$ and $\deg(r_i) + \deg(t_i) < \deg(r_1) + \deg(q_1)$ for $2 \leq i \leq k-1$. For $i = k-1$ we get the desired result. □

**Corollary.** Let $a, b \in K[x]$ be relatively prime, $c \in K[x]^*$ such that $\deg(c) < \deg(a \cdot b)$. Then $c$ can be represented uniquely as $c = u \cdot a + v \cdot b$, where $\deg(u) < \deg(b)$ and $\deg(v) < \deg(a)$.

*Proof.* By Theorem 3.1.2 we can write $1 = u \cdot a + v \cdot b$, where $\deg(u) < \deg(b)$ and $\deg(v) < \deg(a)$.

Obviously $c = (c \cdot u) \cdot a + (c \cdot v) \cdot b$. If $c \cdot u$ or $c \cdot v$ do not satisfy the degree bounds, then we set $u' := \text{rem}(c \cdot u, b)$ and $v' := c \cdot v + \text{quot}(c \cdot u, b) \cdot a$. Now we have $c = u' \cdot a + v' \cdot b$ and $\deg(u') < \deg(b)$. From comparing coefficients of like powers we also see that $\deg(v') < \deg(a)$. This proves the existence of $u$ and $v$.

If $u_1, v_1$ and $u_2, v_2$ are two pairs of linear coefficients satisfying the degree constraints, then $(u_1 - u_2) \cdot a = (v_2 - v_1) \cdot b$. So $a$ divides $v_2 - v_1$. This is only possible if $v_2 - v_1 = 0$. Thus, the linear coefficients $u, v$ are uniquely determined. □

**Theorem 3.1.3.** Let $K$ be a field and $a, b \in K[x]$ with $\deg(a) \geq \deg(b) > 0$. Let $g$ be the greatest common divisor of $a$ and $b$. Then the number of arithmetic operations in $K$ required by E_EUCLID is $\mathcal{O}((\deg(a) - \deg(g) + 1) \cdot \deg(b))$.

*Proof.* Let $k$ be the length of the sequence of remainders computed by E_EUCLID, i.e., $g = r_{k-1}$. The complexity of the body of the "while" loop is dominated by the polynomial division for computing $\text{quot}(r_{i-1}, r_i)$. By Theorem 2.2.1 the number of arithmetic operations in this division is $\mathcal{O}((\deg(r_i) + 1) \cdot (\deg(r_{i-1}) - \deg(r_i) + 1)$. So the complexity of step (2), and also of the whole algorithm, is dominated by

$$\sum_{i=1}^{k-1}(\deg(r_i)+1)\cdot(\deg(r_{i-1})-\deg(r_i)+1)$$
$$\leq (\deg(b)+1)\cdot(\deg(a)-\deg(g)+k-1)\ .$$

Since $k-2 \leq \deg(a)-\deg(g)$, we get $\deg(b)\cdot(\deg(a)-\deg(g)+1)$ as a dominating function for the complexity of E_EUCLID. □

The Chinese remainder algorithm

For the remainder of this section we assume that $D$ is a Euclidean domain. The problem that we want to solve is the following.

*Chinese remainder problem (CRP)*
Given: $r_1, \ldots, r_n \in D$ (remainders)
$m_1, \ldots, m_n \in D^*$ (moduli), pairwise relatively prime
Find: $r \in D$, such that $r \equiv r_i \bmod m_i$ for $1 \leq i \leq n$

We indicate the size of the CRP by an appropriate index, i.e., $\text{CRP}_n$ is a CRP with $n$ remainders and moduli. We describe two solutions of the CRP. The first one is usually associated with the name of J. L. Lagrange. The second one is associated with I. Newton and is a recursive solution.

In the Lagrangian solution one first determines $u_{kj}$ such that

$$1 = u_{kj}\cdot m_k + u_{jk}\cdot m_j, \qquad \text{for } 1 \leq j,k \leq n,\ j \neq k\ .$$

This can obviously be achieved by the extended Euclidean algorithm. Next one considers the elements

$$l_k := \prod_{j=1, j\neq k}^{n} u_{jk}m_j, \qquad \text{for } 1 \leq k \leq n\ .$$

Clearly, $l_k \equiv 0 \bmod m_j$ for all $j \neq k$. On the other hand, $l_k = \prod_{j=1,j\neq k}^{n}(1 - u_{kj}m_k) \equiv 1 \bmod m_k$. So

$$r = \sum_{k=1}^{n} r_k \cdot l_k$$

solves CRP.

The disadvantage of the Lagrangian approach is that it yields a static algorithm, i.e., it is virtually impossible to increase the size of the problem by one more pair $r_{n+1}, m_{n+1}$ without having to recompute everything from the start. This is the reason why we do not investigate this approach further.

So now we consider the recursive Newton approach. Let us first deal with the special case $n = 2$, i.e., with $\text{CRP}_2$. For given remainders $r_1, r_2$ and moduli $m_1, m_2$ we want to find an $r \in D$ such that

1. $r \equiv r_1 \bmod m_1$,
2. $r \equiv r_2 \bmod m_2$.

The solutions of congruence (1) have the form $r_1 + \sigma m_1$ for arbitrary $\sigma \in D$. Moreover, we also have a solution of (2) if $r_1 + \sigma m_1 \equiv r_2 \bmod m_2$, i.e., if $\sigma m_1 \equiv r_2 - r_1 \bmod m_2$. By Theorem 3.1.1 there is a $c \in D$ such that $cm_1 \equiv 1 \bmod m_2$. So for $\sigma = (r_2 - r_1)c$ we also get a solution of (2). Thus, we have shown that $\mathrm{CRP}_2$ always has a solution. The obvious algorithm is called the *Chinese remainder algorithm.*

**Theorem 3.1.4** (Chinese remainder theorem). $\mathrm{CRP}_2$ always has a solution, which can be computed by the algorithm CRA_2.

**Algorithm CRA_2**(in: $r_1, r_2, m_1, m_2$; out: $r$);
[$r_1, r_2, m_1, m_2$ determine a $\mathrm{CRP}_2$ over $D$; $r$ solves the $\mathrm{CRP}_2$]
1. $c := m_1^{-1} \bmod m_2$;
2. $r_1' := r_1 \bmod m_1$;
3. $\sigma := (r_2 - r_1')c \bmod m_2$;
4. $r := r_1' + \sigma m_1$; return.

The general CRP of size $n$ can be solved by reducing it to CRPs of size 2. This reduction is based on the following facts, which will be proved in the Exercises.

**Lemma 3.1.5.** a. Let $m_1, \ldots, m_n \in D^*$ be pairwise relatively prime and let $M = \prod_{i=1}^{n-1} m_i$. Then $m_n$ and $M$ are relatively prime.

b. Let $r, r' \in D$, and $m_1, m_2 \in D^*$ be relatively prime. Then $r \equiv r' \bmod m_1$ and $r \equiv r' \bmod m_2$ if and only if $r \equiv r' \bmod m_1 m_2$.

So now let $R_2$ be a solution of the first two congruences of $\mathrm{CRP}_n$. Then the solutions of the original $\mathrm{CRP}_n$ are the same as the solutions of the following $\mathrm{CRP}_{n-1}$

$$r \equiv R_2 \bmod m_1 m_2 \;,$$
$$r \equiv r_i \bmod m_i, \qquad \text{for } i = 3, \ldots, n \;.$$

Iterating this process we finally arrive at a CRP of size 2 which can be solved by CRA_2.

**Theorem 3.1.6.** A $\mathrm{CRP}_n$ of any size $n$ always has a solution, which can be computed by the algorithm CRA_n.

**Algorithm CRA_n**(in: $r_1, \ldots, r_n, m_1, \ldots, m_n$; out: $r$);
[$r_1, \ldots, r_n, m_1, \ldots, m_n$ determine a $\mathrm{CRP}_n$ over $D$; $r$ solves the $\mathrm{CRP}_n$]
1. $M := m_1$;
2. $r := r_1 \bmod m_1$;
3. for $k = 2$ to $n$ do

$\{r := \text{CRA\_2}(r, r_k, M, m_k);$
$M := M \cdot m_k\};$
return.

*Example 3.1.2.* We want to find an integer $r$ such that
1. $r \equiv 3 \bmod 4$ ,
2. $r \equiv 5 \bmod 7$ ,
3. $r \equiv 2 \bmod 3$ .

We apply CRA_n in the Euclidean domain $\mathbb{Z}$. First, CRA_2 is applied to the congruences (1), (2): $c = 4^{-1} \bmod 7 = 2$, $\sigma = (5-3) \cdot 2 \bmod 7 = 4$, $r = 3 + 4 \cdot 4 = 19$. So $r = 19$ solves (1), (2). Now the problem is reduced to

$$(1,2)\ r \equiv 19 \bmod 28\ , \qquad (3)\ r \equiv 2 \bmod 3\ .$$

By another application of CRA_2 we get $c = 28^{-1} \bmod 3 = 1$, $\sigma = (2-19) \cdot 1 \bmod 3 = 1$, $r = 19 + 1 \cdot 28 = 47$. So $r = 47$ is the least positive solution of the CRP (1), (2), (3).

A CRP of size $n > 2$ could also be reduced to CRPs of size 2 by splitting the remainders and moduli into two groups

$$(r_1, \ldots, r_{\lfloor n/2 \rfloor},\ m_1, \ldots, m_{\lfloor n/2 \rfloor}) \quad \text{and}$$
$$(r_{\lfloor n/2 \rfloor+1}, \ldots, r_n,\ m_{\lfloor n/2 \rfloor+1}, \ldots, m_n)\ ,$$

recursively applying this splitting process to the problems of size $n/2$, solving the resulting CRPs of size 2, and combining the partial results again by solving CRPs of size 2. Such a reduction lends itself very naturally to parallelization. The disadvantage is that in order to add one more remainder $r_{n+1}$ and modulus $m_{n+1}$ to the problem, the separation into the two groups of equal size gets destroyed. That, however, is exactly the pattern of most applications in computer algebra.

The Chinese remainder problem can, in fact, be described in greater generality. Let $R$ be a commutative ring with unity 1.

*Abstract Chinese remainder problem*
Given: $r_1, \ldots, r_n \in R$ (remainders)
$I_1, \ldots, I_n$ ideals in $R$ (moduli), such that $I_i + I_j = R$ for all $i \neq j$
Find: $r \in R$, such that $r \equiv r_i \bmod I_i$ for $1 \leq i \leq n$

The abstract Chinese remainder problem can be treated basically in the same way as the CRP over Euclidean domains. Again there is a Lagrangian and a Newtonian approach and one can show that the problem always has a solution and if $r$ is a solution then the set of all solutions is given by $r + I_1 \cap \ldots \cap I_n$.

That is, the map $\phi$: $r \mapsto (r + I_1, \ldots, r + I_n)$ is a homomorphism from $R$ onto $\prod_{j=1}^{n} R_{/I_j}$ with kernel $I_1 \cap \ldots \cap I_n$. However, in the absence of the Euclidean algorithm it is not possible to compute a solution of the abstract CRP. See Lauer (1983).

A preconditioned Chinese remainder algorithm

If the CRA is applied in a setting where many conversions w.r.t. a fixed set of moduli have to be computed, it is reasonable to precompute all partial results depending on the moduli alone. This idea leads to a preconditioned CRA, as described in Aho et al. (1974).

**Theorem 3.1.7.** Let $r_1, \ldots, r_n$ and $m_1, \ldots, m_n$ be the remainders and moduli, respectively, of a CRP in the Euclidean domain $D$. Let $m$ be the product of all the moduli. Let $c_i = m/m_i$ and $d_i = c_i^{-1} \bmod m_i$ for $1 \leq i \leq n$. Then

$$r = \sum_{i=1}^{n} c_i d_i r_i \bmod m \tag{3.1.1}$$

is a solution to the corresponding CRP.

*Proof.* Since $c_i$ is divisible by $m_j$ for $j \neq i$, we have $c_i d_i r_i \equiv 0 \bmod m_j$ for $j \neq i$. Therefore

$$\sum_{i=1}^{n} c_i d_i r_i \equiv c_j d_j r_j \equiv r_j \bmod m_j, \qquad \text{for all } 1 \leq j \leq n . \qquad \square$$

A more detailed analysis of (3.1.1) reveals many common factors of the expressions $c_i d_i r_i$. Let us assume that $n$ is a power of 2, $n = 2^t$. Obviously, $m_1 \cdot \ldots \cdot m_{n/2}$ is a factor of $c_i d_i r_i$ for all $i > n/2$ and $m_{n/2+1} \cdot \ldots \cdot m_n$ is a factor of $c_i d_i r_i$ for all $i \leq n/2$. So we could write (3.1.1) as

$$r = \left(\sum_{i=1}^{n/2} c_i' d_i r_i\right) \cdot \prod_{i=n/2+1}^{n} m_i + \left(\sum_{i=n/2+1}^{n} c_i'' d_i r_i\right) \cdot \prod_{i=1}^{n/2} m_i , \tag{3.1.2}$$

where $c_i' = (m_1 \cdot \ldots \cdot m_{n/2})/m_i$ and $c_i'' = (m_{n/2+1} \cdot \ldots \cdot m_n)/m_i$. The expression (3.1.2) suggests a divide-and-conquer approach. The quantities we will use are

$$q_{ij} = \prod_{l=i}^{i+2^j-1} m_l \quad \text{and} \quad s_{ij} = \sum_{l=i}^{i+2^j-1} q_{ij} d_l r_l / m_l .$$

For $j = 0$ we have $s_{i0} = d_i r_i$, and for $j > 0$ we can compute $s_{ij}$ by the formula

$$s_{ij} = s_{i,j-1} \cdot q_{i+2^{j-1},j-1} + s_{i+2^{j-1},j-1} \cdot q_{i,j-1} \; .$$

Finally we reach $s_{1t} = r$, i.e., an evaluation of (3.1.2). These considerations lead to CRA_PC, a preconditioned CRA.

**Algorithm CRA_PC**(in: $r_1, \ldots, r_n, m_1, \ldots, m_n, d_1, \ldots, d_n$; out: $r$);
[$r_1, \ldots, r_n, m_1, \ldots, m_n$ determine a CRP$_n$ over $D$,
$d_i = (m/m_i)^{-1} \bmod m_i$ for $1 \leq i \leq n$, where $m = \prod_{j=1}^{n} m_j$,
$n = 2^t$ for some $t \in \mathbb{N}$; $r$ solves the CRP$_n$]
1. [compute the $q_{ij}$'s]
   for $i = 1$ to $n$ do $q_{i0} := m_i$;
   for $j = 1$ to $t$ do
     for $i = 1$ step $2^j$ to $n$ do
     $q_{ij} := q_{i,j-1} \cdot q_{i+2^{j-1},j-1}$;
2. [compute the $s_{ij}$'s]
   for $i = 1$ to $n$ do $s_{i0} := d_i \cdot r_i$;
   for $j = 1$ to $t$ do
     for $i = 1$ step $2^j$ to $n$ do
     $s_{ij} := s_{i,j-1} \cdot q_{i+2^{j-1},j-1} + s_{i+2^{j-1},j-1} \cdot q_{i,j-1}$;
3. $r := s_{1t}$; return.

A correctness proof for CRA_PC can be derived easily by induction on $j$.

*Example 3.1.3.* We execute CRA_PC over the integers for $(r_1, r_2, r_3, r_4) = (1, 2, 4, 3)$ and $(m_1, m_2, m_3, m_4) = (2, 3, 5, 7)$. The corresponding inverses are $(d_1, d_2, d_3, d_4) = (1, 1, 3, 4)$. We start by computing the values of the $q_{ij}$'s in step (1):

$$q_{10} = m_1 = 2, \quad q_{20} = m_2 = 3, \quad q_{30} = m_3 = 5, \quad q_{40} = m_4 = 7$$
$$q_{11} = m_1 m_2 = 6, \quad q_{31} = m_3 m_4 = 35$$
$$q_{12} = m_1 m_2 m_3 m_4 = 210$$

In step (2) the computation of the $s_{ij}$'s yields

$$s_{10} = 1, \quad s_{20} = 2, \quad s_{30} = 12, \quad s_{40} = 12$$
$$s_{11} = 7, \quad s_{31} = 144$$
$$s_{12} = 1109$$

So the result returned in step (3) is 1109, which is congruent to 59 modulo 210. In fact, we could reach the minimal solution 59 if in step (2) we took every $s_{ij}$ modulo $q_{ij}$.

CRA over the integers

When we apply the Chinese remainder algorithm CRA to integers, we can always assume that the moduli are positive and that the function mod $m$ returns the smallest positive remainder. The following is based on these assumptions.

**Theorem 3.1.8.** Over the integers the algorithm CRA_n computes the unique solution of the CRP satisfying $0 \leq r < \prod_{k=1}^{n} m_k$.

*Proof.* By inspection of CRA_2 we find that $0 \leq r_1' < m_1$ and $0 \leq \sigma < m_2$. So the result computed by CRA_2 satisfies $0 \leq r < m_1 m_2$. Since CRA_n is simply a recursive application of CRA_n we get the bounds for $r$.

If $r, r'$ are two solutions of the CRP within the bounds, then $\prod_{k=1}^{n} m_k | r - r'$, so they have to be equal. □

In a typical application of CRA_n on a computer, the remainders $r_k$ and the moduli $m_k$ will be single digit numbers, i.e., they will fit into single computer words. So when CRA_2 is applied in the "for" loop of CRA_n, $r$ and $M$ are of length $k-1$ and $r_k, m_k$ are of length 1. In computing $M^{-1} \bmod m_k$ one divides $M$ by $m_k$ and afterwards applies the Euclidean algorithm to $\text{rem}(M, m_k)$ and $m_k$. The division takes time $\mathcal{O}(k)$ and the Euclidean algorithm takes some constant time (inputs are single digits). Steps (2) and (4) in CRA_2 take constant time. Step (3) in CRA_2 takes time $\mathcal{O}(k)$. So the complexity of the call to CRA_2 in the $k$-th iteration in CRA_n is $\mathcal{O}(k)$. Also the multiplication $M \cdot m_k$ takes time $\mathcal{O}(k)$. So the whole $k$-th iteration in CRA_n takes time $\mathcal{O}(k)$. By summation over all iterations we get that the complexity of CRA_n is $\mathcal{O}(n^2)$.

**Theorem 3.1.9.** If all the remainders and moduli are positive integers with $0 \leq r_i < m_i$ and $L(m_i) = 1$ for $1 \leq i \leq n$, then the complexity of CRA_n is $\mathcal{O}(n^2)$.

**Theorem 3.1.10.** Let $n$ be a power of 2 and let $m_1, \ldots, m_n, r_1, \ldots, r_n$ determine a CRP over the integers. Assume that every one of the moduli and remainders has length less than or equal to the positive integer $b$. Let $M(l)$ denote the complexity function for multiplying two integers of length $l$. Then CRA_PC (where in step (2) all $s_{ij}$ are taken modulo $q_{ij}$) takes time proportional to $M(bn) \cdot \log n$.

A proof of Theorem 3.1.10 can be found in Aho et al. (1974: theorem 8.12).

CRA in polynomial rings

Let us now consider the Euclidean domain $K[x]$, where $K$ is a field. As for the case of the integers we investigate the solution of CRP by CRA_n and we give a complexity analysis of CRA_n. We also consider the case of linear moduli, which leads to Newton's scheme of interpolation.

**Theorem 3.1.11.** In $K[x]$ the algorithm CRA_n computes the unique solution $r$ of the CRP satisfying $\deg(r) < \sum_{i=1}^{n} \deg(m_i)$.

*Proof.* Investigating the solution $r$ computed by CRA_2, we find that $\deg(r) = \deg(r_1' + \sigma m_1) \leq \max\{\deg(m_1) - 1, \deg(m_2) - 1 + \deg(m_1)\} < \deg(m_1) + \deg(m_2)$. Induction yields the degree bound for the output of CRA_n. The uniqueness can be shown by an argument analogous to the one in Theorem 3.1.8 and is left to the reader. □

**Theorem 3.1.12.** If all the remainders $r_i$ have degree less than the corresponding moduli $m_i$ and $\deg(m_i) \leq d$ for $1 \leq i \leq n$, then the number of arithmetic operations in $K$ required by CRA_n on $K[x]$ is $\mathcal{O}(d^2 n^2)$.

*Proof.* In the $k$-th iteration, the degrees of $r$ and $M$ are bounded by $(k-1)d$. So the division of $M$ by $m_k$ takes $(k-1)d^2$ arithmetical operations, and the subsequent application of E_EUCLID takes $d^2$ operations. Steps (2) to (4) of CRA_2 are also bounded by $(k-1)d^2$. So the number of arithmetic operations in CRA_n is of the order $\sum_{k=2}^{n}(k-1)d^2 \sim d^2 n^2$. □

A special case of the CRP in $K[x]$ is the interpolation problem. All the moduli $m_i$ are linear polynomials of the form $x - \beta_i$.

*Interpolation problem IP*
Given: $\alpha_1, \ldots, \alpha_n \in K$,
$\beta_1, \ldots, \beta_n \in K$, such that $\beta_i \neq \beta_j$ for $i \neq j$,
Find: $u(x) \in K[x]$, such that $u(\beta_i) = \alpha_i$ for $1 \leq i \leq n$.

Since $p(x) \bmod (x - \beta) = p(\beta)$ for $\beta \in K$, the interpolation problem is a special case of the CRP. The inverse of $p(x)$ in $K[x]_{/\langle x-\beta\rangle}$ is $p(\beta)^{-1}$. So CRA_n yields a solution algorithm for IP, namely the Newton interpolation algorithm. By applying Theorem 3.1.12 we see that the number of arithmetic operations in the Newton interpolation algorithm is of the order $n^2$.

The modular method

Let $\mathcal{P}$ be a problem whose input and output are from a domain $D$. The basic idea of the modular method in computer algebra consists of applying homomorphisms $\varphi_1, \ldots, \varphi_m$ to $D$, $\varphi_i$: $D \mapsto D_i$, such that the corresponding problem $\mathcal{P}_i$ can be more easily solved in $D_i$ and the solutions of the problems in the image domains can be combined to yield the solution of the original problem $\mathcal{P}$ in $D$. So usually we need a criterion for detecting whether the following diagram commutes, i.e., whether $\varphi_i$ is a "good" or "lucky" homomorphism,

$$\begin{array}{ccc} D & \overset{\mathcal{P}}{\longmapsto} & D \\ \varphi_i \downarrow & \text{comm.} & \downarrow \varphi_i \\ D_i & \overset{\mathcal{P}_i}{\longmapsto} & D_i \end{array}$$

and how many homomorphisms we need for reconstructing the actual solution of $\mathcal{P}$ in $D$.

The constructive solvability of Chinese remainder problems in Euclidean domains such as $\mathbb{Z}$ or $K[x]$, $K$ a field, enables us to solve many problems in computer algebra by the modular method. The basic idea consists in solving a problem $\mathcal{P}$ over the ED $D$ by reducing it to several problems modulo different primes, solving the simpler problems modulo the primes, and then combining the partial solutions by the Chinese remainder algorithm. In many situations, such as greatest common divisors or resultants of polynomials, this modular method results in algorithms with extremely good complexity behaviour.

### Exercises

1. Let $D$ be an ED, $a, b \in D^*$. Prove that if $b$ is a proper divisor of $a$, then $\deg(b) < \deg(a)$.
2. Prove that in an ED $D$ every ideal is principal.
3. Prove that in an ED every non-zero element can be factored uniquely into a product of finitely many irreducible elements, up to units and reordering of factors.
4. Prove that the ring $\mathbb{Z}[i]$, $i = \sqrt{-1}$, of Gaussian integers is an ED and compute the gcd of $5 - 8i$ and $7 + 3i$.
5. Let $f(x) = x^3 - x^2 + 2$, $g(x) = x^2 + x + 1$ be polynomials over $\mathbb{Q}$. Compute a representation of $h(x) = x^4 + 2x$ as $h(x) = p(x)f(x) + q(x)g(x)$, where $\deg(p) < 2$ and $\deg(q) < 3$.
6. Compute the polynomial $r(x) \in \mathbb{Q}[x]$ of least degree satisfying
$$r(x) \equiv 2x^2 + 1 \mod x^3 + x^2 - 1$$
$$r(x) \equiv x + 2 \mod x^2 + 2x + 2.$$
7. Prove Lemma 3.1.5.
8. Consider an algorithm for solving a $\mathrm{CRP}_n$ by splitting it into two $\mathrm{CRP}_{n/2}$ recursively, as long as $n > 2$. What is the time complexity of such an algorithm over the integers?
9. Let $R$ be a commutative ring with 1. Let $I, I_1, \ldots, I_n$ be ideals in $R$ satisfying $I + I_k = R$ for $1 \le k \le n$. Prove that $I + I_1 \cap \ldots \cap I_n = R$.
10. Let $R$ be a commutative ring with 1. Let $I_1, \ldots, I_n$ be ideals in $R$ satisfying $I_i + I_j = R$ for $i \neq j$. Then $R_{/I_1 \cap \ldots \cap I_n} \cong \prod_{i=1}^n R_{/I_i}$.
11. Generalize the Corollary to Theorem 3.1.2, i.e., show that for $a_1, \ldots, a_r \in K[x]$ pairwise relatively prime and $c \in K[x]$ with $\deg(c) < \deg(a_1) + \ldots + \deg(a_r) = n$ there exist $u_1, \ldots, u_r \in K[x]$ with $\deg(u_i) < \deg(a_i)$ for $1 \le i \le r$, such that $c = \sum_{i=1}^r (u_i \prod_{j=1, j\neq i}^r a_j)$.

## 3.2 $p$-adic approximation

In contrast to the modular method, which solves a problem by solving various homomorphic images of the problem and combining the solutions in the image domains by the Chinese remainder algorithm, in this section we consider a single modulus $p$. The approach consists of solving a problem modulo $p$ and then lifting this solution to a solution modulo $p^k$ for suitable $k$. Newton's approximation algorithm for determining the roots of equations can be adapted to the problem of lifting modular solutions. Throughout this section let $R$ be a commutative ring with identity 1 and $I$ an ideal in $R$. Before we state and prove the main theorem on this lifting approach we start with a technical lemma on Taylor expansion.

**Lemma 3.2.1.** Let $f \in R[x_1, \dots, x_r]$, $r \geq 1$, $y_1, \dots, y_r$ new polynomial variables. Then

$$f(x_1 + y_1, \dots, x_r + y_r) = f(x_1, \dots, x_r) + \sum_{j=1}^{r} \frac{\partial f}{\partial x_j} \cdot y_j + h ,$$

where $h \in R[x_1, \dots, x_r, y_1, \dots, y_r]$ and $h \equiv 0 \pmod{\langle y_1, \dots, y_r \rangle^2}$.

*Proof.* Obviously

$$\begin{aligned} f(x_1 + y_1, \dots, x_r + y_r) = g(x_1, \dots, x_r) &+ \sum_{j=1}^{r} g_j(x_1, \dots, x_r) \cdot y_j \\ &+ h(x_1, \dots, x_r, y_1, \dots, y_r) \end{aligned} \qquad (*)$$

for some polynomials $g, g_1, \dots, g_r, h$ with $h \equiv 0 \pmod{\langle y_1, \dots, y_r \rangle^2}$. Substituting $(0, \dots, 0)$ for $(y_1, \dots, y_r)$ in $(*)$ yields $g(x_1, \dots, x_r) = f(x_1, \dots, x_r)$. Derivation of $(*)$ w.r.t. $y_j$ by the chain rule yields

$$\frac{\partial f}{\partial x_j}(x_1 + y_1, \dots, x_r + y_r) = g_j(x_1, \dots, x_r) + \frac{\partial h}{\partial y_j}(x_1, \dots, x_r, y_1, \dots, y_r) .$$

Substituting $(0, \dots, 0)$ for $(y_1, \dots, y_r)$ and using the fact that $\partial h / \partial y_j \equiv 0 \pmod{\langle y_1, \dots, y_r \rangle}$, we get $\partial f / \partial x_j = g_j$. □

**Theorem 3.2.2** (Lifting theorem). Let $I$ be the ideal generated by $p_1, \dots, p_l$ in $R$, $f_1, \dots, f_n \in R[x_1, \dots, x_r]$, $r \geq 1$, and $a_1, \dots, a_r \in R$ such that

$$f_i(a_1, \dots, a_r) \equiv 0 \pmod{I} \quad \text{for } i = 1, \dots, n .$$

Let $U$ be the Jacobian matrix of $f_1, \dots, f_n$ evaluated at $(a_1, \dots, a_r)$, i.e.,

$$U = (u_{ij})_{\substack{i=1,\dots,n \\ j=1,\dots,r}} , \qquad \text{where } u_{ij} = \frac{\partial f_i}{\partial x_j}(a_1, \dots, a_r) .$$

Assume that $U$ is right-invertible modulo $I$, i.e., there is an $r \times n$ matrix $W = (w_{jl})$ such that $U \cdot W \equiv E_n \pmod{I}$ ($E_n$ is the $n \times n$ identity matrix).

Then for every $t \in \mathbb{N}$ there exist $a_1^{(t)}, \ldots, a_r^{(t)} \in R$ such that

$$f_i(a_1^{(t)}, \ldots, a_r^{(t)}) \equiv 0 \pmod{I^t} \quad \text{for } 1 \le i \le n ,$$

and

$$a_j^{(t)} \equiv a_j \pmod{I} \quad \text{for } 1 \le j \le r .$$

*Proof.* We proceed by induction on $t$. For $t = 1$ the statement is satisfied by $a_j^{(1)} = a_j$ for $1 \le j \le r$.

So now let $t \ge 1$ and assume that the statement of the theorem holds for $t$. We show that it also holds for $t+1$. Let $q_1, \ldots, q_m$ be the generators of the ideal $I^t$ (a possible choice for the $q$'s are the products $p_1^{e_1} \cdots p_l^{e_l}$ with $e_1 + \ldots + e_l = t$). By the induction hypothesis there exist $v_{ik} \in R, i = 1, \ldots, n, k = 1, \ldots, m$ such that

$$f_i(a_1^{(t)}, \ldots, a_r^{(t)}) = \sum_{k=1}^{m} v_{ik} q_k . \tag{3.2.1}$$

We set

$$a_j^{(t+1)} = a_j^{(t)} + B_j \quad \text{for some } B_j \in I^t, \quad j = 1, \ldots, r$$

(this guarantees $a_j^{(t+1)} \equiv a_j \pmod{I}$) and determine the coefficients $b_{jk}$ in $B_j = \sum_{k=1}^m b_{jk} q_k$ so that $f_i(a_1^{(t+1)}, \ldots, a_r^{(t+1)}) \equiv 0 \pmod{I^{t+1}}$.

Let $u_{ij}^{(t)} = \frac{\partial f_i}{\partial x_j}(a_1^{(t)}, \ldots, a_r^{(t)})$ for $i = 1, \ldots, n$, $j = 1, \ldots, r$. Then by Lemma 3.2.1

$$\begin{aligned} f_i(a_1^{(t+1)}, \ldots, a_r^{(t+1)}) &= f_i(a_1^{(t)} + B_1, \ldots, a_r^{(t)} + B_r) \\ &\equiv f_i(a_1^{(t)}, \ldots, a_r^{(t)}) + \sum_{j=1}^{r} u_{ij}^{(t)} B_j \pmod{I^{t+1}} \\ &= \sum_{k=1}^{m} v_{ik} q_k + \sum_{j=1}^{r} u_{ij}^{(t)} \cdot \sum_{k=1}^{m} b_{jk} q_k \\ &= \sum_{k=1}^{m} (v_{ik} + \sum_{j=1}^{r} u_{ij}^{(t)} b_{jk}) q_k \quad \text{for } 1 \le i \le n . \end{aligned} \tag{3.2.2}$$

So $f_i(a_1^{(t+1)}, \ldots, a_r^{(t+1)}) \equiv 0 \pmod{I^{t+1}}$ if and only if

$$v_{ik} + \sum_{j=1}^{r} u_{ij}^{(t)} b_{jk} \equiv 0 \pmod{I} \quad \text{for } i = 1, \ldots, n, \quad k = 1, \ldots, m . \tag{3.2.3}$$

The conditions (3.2.3) determine $m$ systems of linear equations with the common coefficient matrix $(u_{ij}^{(t)})$. But $a_j^{(t)} \equiv a_j \pmod{I}$, so $u_{ij}^{(t)} \equiv u_{ij} \pmod{I}$. Thus, in order to solve (3.2.3) it suffices to solve

$$v_{ik} + \sum_{j=1}^{r} u_{ij} b_{jk} \equiv 0 \pmod{I} \quad \text{for } i = 1, \ldots, n, \quad k = 1, \ldots, m \,. \quad (3.2.4)$$

Since $W = (w_{jl})$ is a right-inverse of $U$, the solution of (3.2.4) is

$$b_{jk} = -\sum_{l=1}^{n} w_{jl} v_{lk} \bmod I \quad \text{for } j = 1, \ldots, r, \quad k = 1, \ldots, m \,. \qquad \square$$

The proof of the lifting theorem is constructive, so we can immediately extract an algorithm LIFT for lifting modular solutions of systems of polynomial equations.

**Algorithm LIFT**(in: $(p_1, \ldots, p_l)$, $(f_1, \ldots, f_n)$, $(a_1, \ldots, a_r)$, $W, t$; out: $(\tilde{a}_1, \ldots, \tilde{a}_r)$);
[$p_1, \ldots, p_l$ generate an ideal $I$ in $R$, $f_1, \ldots, f_n \in R[x_1, \ldots, x_r]$, $a_1, \ldots, a_r \in R$, such that $f_i(a_1, \ldots, a_r) \equiv 0 \pmod{I}$ for $i = 1, \ldots, n$, $W$ is an $r \times n$ matrix over $R$ such that $U \cdot W \equiv E_n \pmod{I}$, where $U = (u_{ij})$ with $u_{ij} = \frac{\partial f_i}{\partial x_j}(a_1, \ldots, a_r)$ for $i = 1, \ldots, n$, $j = 1, \ldots, r$, and $t$ is a positive integer;
the output $\tilde{a}_1, \ldots, \tilde{a}_r$ are elements of $R$ such that $f_i(\tilde{a}_1, \ldots, \tilde{a}_r) \equiv 0 \pmod{I^t}$ for $i = 1, \ldots, n$, and $\tilde{a}_j \equiv a_j \pmod{I}$ for $j = 1, \ldots, r$]
1. for $j = 1$ to $r$ do $a_j^{(1)} := a_j$;
2. for $s = 1$ to $t - 1$ do
   {compute $q_1, \ldots, q_m \in R$ such that $I^s = \langle q_1, \ldots, q_m \rangle$;
   compute $v_{ik} \in R$, $i = 1, \ldots, n$, $k = 1, \ldots, m$ such that
   $f_i(a_1^{(s)}, \ldots, a_r^{(s)}) = \sum_{k=1}^{m} v_{ik} q_k$ for $i = 1, \ldots, n$;
   $b_{jk} := -\sum_{l=1}^{n} w_{jl} v_{lk} \bmod I$ for $j = 1, \ldots, r$, $k = 1, \ldots, m$;
   for $j = 1$ to $r$ do $a_j^{(s+1)} := a_j^{(s)} + \sum_{k=1}^{m} b_{jk} q_k$};
3. for $j = 1$ to $r$ do $\tilde{a}_j := a_j^{(t)}$; return.

*Example 3.2.1.* Let $R = \mathbb{Z}$, and consider the two polynomial equations

$$f_1(x_1, x_2) = x_1 x_2 - x_2^2 - 10 = 0 \,,$$
$$f_2(x_1, x_2) = x_1^2 - 4x_1 x_2 + x_1 = 0 \,.$$

As the ideal $I$ we choose $I = \langle 3 \rangle$. An initial solution modulo $I$ is $a_1 = 1, a_2 = -1$. So as the evaluated Jacobian we get

$$U = \begin{pmatrix} -1 & 3 \\ 7 & -4 \end{pmatrix} .$$

A right-inverse of $U$ modulo $I$ is

$$W = \begin{pmatrix} 2 & 0 \\ 2 & 2 \end{pmatrix} .$$

Let us carry out the algorithm LIFT with these input values.

For $s = 1$ we get $q_1 = 3$ and $a_1^{(1)} = 1, a_2^{(1)} = -1$. $f_1(1, -1) = -4 \cdot 3$, so $v_{11} = -4$. $f_2(1, -1) = 2 \cdot 3$, so $v_{21} = 2$. $-W \cdot (-4, 2)^T \equiv (2, 1)^T \pmod 3$, so we get $b_{11} = 2, b_{21} = 1$ and $a_1^{(2)} = 7, a_2^{(2)} = 2$.

All the following iterations yield $(b_{11}, b_{21}) = (0, 0)$ and in fact we have reached an integral solution $(7, 2)$ of the system of equations.

Now suppose that we just consider the single equation

$$f_1(x_1, x_2) = x_1 x_2 - x_2^2 - 10 = 0$$

and also start with $(a_1^{(1)}, a_2^{(1)}) = (1, -1)$. As the inverse modulo 3 we take $W = (-1\ 1)^T$. With these inputs the algorithm LIFT produces the sequence of modular solutions $(a_1^{(1)}, a_2^{(1)}) = (1, -1)$, $(a_1^{(2)}, a_2^{(2)}) = (-2, 2)$, $(a_1^{(3)}, a_2^{(3)}) = (-20, 20)$, etc. There are exactly 8 integral solutions of $f_1(x_1, x_2) = 0$, namely $(\pm 11, \pm 1)$, $(\pm 7, \pm 2)$, $(\pm 7, \pm 5)$, $(\pm 11, \pm 10)$. Obviously, none of them is approximated by the lifting process. What we get is an approximation of a 3-adic solution. However, choosing $W = (2\ 1)^T$ as the right-inverse of the Jacobian, and starting from the same initial solution leads to the integral solution $(7, 2)$.

As we have seen in Example 3.2.1, the algorithm LIFT does not necessarily converge to a solution of the system of polynomial equations in $R$. For $R = \mathbb{Z}$ and $I = \langle p \rangle$, $p$ a prime number, we get a so-called $p$-adic solution.

A brief exposition of what $p$-adic numbers are might be helpful. Let $p$ be a prime number. For any nonzero integer $a$, let the *order of $a$ w.r.t. $p$*, $\mathrm{ord}_p a$, be the highest power of $p$ which divides $a$, i.e., the greatest $m$ such that $a \equiv 0 \pmod{p^m}$. The function $\mathrm{ord}_p$ can be extended to rational numbers $x = a/b$ by setting $\mathrm{ord}_p x := \mathrm{ord}_p a - \mathrm{ord}_p b$. Using the order function we now define a norm $|\,.\,|_p$ on $\mathbb{Q}$ by

$$|x|_p := \begin{cases} \frac{1}{p^{\mathrm{ord}_p x}}, & \text{if } x \neq 0; \\ 0, & \text{if } x = 0. \end{cases}$$

$|\,.\,|_p$ is non-Archimedean. Two Cauchy sequences $\{a_i\}$, $\{b_i\}$ of rational numbers w.r.t. $|\,.\,|_p$ are called equivalent if $|a_i - b_i|_p \to 0$ as $i \to \infty$. Now we define $\mathbb{Q}_p$ to be the set of equivalence classes of Cauchy sequences. We define the norm $|\,.\,|_p$ of an equivalence class $a$ to be $\lim_{i \to \infty} |a_i|_p$, where $\{a_i\}$ is any representative of $a$.

Arithmetic operations can be defined on $\mathbb{Q}_p$ in a natural way, so $\mathbb{Q}_p$ is a field, the *field of p-adic numbers*. The rational numbers $\mathbb{Q}$ can be identified with the subfield of $\mathbb{Q}_p$ consisting of equivalence classes containing a constant Cauchy sequence. $\mathbb{Q}_p$ is complete w.r.t. the norm $|\,.\,|_p$, i.e., every Cauchy sequence has a limit. For a thorough introduction to $p$-adic numbers we refer to Mahler (1973) and Koblitz (1977).

$p$-adic lifting will play an important role in the factorization of polynomials with integer coefficients. The special case of the lifting theorem in the context of factorization is the Hensel lemma. H. Zassenhaus (1969) proposed a "quadratic Hensel construction." This idea can be generalized to the lifting theorem in the following way. Analyzing the proof of the lifting theorem we see that Eq. (3.2.2) does not only hold modulo $I^{t+1}$, but in fact modulo $I^{2t}$. So we aim to determine $a_j^{(t+1)}$ such that

$$f_i(a_1^{(t+1)}, \ldots, a_r^{(t+1)}) \equiv 0 \pmod{I^{2t}} .$$

This is the case if

$$v_{ik} + \sum_{j=1}^{r} u_{ij}^{(t)} b_{jk} \equiv 0 \pmod{I^t} \quad \text{for } i = 1, \ldots, n, \quad k = 1, \ldots, m . \quad (3.2.5)$$

The system is solvable if

$$U^{(t)} = \left(u_{ij}^{(t)}\right)_{\substack{i=1,\ldots,n \\ j=1,\ldots,r}}$$

is invertible modulo $I^t$, i.e., if there is a matrix $W^{(t)}$ such that

$$U^{(t)} \cdot W^{(t)} \equiv E_n \pmod{I^t} . \quad (3.2.6)$$

$U^{(t)} \equiv U \pmod{I}$ and $U$ is invertible modulo $I$. So there exists an initial solution to the system of equations (3.2.6), i.e., a matrix $W$ such that $U^{(t)} \cdot W \equiv E_n \pmod{I}$. The Jacobi matrix of the system (3.2.6) is $U$. Thus, by the lifting theorem there is a matrix $W^{(t)}$ satisfying (3.2.6).

The algorithm LIFT_Q lifts both the modular solution and the matrix $W$ in a quadratic manner, i.e., in every step a solution modulo $I^t$ is lifted to one modulo $I^{2t}$.

**Algorithm LIFT_Q**(in: $(p_1, \ldots, p_l)$, $(f_1, \ldots, f_n)$, $(a_1, \ldots, a_r)$, $W, t$; out: $(\tilde{a}_1, \ldots, \tilde{a}_r)$);
[$p_1, \ldots, p_l$ generate an ideal $I$ in $R$, $f_1, \ldots, f_n \in R[x_1, \ldots, x_r]$, $a_1, \ldots, a_r \in R$, such that $f_i(a_1, \ldots, a_r) \equiv 0 \pmod{I}$ for $i = 1, \ldots, n$, $W$ is an $r \times n$ matrix over $R$ such that $U \cdot W \equiv E_n \pmod{I}$, where $U = (u_{ij})$ with $u_{ij} = \frac{\partial f_i}{\partial x_j}(a_1, \ldots, a_r)$ for $i = 1, \ldots, n$, $j = 1, \ldots, r$, and $t$ is a positive integer;
the output $\tilde{a}_1, \ldots, \tilde{a}_r$ are elements of $R$ such that $f_i(\tilde{a}_1, \ldots, \tilde{a}_r) \equiv 0 \pmod{I^{2^t}}$ for $i = 1, \ldots, n$, and $\tilde{a}_j \equiv a_j \pmod{I}$ for $j = 1, \ldots, r$]
1. $s := 0$;
   for $j = 1$ to $r$ do $a_j^{(0)} := a_j$;
   $W^{(0)} := W$;
2. while $s < t$ do
   {[determine ideal basis]

compute $q_1, \ldots, q_m \in R$ such that $I^{2^s} = \langle q_1, \ldots, q_m \rangle$;
[lift $a_j^{(s)}$]
compute $v_{ik} \in R$, $i = 1, \ldots, n$, $k = 1, \ldots, m$ such that
$f_i(a_1^{(s)}, \ldots, a_r^{(s)}) = \sum_{k=1}^{m} v_{ik} q_k$ for $i = 1, \ldots, n$;
$b_{jk} := -\sum_{l=1}^{n} w_{jl}^{(s)} v_{lk}$ for $j = 1, \ldots, r$, $k = 1, \ldots, m$;
for $j = 1$ to $r$ do $a_j^{(s+1)} := a_j^{(s)} + \sum_{k=1}^{m} b_{jk} q_k$;
[lift $W^{(s)}$]
$u_{ij}^{(s+1)} := \frac{\partial f_i}{\partial x_j}(a_1^{(s+1)}, \ldots, a_r^{(s+1)})$ for $i = 1, \ldots, n$, $j = 1, \ldots, r$;
for $k = 1, \ldots, m$ compute an $n \times n$ matrix $D_k$ over $R$ such that
$U^{(s+1)} \cdot W^{(s)} = E_n + \sum_{k=1}^{m} D_k q_k$;
for $k = 1$ to $m$ do $Z_k := -W^{(s)} D_k$;
$W^{(s+1)} := W^{(s)} + \sum_{k=1}^{m} Z_k q_k$;
[increase exponent]
$s := s + 1$};
3. for $j = 1$ to $r$ do $\tilde{a}_j := a_j^{(t)}$; return.

*Example 3.2.2.* As in the previous example let $R = \mathbb{Z}$, $I = \langle 3 \rangle$,

$$f_1(x_1, x_2) = x_1 x_2 - x_2^2 - 10 = 0 ,$$

$W = (-1\ 1)^T$ and $a_1^{(0)} = 1, a_2^{(0)} = -1$ the initial solution modulo 3.

Applying the quadratic lifting algorithm LIFT_Q we get the approximation $(a_1^{(1)}, a_2^{(1)}) = (-2, 2)$, $(a_1^{(2)}, a_2^{(2)}) = (-20, 20)$, $(a_1^{(3)}, a_2^{(3)}) = (-830, 830), \ldots$ of a 3-adic solution of $f_1 = 0$.

$f_1(a_1^{(3)}, a_2^{(3)}) = 1377810 \equiv 0 \pmod{3^{2^3} = 6561}$.

A quadratic lifting scheme obviously takes fewer lifting steps for reaching a certain $p$-adic approximation. On the other hand, every single lifting step requires the lifting of the inverse of the Jacobian matrix.

### Exercises

1. Define arithmetic operations such that $\mathbb{Q}_p$ is a field and the rational numbers $\mathbb{Q}$ are a natural subfield of $\mathbb{Q}_p$.
2. Work out the details in Example 3.2.2.

## 3.3 The fast Fourier transform

The Fourier transform has innumerable applications in science and technology, where it is generally used for constructing fast algorithms. In this section we will introduce and analyze the discrete Fourier transform, its inverse, and applications to convolutions and products. An efficient algorithm for the computation of the discrete Fourier transform, the *fast Fourier transform* (FFT), will be developed. Finally, we get the currently fastest known multiplication algorithm for integers.

Throughout this section let $K$ be a computable field and let $n = 1 + \cdots + 1$ ($n$ times) be invertible in $K$, i.e., $\text{char}(K) \not| n$. A vector $a = (a_0, \ldots, a_{n-1})$ over $K$ determines a polynomial

$$a(x) = \sum_{k=0}^{n-1} a_k x^k . \tag{3.3.1}$$

This polynomial can also be represented by its values at $n$ different evaluation points in $K$. The discrete Fourier transform of the vector $a$ is exactly such an evaluation at certain well-chosen points, which are called Fourier points.

*Definition 3.3.1.* An element $\omega$ in $K$ is a *primitive n-th root of unity* iff $\omega^n = 1$ and $\omega^j \neq 1$ for $0 < j < n$.

So, for example, $e^{2\pi i/n}$ is a primitive $n$-th root of unity in the field of complex numbers.

*Definition 3.3.2.* Let $\omega$ be a primitive $n$-th root of unity in $K$. Let $A = (A_{ij})_{0 \leq i,j < n}$ be the $(n \times n)$ matrix with $A_{ij} = \omega^{ij}$. Let $a = (a_0, a_1, \ldots, a_{n-1})^T$ be a vector of length $n$ over $K$. The vector $F_\omega(a) = A \cdot a$, whose $i$-th component $(0 \leq i < n)$ is $\sum_{k=0}^{n-1} a_k \omega^{ik}$, is the *discrete Fourier transform* (DFT) of $a$ (w.r.t. $\omega$).

Obviously the $i$-th component $\sum_{k=0}^{n-1} a_k \omega^{ik}$ of $F_\omega(a)$ is the evaluation of the polynomial in (3.3.1) corresponding to the vector $a$ at $x = \omega^i$. We can write

$$F_\omega(a) = \begin{pmatrix} F_\omega(a)_0 \\ F_\omega(a)_1 \\ \vdots \\ F_\omega(a)_{n-1} \end{pmatrix}$$

$$= \begin{pmatrix} 1 & 1 & 1 & \cdots & 1 \\ 1 & \omega & \omega^2 & \cdots & \omega^{n-1} \\ \vdots & \vdots & \vdots & & \vdots \\ 1 & \omega^{n-1} & \omega^{2(n-1)} & \cdots & \omega^{(n-1)^2} \end{pmatrix} \cdot \begin{pmatrix} a_0 \\ a_1 \\ \vdots \\ a_{n-1} \end{pmatrix} = A \cdot a .$$

The matrix $A$ is the evaluation of the Vandermonde matrix at $\omega$. We call $A$ the *Vandermonde matrix at $\omega$ of order $n$*.

Sometimes we think of the argument $a$ to the Fourier transform as a vector, and at other times as a polynomial. We will freely use whatever notion is more convenient. Moreover, we allow ourselves to write the vector $a$ sometimes as a column vector and at other times as a row vector.

**Lemma 3.3.1.** Let $\omega$ be a primitive $n$-th root of unity in $K$, and let $A$ be the Vandermonde matrix at $\omega$ of order $n$. Then $A$ is regular and $A^{-1} = \frac{1}{n}(\omega^{-ij})_{0\leq i,j<n}$.

*Proof.* Let $A^{-1}$ be defined as above. The element in the $i$-th row and $j$-th column of $A \cdot A^{-1}$ is

$$\frac{1}{n} w_{ij} \doteq \frac{1}{n}\sum_{k=0}^{n-1} \omega^{ik}\omega^{-kj} .$$

Obviously $w_{ii} = n$ for $0 \leq i < n$. Now let $i \neq j$. Then $w_{ij} = \sum_{k=0}^{n-1} \omega^{(i-j)k}$. Since $0 <| i - j |< n$, we have $\omega^{i-j} \neq 1$. Using the summation formula for geometric series we get

$$w_{ij} = \frac{(\omega^{i-j})^n - 1}{\omega^{i-j} - 1} = \frac{(\omega^n)^{i-j} - 1}{\omega^{i-j} - 1} = 0 . \qquad \square$$

*Definition 3.3.3.* Let $n$, $\omega$, $A$ and $a$ be as in Definition 3.3.2. The vector $F_\omega^{-1}(a) = A^{-1} \cdot a$, whose $i$-th component $(0 \leq i < n)$ is $\frac{1}{n}\sum_{k=0}^{n-1} a_k\omega^{-ik}$, is the *inverse discrete Fourier transform* (iDFT) of $a$ (w.r.t. $\omega$).

In order to speed up the evaluation of the polynomial (3.3.1), we will use the fact that the set of Fourier points $\{\omega^i \mid 0 \leq i < n\}$ contains an additive inverse of every one of its elements for properly chosen $n$.

**Lemma 3.3.2.** Let $n = 2m$, $\omega$ a primitive $n$-th root of unity in $K$. Then
a. $\omega^{m+j} = -\omega^j$ for $0 \leq j < m$,
b. $\omega^2$ is a primitive $m$-th root of unity.

*Proof.* a. Since $\omega$ is an $n$-th root of unity, we have

$$(\omega^{m+j})^2 = (\omega^j)^2\omega^n = (\omega^j)^2 .$$

Any solution of $x^2 - (\omega^j)^2 = (x + \omega^j)(x - \omega^j)$ must be either $\omega^j$ or $-\omega^j$. But $\omega^{m+j} \neq \omega^j$, so we must have $\omega^{m+j} = -\omega^j$.

b. Clearly $\omega^2$ is an $m$-th root of unity. If $(\omega^2)^j = 1$ for some $0 < j < m$, then $\omega^k = 1$ for some $0 < k < n$, in contradiction to $\omega$ being primitive. $\square$

**Theorem 3.3.3.** Let $n = 2m$, $\omega$ a primitive $n$-th root of unity in $K$. With the notation

$$a(x) = a^{(e)}(x^2) + x \cdot a^{(o)}(x^2), \qquad a^{(e)}(y) = \sum_{j=0}^{m-1} a_{2j}y^j, \quad a^{(o)}(y) = \sum_{j=0}^{m-1} a_{2j+1}y^j$$

we can express $F_\omega(a)$ as

$$F_\omega(a) = \begin{pmatrix} a(1) \\ a(\omega) \\ \vdots \\ a(\omega^{m-1}) \\ a(\omega^m) \\ a(\omega^{m+1}) \\ \vdots \\ a(\omega^{2m-1}) \end{pmatrix} = \begin{pmatrix} a^{(e)}(1) + a^{(o)}(1) \\ a^{(e)}(\omega^2) + \omega a^{(o)}(\omega^2) \\ \vdots \\ a^{(e)}((\omega^2)^{m-1}) + \omega^{m-1}a^{(o)}((\omega^2)^{m-1}) \\ a^{(e)}(1) - a^{(o)}(1) \\ a^{(e)}(\omega^2) - \omega a^{(o)}(\omega^2) \\ \vdots \\ a^{(e)}((\omega^2)^{m-1}) - \omega^{m-1}a^{(o)}((\omega^2)^{m-1}) \end{pmatrix}.$$

*Proof.* The first $m$ components of $F_\omega(a)$ are obvious. For $0 \leq j < m$ we have $\omega^{m+j} = -\omega^j$ by Lemma 3.3.2. So $a(\omega^{m+j}) = a(-\omega^j) = a^{(e)}((\omega^j)^2) - \omega^j a^{(o)}((\omega^j)^2)$. □

From Theorem 3.3.3 and Lemma 3.3.2 (b) we see that $F_\omega(a)$ can be evaluated very efficiently if $n$ is a power of 2. This observation is the basis for the fast Fourier transform.

**Algorithm FFT**(in: $n$,$\omega$,$a$; out: $b$);
[the integer $n$ is a power of 2 and invertible in $K$, $\omega$ is a primitive $n$-th root of unity in $K$, $a = (a_0, \ldots, a_{n-1})^T$ is a vector of length $n$ over $K$, which is also interpreted as a polynomial $a(x) = \sum_{i=0}^{n-1} a_i x^i$; $b$ is the Fourier transform $F_\omega(a)$ of $a$.
We assume that the powers $\omega^i$, $0 \leq i < n$, are precomputed.]
1. if $n = 1$ then $\{b := a$; return$\}$;
2. $m := n/2$;
   $a^{(e)} := (a_{2i})_{i=0,\ldots,m-1}$; $a^{(o)} := (a_{2i+1})_{i=0,\ldots,m-1}$;
3. [recursive calls]
   $c^{(e)} := \mathrm{FFT}(m, \omega^2, a^{(e)})$;
   $c^{(o)} := \mathrm{FFT}(m, \omega^2, a^{(o)})$;
4. [combination]
   for $j := 0$ to $m - 1$ do
   $\{b_j := c_j^{(e)} + \omega^j \cdot c_j^{(o)}$;
   $b_{m+j} := c_j^{(e)} - \omega^j \cdot c_j^{(o)}\}$;
   return.

*Example 3.3.1.* Let $K = \mathbb{Z}_{17}$, $n = 8$, $\omega = 2$. Then $n$ is invertible in $K$ and $\omega$ is a primitive $n$-th root of unity in $K$. We compute the Fourier transform of the vector $a = (2, 3, 5, 1, 4, 6, 1, 2)^T$ by the algorithm FFT. $a$ is decomposed into its even and odd parts $a^{(e)} = (2, 5, 4, 1)^T$, $a^{(o)} = (3, 1, 6, 2)^T$. Recursive application of FFT to $4, 4, a^{(e)}$ yields the vector $c^{(e)} = (12, 14, 0, 16)^T$. Recursive application of FFT to 4, 4, $a^{(o)}$ yields the vector $c^{(o)} = (12, 10, 6, 1)^T$. So in the combination step we get $b = (7, 0, 7, 7, 0, 11, 10, 8)^T$ as the Fourier transform of $a$.

**Theorem 3.3.4.** The number of field operations in the algorithm FFT is $\frac{3}{2}n \log_2 n$.

*Proof.* The time (number of field operations) $T(n)$ of executing FFT on an input of length $n$ satisfies the equation

$$T(n) = 2T(n/2) + \tfrac{3}{2}n \ .$$

Iterating this formula $(\log_2 n \ - \ 1)$ times gives

$$T(n) = 3(\log_2 n)2^{(\log_2 n)-1} + nT(1) \ .$$

But $T(1) = 0$, so we get $\frac{3}{2}n \log_2 n$ as the total number of field operations. □

For executing FFT over finite fields we need to know whether for a given $n$ we can find an appropriate finite field $\mathbb{Z}_p$ and a primitive $n$-th root of unity in it. Fortunately this is the case.

**Theorem 3.3.5.** $\mathbb{Z}_p$ has a primitive $n$-th root of unity if and only if $n \mid (p-1)$.

*Proof.* By Lagrange's theorem, the order of a group element divides the order of the group. So $n$ must divide the order $p - 1$ of the multiplicative group $\mathbb{Z}_p^*$.

Now suppose $n \mid (p - 1)$. The multiplicative group of $\mathbb{Z}_p$ is cyclic, so it contains a primitive element $\alpha$. Thus, $\beta = \alpha^{(p-1)/n}$ has order $n$ in $\mathbb{Z}_p^*$, i.e., $\beta$ is a primitive $n$-th root of unity in $\mathbb{Z}_p$. □

So for $n = 2^k$ there is a primitive $n$-th root of unity in $\mathbb{Z}_p$ if and only if $p$ is of the form

$$p = 2^k q + 1 \ .$$

The generalized prime number theorem states that for relatively prime integers $a, b$ the number of primes less or equal to $x$ in the arithmetic progression $a \cdot q + b$ $(q = 1, 2, \ldots)$ is approximately

$$\frac{x}{\log x}\phi(a) \ ,$$

where $\phi$ is Euler's phi function. As a consequence we get that the number of primes $p = 2^k q + 1 \le x$ is approximately

$$\frac{x}{\log x}2^{k-1} \ .$$

For instance, there are approximately 180 primes $p = 2^k q + 1$ ($q$ odd) with exponent $k \geq 20$ below $x = 2^{31}$. Any of these primes could be used to compute FFTs of size $2^{20}$.

### Polynomial multiplication and convolution

There exists a very tight connection between the Fourier transform and the evaluation of polynomials. Let

$$a(x) = \sum_{i=0}^{n-1} a_i x^i$$

be a polynomial of degree $n-1$. We can represent $a$ by its vector of coefficients $(a_0, \ldots, a_{n-1})$, or by its values at $n$ distinct evaluation points $x_0, \ldots, x_{n-1}$. Computing the coefficient representation of a polynomial from the list of values at evaluation points is interpolation. On the other hand, the DFT of the coefficient vector is the representation of $a$ as the values at the evaluation points $\omega^0, \omega, \ldots, \omega^{n-1}$. So the inverse DFT is just a particular way of interpolating, namely w.r.t. interpolation points which are the powers of a primitive root of unity.

*Definition 3.3.4.* Let $a = (a_0, \ldots, a_{n-1})$, $b = (b_0, \ldots, b_{n-1})$ be vectors over $K$. The *convolution* of $a$ and $b$, written as $a \odot b$, is the vector $c = (c_0, \ldots, c_{2n-1})$, with $c_i = \sum_{j=0}^{n-1} a_j b_{i-j}$, where $a_k = b_k = 0$ for $k < 0$ or $k \leq n$.

The *positive wrapped convolution* of $a$ and $b$ is the vector $c = (c_0, \ldots, c_{n-1})$ with $c_i = \sum_{j=0}^{i} a_j b_{i-j} + \sum_{j=i+1}^{n-1} a_j b_{n+i-j}$.

The *negative wrapped convolution* of $a$ and $b$ is the vector $c = (c_0, \ldots, c_{n-1})$ with $c_i = \sum_{j=0}^{i} a_j b_{i-j} - \sum_{j=i+1}^{n-1} a_j b_{n+i-j}$.

The motivation for considering convolutions comes from the multiplication of polynomials. The coefficients of the product of two polynomials of degree $n-1$ are exactly the components of the convolution of their coefficient vectors. On the other hand, if the polynomials are represented as their values at $2n$ evaluation points, the representation of their product can be determined by pairwise multiplication of the values.

**Theorem 3.3.6** (Convolution theorem). Let $\omega$ be a primitive $2n$-th root of unity in $K$. Let $a = (a_0, \ldots, a_{n-1})$ and $b = (b_0, \ldots, b_{n-1})$ be vectors of length $n$ over $K$, and $\bar{a} = (a_0, \ldots, a_{n-1}, 0, \ldots, 0)$, $\bar{b} = (b_0, \ldots, b_{n-1}, 0, \ldots, 0)$ the corresponding vectors of length $2n$, where the trailing components have been filled with 0's. Then $a \odot b = F_\omega^{-1}(F_\omega(\bar{a}) \cdot F_\omega(\bar{b}))$, where "$\cdot$" means componentwise multiplication.

*Proof.* Obviously the vector $c = a \odot b$ consists of the coefficients of $a \cdot b$. So if we can show that $F_\omega(c) = F_\omega(\bar{a}) \cdot F_\omega(\bar{b})$ then we are done. But this is clear, since every line is of the form

$$h_\omega^{(i)}(a \cdot b) = h_\omega^{(i)}(a) \cdot h_\omega^{(i)}(b) ,$$

where $h_\omega^{(i)}$ is the evaluation homomorphism w.r.t. $\omega^i$. □

*Example 3.3.2.* Let us suppose that we are given the polynomial $a(x) = x + 10$ and we want to compute $(x + 10)^3$. Making use of the convolution theorem we represent $a$ as the vector $(10, 1, 0, 0)$, apply the Fourier transform $F_\omega$ w.r.t. some 4th root of unity $\omega$, $F_\omega(a) = (\hat{a}_0, \hat{a}_1, \hat{a}_2, \hat{a}_3)$, raise all the components of the transform to the third power, and apply the inverse Fourier transform, getting $F_\omega^{-1}(\hat{a}_0^3, \hat{a}_1^3, \hat{a}_2^3, \hat{a}_3^3) = (1000, 300, 30, 1)$, representing the polynomial $x^3 + 30x^2 + 300x + 1000 = a(x)^3$.

Quite clearly this approach will be a slow-down compared to the usual method of multiplication. The same is true for higher degree but sparse polynomials. However, if the polynomials to be multiplied are of high degree and dense, then the Fourier transform might be a feasible alternative.

The addition of trailing 0's can be avoided by using the wrapped convolutions. In describing the Schönhage–Strassen multiplication algorithm we will make use of the wrapped convolutions. Evaluating two polynomials of degree $n - 1$ at $n$ points, multiplying these values componentwise, and applying the inverse transformation, we get exactly the components of the positive wrapped convolution.

**Theorem 3.3.7.** Let $\omega$ be a primitive $n$-th root of unity in $K$, $\psi^2 = \omega$. Let $a = (a_0, \ldots, a_{n-1})$ and $b = (b_0, \ldots, b_{n-1})$ be vectors of length $n$ over $K$.

a. The positive wrapped convolution of $a$ and $b$ is $F_\omega^{-1}(F_\omega(a) \cdot F_\omega(b))$.
b. Let $d = (d_0, \ldots, d_{n-1})$ be the negative wrapped convolution of $a$ and $b$. Let $\hat{a}, \hat{b}$ and $\hat{d}$ be defined as $(a_0, \psi a_1, \ldots, \psi^{n-1} a_{n-1})$, $(b_0, \psi b_1, \ldots, \psi^{n-1} b_{n-1})$, and $(d_0, \psi d_1, \ldots, \psi^{n-1} d_{n-1})$. Then $\hat{d} = F_\omega^{-1}(F_\omega(\hat{a}) \cdot F_\omega(\hat{b}))$.

*Proof.* We consider only part (b). Observe that

$$d_p = \sum_{j=0}^{p} a_j \cdot b_{p-j} - \sum_{j=p+1}^{n-1} a_j \cdot b_{n+p-j} .$$

Let $F(\hat{d}) = (d'_0, \ldots, d'_{n-1})$, i.e.,

$$d'_l = \sum_{p=0}^{n-1} \psi^p \cdot d_p \cdot \omega^{lp} =$$

$$= \sum_{p=0}^{n-1}\sum_{j=0}^{p} \psi^p \cdot a_j \cdot b_{p-j} \cdot \omega^{lp} - \sum_{p=0}^{n-1}\sum_{j=p+1}^{n-1} \psi^p \cdot a_j \cdot b_{n+p-j} \cdot \omega^{lp}$$

$$= \sum_{p=0}^{n-1}\sum_{j=0}^{p} \psi^p \cdot a_j \cdot b_{p-j} \cdot \omega^{lp} + \sum_{p=0}^{n-1}\sum_{j=p+1}^{n-1} \psi^{n+p} \cdot a_j \cdot b_{n+p-j} \cdot \omega^{lp} . \quad (3.3.2)$$

On the other hand, let

$$F(\hat{a}) = (a'_0, \ldots, a'_{n-1}), \quad a'_l = \sum_{q=0}^{n-1} \psi^q \cdot a_q \cdot \omega^{lq} ,$$

$$F(\hat{b}) = (b'_0, \ldots, b'_{n-1}), \quad b'_l = \sum_{r=0}^{n-1} \psi^r \cdot b_r \cdot \omega^{lr} ,$$

i.e.,

$$a'_l \cdot b'_l = \sum_{q=0}^{n-1}\sum_{r=0}^{n-1} \psi^{q+r} \cdot a_q \cdot b_r \cdot \omega^{l(q+r)}$$

$$= \sum_{s=0}^{n-1}\sum_{t=0}^{s} \psi^s \cdot a_t \cdot b_{s-t} \cdot \omega^{ls} + \sum_{s=0}^{n-1}\sum_{t=p+1}^{n-1} \psi^{n+s} \cdot a_t \cdot b_{n+s-t} \cdot \omega^{ls} . \quad (3.3.3)$$

The statement follows from (3.3.2) = (3.3.3). □

**Theorem 3.3.8.** Let $\omega$ be a primitive $n$-th root of unity in $K$, $\psi^2 = \omega$, $\psi^n = -1$. Let $a = (a_0, \ldots, a_{n-1})$ and $b = (b_0, \ldots, b_{n-1})$ be vectors of length $n$ over $K$. The convolution $a \odot b$ as well as the wrapped convolutions of $a$ and $b$ can be computed in $\mathcal{O}(n \log n)$ arithmetic operations.

*Proof.* Combine Theorems 3.3.4, 3.3.6, and 3.3.7. □

FFT in the integers modulo $m$

With only slight modifications of the proofs, the whole theory of the fast Fourier transform can be generalized to commutative rings with identity $R$, see for instance Aho et al. (1974). The complexity bounds remain the same. Fourier transforms in rings $\mathbb{Z}_m$ are particularly important in the fast Schönhage–Strassen integer multiplication algorithm. For given $n$ and $\omega$, powers of 2, we will need to explicitly find a ring $\mathbb{Z}_m$ in which we can execute the Fourier transform. The existence and form of such an $m$ is given by the following theorem.

**Theorem 3.3.9.** Let $n$ and $\omega$ be powers of 2 with positive exponents and let $m = \omega^{n/2} + 1$. In $\mathbb{Z}_m$ there is a multiplicative inverse of $n$ and $\omega$ is a primitive $n$-th root of unity.

*Proof.* $m$ and $n$ are relatively prime, so $n$ is invertible in $\mathbb{Z}_m$. We have $\omega^n = \omega^{n/2} \cdot \omega^{n/2} \equiv (-1) \cdot (-1) \equiv 1 \bmod m$. So $\omega$ is an $n$-th root of unity in $\mathbb{Z}_m$. Finally we have to show that $\omega^j \neq 1$ for $0 < j < n$. For $0 < j < n/2$ we have $1 < \omega^j < m - 1$, so $\omega^j \neq \pm 1$ modulo $m$. For $j = n/2$ we have $\omega^j = -1$ modulo $m$. For $n/2 < j < n$ we have $\omega^j = \omega^{n/2}\omega^{j-n/2} = -\omega^{j-n/2} \neq \pm 1$ modulo $m$. □

Now let us determine the bit complexity of the Fourier transform in rings $\mathbb{Z}_m$, $m = \omega^{n/2} + 1$.

**Theorem 3.3.10.** Let $\omega$ and $n$ be powers of 2 and $m = \omega^{n/2} + 1$. Let $a = (a_0, a_1, \ldots, a_{n-1})$ be a vector over $\mathbb{Z}$ and $0 \leq a_i < m$ for each $i$. Then the DFT of $a$ and the iDFT of $a$ modulo $m$ can be computed in time dominated by $n^2 \log n \log \omega$.

*Proof.* We apply the algorithm FFT. The time $T(n, \omega)$ for executing FFT on inputs of size $(n, \omega)$ is 2 times the time for executing FFTs of size $(n/2, \omega^2)$ (from step (3)) plus $2(n/2)$ times the time for executing the multiplication by $\omega^j$ and the subsequent addition or subtraction in step (4) for computing $b_j$ or $b_{m+j}$. Multiplication by $\omega^j$ is a left-shift, and the result is less than $\omega^n$. So we can write $\omega^j \cdot c_j^{(o)} = z_0 + z_1\omega^{n/2} \equiv z_0 - z_1 \bmod m$, where $0 \leq z_i < \omega^{n/2}$, and therefore the computation of $b_j$ takes time dominated by $\log m \sim n \log \omega$. Thus, we get

$$T(n, \omega) \preceq 2T\left(\frac{n}{2}, \omega^2\right) + n^2 \log \omega \; .$$

Iterating this equation $\log n$ times, we get the complexity of the DFT as

$$T(n, \omega) \preceq n^2 \log n \log \omega \; .$$

For computing the iDFT we substitute $\omega^{-1}$ for $\omega$ in FFT. Multiplication by $\omega^{-p}$ is the same as multiplication by $\omega^{n-p}$. This means that the multiplication in step (4) is again a left-shift yielding a result bounded by $\omega^{(3/2)n}$. Therefore either $\omega^{-j} \cdot c_j^{(o)} \equiv -1 \bmod m$ or $\omega^{-j} \cdot c_j^{(o)} = z_0 + z_1\omega^{n/2} + z_2\omega^n \equiv z_0 - z_1 + z_2 \bmod m$. So the computation of $b_j$ takes time dominated by $\log m \sim n \log \omega$ and we get the same complexity bound as above. The only complication is that now we need to multiply the result by $1/n$. If $n = 2^k$, then

$$2^k \cdot 2^{n \log \omega - k} \equiv 2^{n \log \omega} \equiv \omega^n \equiv \omega^n + 2\omega^{n/2} + 2 \equiv m \cdot m + 1 \equiv 1 \bmod m \; ,$$

so multiplication by $n^{-1}$ can be computed as a left-shift by $n \log \omega - k$ positions. The result is again bounded by $\omega^{(3/2)n}$, so its remainder modulo $m$ can be computed in time dominated by $n \log \omega$. Thus, also the time for computing iDFT is dominated by $n^2 \log n \log \omega$. □

The Schönhage–Strassen integer multiplication algorithm

In Sect. 2.1 we have introduced the Karatsuba algorithm INT_MULTK for multiplying two integers of length $n$ by partitioning them into integers of length $n/2$. The complexity of INT_MULTK is proportional to $n^{\log_2 3}$. This approach is generalizable and leads to the so-called Schönhage–Strassen integer multiplication algorithm (Schönhage and Strassen 1971).

We represent the inputs as $b$ blocks of $l$ bits each. These $b$ blocks are regarded as the coefficients of a polynomial. In order to get the coefficients of the product of these polynomials, we evaluate them at suitable points, multiply these values, and interpolate. Choosing the $n$-th roots of unity as evaluation points, we can apply the algorithm FFT and the convolution theorem. Recursive application of this process leads to a multiplication algorithm for integers of length $n$ with complexity $n \log n \log\log n$.

For simplifying the analysis we will assume that $n$ is a power of 2. This can always be achieved by adding leading 0's and the complexity function will remain unchanged (only the constant factor is increased). Actually we will compute the product of two integers of length $n$ modulo $2^n + 1$. If we want the exact product of two integers of length $n$, we must again add leading 0's and compute the product of integers of length $2n$ modulo $2^{2n} + 1$. Again, the complexity function remains unchanged.

So now let $u$ and $v$ be binary integers in the range $0 \le u, v \le 2^n$ which should be multiplied modulo $2^n + 1$. If either $u$ or $v$ is equal to $2^n$, we use the special symbol $-1$ to represent it and we treat this situation as an easy special case, e.g., $u2^n \equiv -u \equiv 2^n + 1 - u \bmod 2^n + 1$.

Now for $n = 2^k$ we set $b = 2^{k/2}$ for $k$ even and $b = 2^{(k-1)/2}$ for $k$ odd. Furthermore, let $l = n/b$. Then $l \ge b$ and $b \mid l$. Both $u$ and $v$ are decomposed into $b$ blocks of $l$ bits each. So

| $u_{b-1}$ | $\cdots$ | $u_0$ |
|---|---|---|
| $l$ bits | | $l$ bits |

| $v_{b-1}$ | $\cdots$ | $v_0$ |
|---|---|---|
| $l$ bits | | $l$ bits |

$$u = u_{b-1}2^{(b-1)l} + \cdots + u_1 2^l + u_0 \quad \text{and} \quad v = v_{b-1}2^{(b-1)l} + \cdots + v_1 2^l + v_0 .$$

The product of $u$ and $v$ is

$$uv = y_{2b-1}2^{(2b-1)l} + \cdots + y_1 2^l + y_0 , \qquad (3.3.4)$$

where

$$y_i = \sum_{j=0}^{b-1} u_j v_{i-j}, \qquad 0 \le i < 2b .$$

We assume that $u_j = v_j = 0$ for $j < 0$ or $j > b - 1$. The term $y_{2b-1}$ is 0 and is only present for reasons of symmetry.

The product $uv$ could be computed by application of the convolution theorem. This would mean that we need $2b$ multiplications and shifts. However, if we use wrapped convolutions, we need only $b$ multiplications. This is the reason for computing $uv$ modulo $2^n + 1$. Because of $2^{bl} + 1 = 2^n + 1$ we have

$$y_i 2^{il} + y_{b+i} 2^{(b+i)l} \equiv (y_i - y_{b+i}) 2^{il} \bmod 2^n + 1 .$$

So (3.3.4) is transformed to

$$uv \equiv w_{b-1} 2^{(b-1)l} + \cdots + w_1 2^l + w_0 \bmod 2^n + 1 ,$$

where

$$w_i = y_i - y_{b+i}, \quad 0 \leq i < b .$$

Since the product of two binary numbers of length $l$ is less than $2^{2l}$ and since $y_i$ and $y_{b+i}$ are sums of $i + 1$ and $b - (i + 1)$ of such products, respectively, we get the bounds $-(b - 1 - i)2^{2l} < w_i < (i + 1)2^{2l}$. So $w_i$ has at most $b2^{2l}$ possible values. If we can compute the $w_i$'s modulo $b2^{2l}$ then we can compute $uv$ modulo $2^n + 1$ in $\mathcal{O}(b \log(b2^{2l}))$ additional steps by adding the $w_i$'s after appropriate shifts.

For computing the $w_i$'s modulo $b2^{2l}$ we compute them both modulo $b$ and modulo $2^{2l} + 1$. Let $w_i' = w_i \bmod b$ and $w_i'' = w_i \bmod (2^{2l} + 1)$. $b$ is a power of 2 and $2^{2l} + 1$ is odd, so $b$ and $2^{2l} + 1$ are relatively prime. $b = 2^p$ divides $l$, so it also divides $2^{2l}$, and therefore we have $2^{2l} + 1 \equiv 1 \bmod b$. As in the proof of the Chinese remainder theorem we get

$$w_i = (2^{2l} + 1)((w_i' - w_i'') \bmod b) + w_i'' ,$$

and $-(b - 1 - i)2^{2l} < w_i < (i + 1)2^{2l}$. The complexity of computing $w_i$ from $w_i'$ and $w_i''$ is $\mathcal{O}(l + \log b)$ for each $i$ ($\mathcal{O}(\log b)$ for $w_i' - w_i'' \bmod b$, $\mathcal{O}(l)$ for the shift by $2l$ positions, $\mathcal{O}(l + \log b)$ for subsequent additions). So in total the complexity is $\mathcal{O}(bl + b \log b)$ or $\mathcal{O}(n)$.

The $w_i$'s are computed modulo $b$ by setting $u_i' = u_i \bmod b$ and $v_i' = v_i \bmod b$ and forming the binary numbers

$$\hat{u} = u_{b-1}' 00 \ldots 0 u_{b-2}' 00 \ldots 0 \ldots 00 \ldots 0 u_0' ,$$
$$\hat{v} = v_{b-1}' 00 \ldots 0 v_{b-2}' 00 \ldots 0 \ldots 00 \ldots 0 v_0'$$

of length $3b \log b$. Every block of 0's has the length $2 \log b$. The computation of the product $\hat{u}\hat{v}$ by the Karatsuba algorithm takes time $\mathcal{O}((3b \log b)^{1.6})$, i.e., less than $\mathcal{O}(n)$. $\hat{u}\hat{v} = \sum_{i=0}^{2b-1} y_i' 2^{(3 \log b)i}$, where $y_i' = \sum_{j=0}^{2b-1} u_j' v_{i-j}'$. Furthermore, $y_i' < 2^{3 \log b}$. So the $y_i$'s can easily be extracted from the product $\hat{u}\hat{v}$. Then the values of the $w_i$'s modulo $b$ are simply $y_i' - y_{b+i}' \bmod b$.

The $w_i$'s modulo $2^{2l}+1$ are computed via a wrapped convolution. This means we have to compute a DFT, multiply the resulting vectors componentwise, and compute an iDFT. Let $\omega = 2^{4l/b}$ and $m = 2^{2l} + 1$. By Theorem 3.3.9 $b$ has a multiplicative inverse in $\mathbb{Z}_m$ and $\omega$ is a primitive $b$-th root of unity. So by Theorem 3.3.7 the negative wrapped convolution of $[u_0, \psi u_1, \ldots, \psi^{b-1}u_{b-1}]$ and $[v_0, \psi v_1, \ldots, \psi^{b-1}v_{b-1}]$, where $\psi = 2^{2l/b}$, is of the form

$$[(y_0 - y_b), \psi(y_1 - y_{b+1}), \ldots, \psi^{b-1}(y_{b-1} - y_{2b-1})] \bmod 2^{2l} + 1 ,$$

where $y_i = \sum_{j=0}^{b-1} u_j v_{i-j}$ for $0 \le i \le 2b - 1$. Now the $w_i$'s modulo $2^{2l} + 1$ can be computed by appropriate shifts.

**Algorithm INT_MULTSS**(in: $u, v, n$; out: $w$);
[$u, v$ are binary integers of length $n$, $n = 2^k$ for some $k \in \mathbb{N}$;
$w$ is a binary integer such that $w = uv \bmod 2^n + 1$.]

0. For small $n$ apply one of the usual multiplication algorithms.
   For big $n$ ($n > 3$ at least) set $b = 2^{k/2}$ if $k$ is even and $b = 2^{(k-1)/2}$ if $k$ is odd, and $l = n/b$.
   Let $u = \sum_{i=0}^{b-1} u_i 2^{li}$ and $v = \sum_{i=0}^{b-1} v_i 2^{li}$, where $0 \le u_i, v_i \le 2^l - 1$, be the representations of $u, v$ in the positional number system with base $2^l$.
1. Call FFT for computing $F_\omega[u_0, \psi u_1, \ldots, \psi^{b-1}u_{b-1}]$ modulo $2^{2l} + 1$ and $F_\omega[v_0, \psi v_1, \ldots, \psi^{b-1}v_{b-1}]$ modulo $2^{2l}+1$, where $\omega = 2^{4l/b}$ and $\psi = 2^{2l/b}$.
2. Apply INT_MULTSS recursively for computing the pairwise products of the DFTs of step (1) modulo $2^{2l}+1$. The case that one of the components is $2^{2l}$ is treated as a special case.
3. Compute the iDFT modulo $2^{2l} + 1$ of the vector of pairwise products of step (2).
   The result is $[w_0, \psi w_1, \ldots, \psi^{b-1}w_{b-1}]$ modulo $2^{2l}+1$, where $w_i$ is the $i$-th component of the negative wrapped convolution of $[u_0, u_1, \ldots, u_{b-1}]$ and $[v_0, v_1, \ldots, v_{b-1}]$.
   Compute $w_i'' = w_i \bmod 2^{2l} + 1$ by multiplication of $\psi^i w_i$ by $\psi^{-i}$ modulo $2^{2l} + 1$.
4. Compute $w_i' = w_i \bmod b$ as follows:
   a. Set $u_i' = u_i \bmod b$ and $v_i' = v_i \bmod b$ for $0 \le i < b$.
   b. Construct the binary numbers $\hat{u}$ and $\hat{v}$ by concatenating the $u_i$'s and $v_i$'s with blocks of $2 \log b$ zeros in between.
      So $\hat{u} = \sum_{i=0}^{b-1} u_i' 2^{(3\log b)i}$ and $\hat{v} = \sum_{i=0}^{b-1} v_i' 2^{(3\log b)i}$.
   c. Compute the product $\hat{u}\hat{v}$ by the Karatsuba algorithm.
   d. The product $\hat{u}\hat{v}$ is $\sum_{i=0}^{2b-1} y_i' 2^{(3\log b)i}$, where $y_i' = \sum_{j=0}^{2b-1} u_j' v_{i-j}'$.
      Set $w_i' = (y_i' - y_{b+i}') \bmod b$, for $0 \le i < b$.
5. Compute the $w_i$'s as $w_i = (2^{2l} + 1)((w_i' - w_i'') \bmod b) + w_i''$, where $-(b - 1 - i)2^{2l} < w_i < (i + 1)2^{2l}$.
6. Set $w = \sum_{i=0}^{b-1} w_i 2^{li} \bmod 2^n + 1$.

**Theorem 3.3.11.** The complexity of the algorithm INT_MULTSS is $\mathcal{O}(n \log n \cdot \log\log n)$.

*Proof.* Let $M(k)$ be the complexity of applying INT_MULTKSS to binary integers of length $k$. Then by Theorem 3.3.10 the complexity of the steps (1) to (3) is

$$\begin{aligned}&\mathcal{O}(\max\{b^2 \cdot \log b \cdot log 2^{2l/b},\ b \cdot M(b \cdot \log 2^{2l/b})\})\\ &= \mathcal{O}(\max\{bl \cdot \log b,\ b \cdot M(2l)\})\\ &= \mathcal{O}(bl \cdot \log b + b \cdot M(2l))\ ,\end{aligned}$$

where the first term in the complexity bound comes from the Fourier transforms and the second term from the multiplications of the components. The length of $\hat{u}$ and $\hat{v}$ in (4) is bounded by $3b \cdot \log b$, so the multiplication by the Karatsuba algorithm takes time $\mathcal{O}((3b \cdot \log b)^{1.6})$. For sufficiently big $b$ we have $(3b \cdot \log b)^{1.6} < b^2$, so that the complexity for step (4) is dominated by the term $\mathcal{O}(b^2 \log b)$ in the complexity bound for (1)–(3). The steps (5) and (6) are of complexity $\mathcal{O}(n)$ and can be neglected.

Using the fact that $n = bl$ and $b \leq \sqrt{n}$ we get the recursive relation

$$M(n) \leq c \cdot n \cdot \log n + b \cdot M(2l) \tag{3.3.5}$$

for a constant $c$ and sufficiently big $n$. Setting $M'(n) = M(n)/n$, we can transform (3.3.5) into

$$M'(n) \leq c \cdot \log n + 2M'(2l)$$

and furthermore by $l \leq 2\sqrt{n}$ into

$$M'(n) \leq c \cdot \log n + 2M'(4\sqrt{n})\ . \tag{3.3.6}$$

Now by induction on $n$ we can show that

$$M'(n) \leq c' \cdot \log n \cdot \log\log n \quad \text{for some } c'\ . \tag{3.3.7}$$

Assume that (3.3.7) hold for all $m < n$. Then by the induction hypothesis

$$\begin{aligned}M'(n) &\leq c \cdot \log n + 2M'(4\sqrt{n})\\ &\leq c \cdot \log n + 2 \cdot c' \cdot \log(4\sqrt{n}) \cdot \log\log(4\sqrt{n})\\ &= c \cdot \log n + 2 \cdot c' \cdot (2 + \tfrac{1}{2}\log n) \cdot \log(2 + \tfrac{1}{2}\log n)\\ &= c \cdot \log n + 4 \cdot c' \cdot \log(\underbrace{2 + \tfrac{1}{2}\log n}_{\leq \frac{2}{3}\log n \text{ for sufficiently big } n}) + c' \cdot \log n \cdot \log(\underbrace{2 + \tfrac{1}{2}\log n})\\ &\leq c \cdot \log n + 4c' \cdot \log \tfrac{2}{3} + 4c' \cdot \log\log n\\ &\quad + c' \cdot \log n \cdot \log \tfrac{2}{3} + c' \cdot \log n \cdot \log\log n\ .\end{aligned}$$

For sufficiently big $n$ and $c'$ the first four terms are dominated by the forth, which is negative. So $M'(n) \leq c' \cdot \log n \cdot \log\log n$. This proves (3.3.7).

Relation (3.3.7) implies

$$M(n) \leq c' \cdot n \cdot \log n \cdot \log\log n \ . \qquad \square$$

*Example 3.3.3.* Let us demonstrate how the algorithm INT_MULTSS works by applying it to $u = 217$ and $v = 145$. We choose $n = 16 = 2^4$, i.e., $k = 4$.

0. $b = 2^{4/2} = 4$ and $l = n/b = 4$. We decompose the binary representations of $u$ and $v$ into blocks:

$$u = \underbrace{0000}_{u_3} \cdot 2^{4\cdot 3} + \underbrace{0000}_{u_2} \cdot 2^{4\cdot 2} + \underbrace{1101}_{u_1} \cdot 2^{4\cdot 1} + \underbrace{1001}_{u_0} \cdot 2^{4\cdot 0},$$

$$v = \underbrace{0000}_{v_3} \cdot 2^{4\cdot 3} + \underbrace{0000}_{v_2} \cdot 2^{4\cdot 2} + \underbrace{1001}_{v_1} \cdot 2^{4\cdot 1} + \underbrace{0001}_{v_0} \cdot 2^{4\cdot 0}.$$

1. $\psi = 2^{2\cdot 4/4} = 4$; $\omega = \psi^2 = 16$ is a primitive 4th root of unity modulo $2^8+1$. As the DFT modulo $2^8 + 1$ of $[u_0, \psi u_1, \psi^2 u_2, \psi^3 u_3] = [9, 52, 0, 0]$ we get $[61, 70, 214, 205]$. As the DFT modulo $2^8 + 1$ von $[v_0, \psi v_1, \psi^2 v_2, \psi^3 v_3] = [1, 36, 0, 0]$ we get $[37, 63, 222, 196]$.
2. The pairwise product modulo $2^8 + 1$ is $p = [201, 41, 220, 88]$.
3. Now we have to compute the inverse DFT modulo $2^8+1$ of $p$, i.e., $F_\omega^{-1}(p) = \frac{1}{4}F_{\omega^{-1}}(p)$, where $\omega^{-1} = 241$. We get $F_\omega^{-1}(p) = \frac{1}{4}[36, 219, 35, 0] = [w_0, \psi w_1, \psi^2 w_2, \psi^3 w_3] \bmod 2^8 + 1$. So the $w_i''$'s are $w_0'' = 9$, $w_1'' = 94$, $w_2'' = 117$, $w_3'' = 0$.
4. a. $[u_0', u_1', u_2', u_3'] = [1, 1, 0, 0]$, $\quad [v_0', v_1', v_2', v_3'] = [1, 1, 0, 0]$.

   b. $\hat{u} = \underbrace{00}_{u_3'}\ 0000\ \underbrace{00}_{u_2'}\ 0000\ \underbrace{01}_{u_1'}\ 0000\ \underbrace{01}_{u_0'},$

   $\hat{v} = \underbrace{00}_{v_3'}\ 0000\ \underbrace{00}_{v_2'}\ 0000\ \underbrace{01}_{v_1'}\ 0000\ \underbrace{01}_{v_0'}.$

   c. $\hat{u}\hat{v} = 0 \cdots 0\,\underbrace{000001}_{y_2'}\,\underbrace{000010}_{y_1'}\,\underbrace{000001}_{y_0'}.$

   d. $w_0' = y_0' - y_4' \bmod 4 = 1,$
   $w_1' = y_1' - y_5' \bmod 4 = 2,$
   $w_2' = y_2' - y_6' \bmod 4 = 1,$
   $w_3' = y_3' - y_7' \bmod 4 = 0.$
5. $w_0 = (2^8 + 1) \cdot ((1 - 9) \bmod 4) + 9 = 9,$
   $w_1 = (2^8 + 1) \cdot ((2 - 94) \bmod 4) + 94 = 94,$
   $w_2 = (2^8 + 1) \cdot ((1 - 117) \bmod 4) + 117 = 117,$
   $w_3 = (2^8 + 1) \cdot ((0 - 0) \bmod 4) + 0 = 0.$
6. Combining these partial results we finally get $w = w_0 2^0 + w_1 2^4 + w_2 2^8 + w_3 2^{12} = 31465$.

### Exercises

1. Let $\omega$ be a primitive $n$-th root of unity in $K$, and let $n = 2m$. Show that $\omega^2$ is a primitive $m$-th root of unity in $K$.
2. Let $\omega$ be a primitive $n$-th root of unity in $K$, and let $n$ be even. Show that $\omega^{n/2} = -1$.
3. How could you use the Fourier transform to compute the product of $x^{1000} + 1$ and $x^{1000} + x$ over $\mathbb{C}$? How many evaluations points do you need?
4. Prove Theorem 3.3.7 (a).

## 3.4 Bibliographic notes

The modular approach and the technique of $p$-adic lifting are treated in Lauer (1983). For a way of exploiting sparseness in Hensel lifting we refer to Kaltofen (1985b).

The fast Fourier transform has been discovered by J. M. Cooley and J. W. Tukey (1965). The discrete Fourier transform has a long history. In Cooley et al. (1967) the roots of the FFT are traced back to Runge and König (1924). For overviews on FFT we refer to Cooley et al. (1969), Aho et al. (1974), or Lipson (1981).

# 4 Greatest common divisors of polynomials

## 4.1 Polynomial remainder sequences

If $K$ is a field, then $K[x]$ is a Euclidean domain, so $\gcd(f, g)$ for $f, g \in K[x]$ can be computed by the Euclidean algorithm. Often, however, we are given polynomials $f, g$ over a domain such as $\mathbb{Z}$ or $K[x_1, \ldots, x_{n-1}]$ and we need to compute their gcd.

Throughout this section we let $I$ be a unique factorization domain (ufd) and $K$ the quotient field of $I$.

*Definition 4.1.1.* A univariate polynomial $f(x)$ over the ufd $I$ is *primitive* iff there is no prime in $I$ which divides all the coefficients in $f(x)$.

A key fact concerning primitive polynomials has been established by C. F. Gauss.

**Theorem 4.1.1** (Gauss's lemma). Let $f, g$ be primitive polynomials over the ufd $I$. Then also $f \cdot g$ is primitive.

*Proof.* Let $f(x) = \sum_{i=0}^{m} a_i x^i$, $g(x) = \sum_{i=0}^{n} b_i x^i$. For an arbitrary prime $p$ in $I$, let $j$ and $k$ be the minimal indices such that $p$ does not divide $a_j$ and $b_k$, respectively. Then $p$ does not divide the coefficient of $x^{j+k}$ in $f \cdot g$. □

**Corollary.** Gcd's and factorization are basically the same over $I$ and over $K$.

a. If $f_1, f_2 \in I[x]$ are primitive and $g$ is a gcd of $f_1$ and $f_2$ in $I[x]$, then $g$ is also a gcd of $f_1$ and $f_2$ in $K[x]$.
b. If $f \in I[x]$ is primitive and irreducible in $I[x]$, then it is also irreducible in $K[x]$.

*Proof.* a. Clearly every common divisor of $f_1$ and $f_2$ in $I[x]$ is also a common divisor in $K[x]$. Now let $g'$ be a common divisor of $f_1$ and $f_2$ in $K[x]$. Eliminating the common denominator of coefficients in $g'$ and making the result primitive, we get basically the same divisor. So w.l.o.g. we may assume that $g'$ is primitive in $I[x]$. For some primitive $h_1, h_2 \in I[x]$, $a_1, a_2 \in K$ we can write $f_1 = a_1 \cdot h_1 \cdot g'$, $f_2 = a_2 \cdot h_2 \cdot g'$. Since, by Gauss's lemma, $h_1 g'$ and $h_2 g'$ are primitive, $a_1$ and $a_2$ have to be units in $I$. So $g'$ is also a common divisor of $f_1$ and $f_2$ in $I[x]$.

b. Suppose $f = f_1 \cdot f_2$ for some $f_1, f_2 \in K[x] \setminus K$. Then for some primitive

$f_1', f_2' \in I[x] \setminus I$ and $a \in K$ we have $f = a \cdot f_1' \cdot f_2'$. By Gauss's lemma $f_1' \cdot f_2'$ is also primitive, so $a$ has to be a unit. □

By this corollary the computation of gcds in $I[x]$ can be reduced to the computation of gcds in $K[x]$. From a complexity point of view, however, this reduction is not very efficient, since arithmetic in the quotient field is usually much more costly than in the underlying integral domain. In the following we will develop methods for working directly in the ufd $I$.

*Definition 4.1.2.* Up to multiplication by units we can decompose every polynomial $a(x) \in I[x]$ uniquely into

$$a(x) = \mathrm{cont}(a) \cdot \mathrm{pp}(a) \ ,$$

where $\mathrm{cont}(a) \in I$ and $\mathrm{pp}(a)$ is a primitive polynomial in $I[x]$; $\mathrm{cont}(a)$ is the *content* of $a(x)$, $\mathrm{pp}(a)$ is the *primitive part* of $a(x)$.

*Definition 4.1.3.* Two non-zero polynomials $a(x), b(x) \in I[x]$ are *similar* iff there are *similarity coefficients* $\alpha, \beta \in I^*$ such that $\alpha \cdot a(x) = \beta \cdot b(x)$. In this case we write $a(x) \simeq b(x)$. Obviously $a(x) \simeq b(x)$ if and only if $\mathrm{pp}(a) = \mathrm{pp}(b)$. $\simeq$ is an equivalence relation preserving the degree.

Now we are ready to define what we mean by a polynomial remainder sequence.

*Definition 4.1.4.* Let $k$ be a natural number greater than 1, and $f_1, f_2, \ldots, f_{k+1}$ polynomials in $I[x]$. Then $f_1, f_2, \ldots, f_{k+1}$ is a *polynomial remainder sequence* (prs) iff

$$\deg(f_1) \geq \deg(f_2) \ ,$$
$$f_i \neq 0 \ \text{ for } \ 1 \leq i \leq k \quad \text{and} \quad f_{k+1} = 0 \ ,$$
$$f_i \simeq \mathrm{prem}(f_{i-2}, f_{i-1}) \ \text{ for } \ 3 \leq i \leq k+1 \ .$$

**Lemma 4.1.2.** Let $a, b, a', b' \in I[x]^*$, $\deg(a) \geq \deg(b)$, and $r \simeq \mathrm{prem}(a, b)$.
a. If $a \simeq a'$ and $b \simeq b'$ then $\mathrm{prem}(a, b) \simeq \mathrm{prem}(a', b')$.
b. $\gcd(a, b) \simeq \gcd(b, r)$.

*Proof.* a. Let $\alpha a = \alpha' a'$, $\beta b = \beta' b'$, and $m = \deg(a), n = \deg(b)$. By Lemma 2.2.4

$$\begin{aligned} \beta^{m-n+1} \alpha \mathrm{prem}(a, b) &= \mathrm{prem}(\alpha a, \beta b) \\ &= \mathrm{prem}(\alpha' a', \beta' b') = (\beta')^{m-n+1} \alpha' \mathrm{prem}(a', b') \ . \end{aligned}$$

b. Clearly $a, b$ and $b$, $\mathrm{prem}(a, b)$ have the same primitive divisors, so $\gcd(a, b) \simeq \gcd(b, \mathrm{prem}(a, b))$, and by part (a) this is similar to $\gcd(b, r)$. □

**Algorithm GCD_PRS**(in: $a, b$; out: $g$);
$[a, b \in I[x]^*, g = \gcd(a, b)]$
1. if $\deg(a) \geq \deg(b)$
   then $\{f_1 := \mathrm{pp}(a);\ f_2 := \mathrm{pp}(b)\}$
   else $\{f_1 := \mathrm{pp}(b);\ f_2 := \mathrm{pp}(a)\}$;
2. $d := \gcd(\mathrm{cont}(a), \mathrm{cont}(b))$;
3. compute $f_3, \ldots, f_k, f_{k+1} = 0$ such that $f_1, f_2, \ldots, f_k, 0$ is a prs;
4. $g := d \cdot \mathrm{pp}(f_k)$; return.

Therefore, if $f_1, f_2, \ldots, f_k, 0$ is a prs, then

$$\gcd(f_1, f_2) \simeq \gcd(f_2, f_3) \simeq \ldots \simeq \gcd(f_{k-1}, f_k) \simeq f_k .$$

If $f_1$ and $f_2$ are primitive, then by Gauss's lemma also their gcd must be primitive, i.e., $\gcd(f_1, f_2) = \mathrm{pp}(f_k)$. So the gcd of polynomials over the ufd $I$ can be computed by the algorithm GCD_PRS.

Actually GCD_PRS is a family of algorithms, depending on how exactly we choose the elements of the prs in step (3). Starting from primitive polynomials $f_1, f_2$, there are various possibilities for this choice.

In the so-called *generalized Euclidean algorithm* we simply set

$$f_i := \mathrm{prem}(f_{i-2}, f_{i-1}) \quad \text{for } 3 \leq i \leq k+1 .$$

This choice, however, leads to an enormous blow-up of coefficients, as can be seen in the following example.

*Example 4.1.1.* We consider polynomials over $\mathbb{Z}$. Starting from the primitive polynomials

$$f_1 = x^8 + x^6 - 3x^4 - 3x^3 + 8x^2 + 2x - 5 ,$$
$$f_2 = 3x^6 + 5x^4 - 4x^2 - 9x + 21 ,$$

the generalized Euclidean algorithm generates the prs

$$f_3 = -15x^4 + 3x^2 - 9 ,$$
$$f_4 = 15795x^2 + 30375x - 59535 ,$$
$$f_5 = 1254542875143750x - 1654608338437500 ,$$
$$f_6 = 12593338795500743100931141992187500 .$$

So the gcd of $f_1$ and $f_2$ is the primitive part of $f_6$, i.e., 1.

Although the inputs and the output of the algorithm may have extremely short coefficients, the coefficients in the intermediate results may be enormous. In particular, for univariate polynomials over $\mathbb{Z}$ the length of the coefficients grows exponentially at each step (see Knuth 1981: sect. 4.6.1). This effect of intermediate coefficient growth is even more dramatic in the case of multivariate polynomials.

Another possible choice for computing the prs in GCD_PRS is to shorten the coefficients as much as possible, i.e., always eliminate the content of the intermediate results.

$$f_i := \mathrm{pp}(\mathrm{prem}(f_{i-2}, f_{i-1})) .$$

We call such a prs a *primitive prs*.

*Example 4.1.1* (continued). The primitive prs starting from $f_1, f_2$ is

$$\begin{aligned} f_3 &= 5x^4 - x^2 + 3 , \\ f_4 &= 13x^2 + 25x - 49 , \\ f_5 &= 4663x - 6150 , \\ f_6 &= 1 . \end{aligned}$$

Keeping the coefficients always in the shortest form carries a high price. For every intermediate result we have to determine its content, which means doing a lot of gcd computations in the coefficient domain.

The goal, therefore, is to keep the coefficients as short as possible without actually having to compute a lot of gcds in the coefficient domain. So we set

$$\beta_i f_i := \mathrm{prem}(f_{i-2}, f_{i-1}) ,$$

where $\beta_i$, a factor of $\mathrm{cont}(\mathrm{prem}(f_{i-2}, f_{i-1}))$, needs to be determined. The best algorithm of this form known is Collins's *subresultant prs algorithm* (Collins 1967, Brown and Traub 1971).

First we need some notation. Let

$$a(x) = \sum_{i=0}^{m} a_i x^i , \quad b(x) = \sum_{i=0}^{n} b_i x^i$$

be non-constant polynomials in $I[x]$ of degree $m$ and $n$, respectively, where $m \geq n$.

Let $M(a, b)$ be the *Sylvester matrix* of $a$ and $b$, i.e.,

$$M(a,b) = \begin{pmatrix} a_m & a_{m-1} & \cdots & \cdots & \cdots & a_1 & a_0 & 0 & \cdots & \cdots & \cdots & 0 \\ 0 & a_m & a_{m-1} & \cdots & \cdots & \cdots & a_1 & a_0 & 0 & \cdots & \cdots & 0 \\ & & & & \vdots & & & & & & & \\ 0 & \cdots & \cdots & \cdots & 0 & a_m & a_{m-1} & \cdots & \cdots & \cdots & a_1 & a_0 \\ - & - & - & - & - & - & - & - & - & - & - & - \\ b_n & b_{n-1} & \cdots & \cdots & \cdots & b_1 & b_0 & 0 & \cdots & \cdots & \cdots & 0 \\ 0 & b_n & b_{n-1} & \cdots & \cdots & \cdots & b_1 & b_0 & 0 & \cdots & \cdots & 0 \\ & & & & \vdots & & & & & & & \\ 0 & \cdots & \cdots & \cdots & 0 & b_n & b_{n-1} & \cdots & \cdots & \cdots & b_1 & b_0 \end{pmatrix}.$$

The lines of $M(a,b)$ consist of the coefficients of the polynomials $x^{n-1}a(x), \ldots, xa(x), a(x)$ and $x^{m-1}b(x), \ldots, xb(x), b(x)$, i.e., there are $n$ lines of coefficients of $a$ and $m$ lines of coefficients of $b$. The resultant of $a$ and $b$ is the determinant of $M(a,b)$. In order to get the subresultants, we delete certain lines and columns in $M(a,b)$.

By $M(a,b)_{i,j}$ we denote the matrix resulting from $M(a,b)$ by deleting

- the last $j$ rows of coefficients of $a$,
- the last $j$ rows of coefficients of $b$,
- the last $2j+1$ columns except the $(m+n-i-j)$-th,

for $0 \le i \le j \le n-1$.

*Definition 4.1.5.* Let $a(x), b(x) \in I[x]^*$ with $m = \deg(a) \ge \deg(b) = n$. The determinant of $M(a,b)$ is the *resultant* of $a(x)$ and $b(x)$.

For $0 \le j \le n-1$ the polynomial

$$S_j(a,b)(x) = \sum_{i=0}^{j} \det(M(a,b)_{i,j})x^i$$

is the *j-th subresultant* of $a$ and $b$.

Obviously $\deg(S_j(a,b)) \le j$.

*Example 4.1.2.* Let $a(x) = 2x^4 + x^2 - 4$, $b(x) = 3x^2 + 2$ over the integers. We want to compute the first subresultant $S_1(a,b)$ of $a$ and $b$.

$$M(a,b) = \begin{pmatrix} 2 & 0 & 1 & 0 & -4 & 0 \\ 0 & 2 & 0 & 1 & 0 & -4 \\ - & - & - & - & - & - \\ 3 & 0 & 2 & 0 & 0 & 0 \\ 0 & 3 & 0 & 2 & 0 & 0 \\ 0 & 0 & 3 & 0 & 2 & 0 \\ 0 & 0 & 0 & 3 & 0 & 2 \end{pmatrix}.$$

Thus we get

$$S_1(a,b)(x) = \det(M(a,b)_{1,1})x + \det(M(a,b)_{0,1})$$
$$= \det\begin{pmatrix} 2 & 0 & 1 & 0 \\ 3 & 0 & 2 & 0 \\ 0 & 3 & 0 & 2 \\ 0 & 0 & 3 & 0 \end{pmatrix} \cdot x + \det\begin{pmatrix} 2 & 0 & 1 & -4 \\ 3 & 0 & 2 & 0 \\ 0 & 3 & 0 & 0 \\ 0 & 0 & 3 & 2 \end{pmatrix} = 0x + 102 \ .$$

In the following we will give a relation between the chain of subresultants of polynomials $a(x), b(x)$ and the elements of a prs starting from $a$ and $b$. We will use the following notation: for a prs $f_1, f_2, \ldots, f_k, 0$ in $I[x]$,

$$n_i := \deg(f_i) \quad \text{for } 1 \le i \le k \quad (n_1 \ge n_2 > \ldots > n_k \ge 0) \ ,$$
$$\delta_i := n_i - n_{i+1} \quad \text{for } 1 \le i \le k-1 \ .$$

**Theorem 4.1.3.** Let $f_1, f_2 \in I[x]^*$ and $f_1, f_2, \ldots, f_k, f_{k+1} = 0$ be a prs in $I[x]$. Let $\alpha_i := \mathrm{lc}(f_{i-1})^{\delta_{i-2}+1}$ for $3 \le i \le k+1$, and $\beta_i \in I$ such that $\beta_i f_i = \mathrm{prem}(f_{i-2}, f_{i-1})$ for $3 \le i \le k+1$.

Then for $3 \le i \le k$ we have:

$$S_{n_{i-1}-1}(f_1, f_2) = \gamma_i f_i \ ,$$
$$S_j(f_1, f_2) = 0 \quad \text{for } n_{i-1} - 1 > j > n_i \ ,$$
$$S_{n_i}(f_1, f_2) = \theta_i f_i \ ,$$
$$S_j(f_1, f_2) = 0 \quad \text{for } n_k > j \ge 0 \ ,$$

where

$$\gamma_i = (-1)^{\sigma_i} \cdot \mathrm{lc}(f_{i-1})^{1-\delta_{i-1}} \cdot \left(\prod_{l=3}^{i} (\beta_l/\alpha_l)^{n_{l-1}-n_{i-1}+1} \cdot \mathrm{lc}(f_{l-1})^{\delta_{l-2}+\delta_{l-1}}\right) ,$$

$$\theta_i = (-1)^{\tau_i} \cdot \mathrm{lc}(f_i)^{\delta_{i-1}-1} \cdot \left(\prod_{l=3}^{i} (\beta_l/\alpha_l)^{n_{l-1}-n_i} \cdot \mathrm{lc}(f_{l-1})^{\delta_{l-2}+\delta_{l-1}}\right) ,$$

$$\sigma_i = \sum_{l=3}^{i} (n_{l-2} - n_{i-1} + 1)(n_{l-1} - n_{i-1} + 1) \ ,$$

$$\tau_i = \sum_{l=3}^{i} (n_{l-2} - n_i)(n_{l-1} - n_i) \ .$$

*Proof.* For a proof of Theorem 4.1.3 we refer to Brown and Traub (1971). See also Brown (1978). □

In simpler words, Theorem 4.1.3 states that both $S_{n_{i-1}-1}(f_1, f_2)$ and $S_{n_i}(f_1, f_2)$ are similar to $f_i$, and all the subresultants in between vanish.

*Example 4.1.3.* Let us demonstrate the relations stated in Theorem 4.1.3 by considering the polynomials

$$f_1 = x^4 - 2x^3 + 1, \qquad f_2 = 3x^2 + 2x + 3$$

in $\mathbb{Z}[x]$. We choose $\beta_3 = 1$, i.e., $f_3 = \text{prem}(f_1, f_2) = 58x + 6$.
The first subresultant of $f_1, f_2$ is

$$\begin{aligned} S_1(f_1, f_2) &= \sum_{i=0}^{1} \det(M(f_1, f_2)_{i,1})x^i \\ &= \det\begin{pmatrix} 1 & -2 & 0 & 0 \\ 3 & 2 & 3 & 0 \\ 0 & 3 & 2 & 3 \\ 0 & 0 & 3 & 2 \end{pmatrix} \cdot x + \det\begin{pmatrix} 1 & -2 & 0 & 1 \\ 3 & 2 & 3 & 0 \\ 0 & 3 & 2 & 0 \\ 0 & 0 & 3 & 3 \end{pmatrix} \\ &= -58x - 6 = -\text{prem}(f_1, f_2) \ . \end{aligned}$$

The coefficient of similarity $\gamma_3$ in Theorem 4.1.3 is

$$\begin{aligned} \gamma_3 &= (-1)^{\sigma_3} \cdot 3^{1-\delta_2} \cdot (\beta_3/\alpha_3)^{n_2-n_2+1} \cdot 3^{\delta_1+\delta_2} \\ &= (-1)^3 \cdot 3^0 \cdot (1/27) \cdot 3^3 = -1 \ . \end{aligned}$$

*Example 4.1.4.* Suppose $f_1, f_2, f_3, f_4, f_5, f_6 = 0$ is a prs in $I[x]$ with $n_1 = 10$, $n_2 = 9$, $n_3 = 6$, $n_4 = 5$, $n_5 = 1$. Then we must have the following relations

$$\begin{aligned} &S_{n_{3-1}-1} = S_8 \simeq f_3 \ , \\ &S_7 = 0 \ , \\ &S_{n_3} = S_6 \simeq f_3 \ , \\ &S_{n_{4-1}-1} = S_5 \simeq f_4 \ , \\ &S_{n_4} = S_5 \simeq f_4 \ , \\ &S_{n_{5-1}-1} = S_4 \simeq f_5 \ , \\ &S_3 = S_2 = 0 \ , \\ &S_{n_5} = S_1 \simeq f_5 \ , \\ &S_0 = 0 \ . \end{aligned}$$

So the sequence of subresultants of $f_1, f_2$ basically agrees with a prs starting from $f_1, f_2$. Moreover, this particular prs eliminates much of the content of the intermediate results without ever having to compute a gcd of coefficients. In fact, there is a very efficient way of determining the sequence of subresultants, see Brown (1978).

**Theorem 4.1.4.** Let $f_1, f_2, \ldots, f_{k+1}$ be as in Theorem 4.1.3. In order to get $\gamma_i = 1$ for $3 \leq i \leq k+1$, we have to set $\beta_i$ as follows:

$$\beta_3 = (-1)^{\delta_1+1} ,$$
$$\beta_i = (-1)^{\delta_{i-2}+1} \cdot \mathrm{lc}(f_{i-2}) \cdot h_{i-2}^{\delta_{i-2}} \quad \text{for } i = 4, \ldots, k+1 ,$$

where

$$h_2 = \mathrm{lc}(f_2)^{\delta_1} ,$$
$$h_i = \mathrm{lc}(f_i)^{\delta_{i-1}} \cdot h_{i-1}^{1-\delta_{i-1}} \quad \text{for } i = 3, \ldots, k .$$

By choosing the similarity coefficient $\beta_i$ as in Theorem 4.1.4, we get the so-called *subresultant prs*:

$$f_3 = (-1)^{\delta_1+1} \cdot \mathrm{prem}(f_1, f_2) ,$$
$$f_i = \frac{(-1)^{\delta_{i-2}+1}}{\mathrm{lc}(f_{i-2}) \cdot h_{i-2}^{\delta_{i-2}}} \cdot \mathrm{prem}(f_{i-2}, f_{i-1}) \quad \text{for } i = 4, \ldots, k .$$

This subresultant prs is computed by the algorithm PRS_SR.

**Algorithm PRS_SR**(in: $f_1, f_2$; out: $F = [f_1, f_2, \ldots, f_k]$);
[$f_1, f_2 \in I[x]$, $\deg(f_1) \geq \deg(f_2)$, $f_1, f_2, \ldots, f_k, 0$ are the subresultant prs for $f_1, f_2$ up to sign]
1. $F := [f_2, f_1]$;
   $g := 1$; $h := 1$; $f' := f_2$; $i := 3$;
2. while $f' \neq 0$ and $\deg(f') > 0$ do
   {$\delta := \deg(f_{i-2}) - \deg(f_{i-1})$;
   $f' := \mathrm{prem}(f_{i-2}, f_{i-1})$;
   if $f' \neq 0$
   then {$f_i := f'/(g \cdot h^\delta)$; $F := \mathrm{CONS}(f_i, F)$;
   $g := \mathrm{lc}(f_{i-1})$; $h := h^{1-\delta} \cdot g^\delta$;
   $i := i+1$}};
   $F := \mathrm{INV}(F)$; return.

*Example 4.1.1* (continued). The subresultant prs (up to sign) of $f_1$ and $f_2$ computed by PRS_SR is

$$f_3 = -15x^4 + 3x^2 - 9 ,$$
$$f_4 = 65x^2 + 125x - 245 ,$$
$$f_5 = -9326x + 12300 ,$$
$$f_6 = 260708 .$$

If we apply PRS_SR to univariate polynomials over the integers, we can give a bound on the length of the coefficients that could appear. Let $f_1(x), f_2(x) \in \mathbb{Z}[x]$ of degree $m$ and $n$, respectively, and let the absolute values of all the coefficients be bounded by $d$. We use Hadamard's bound for the determinant of a $p \times p$-matrix $A = (a_{ij})$

$$\det(A) \leq \prod_{i=1}^{p} \sqrt{\sum_{j=1}^{p} a_{ij}^2} .$$

Applying this to the resultant of $f_1$ and $f_2$ we obtain

$$\left(\sqrt{(m+1)d^2}\right)^n \left(\sqrt{(n+1)d^2}\right)^m = d^{m+n}(m+1)^{n/2}(n+1)^{m/2}$$

as a bound for the coefficients of the subresultant prs.

For $\mathbb{Z}[x_1, \ldots, x_r]$, $n$ the maximal degree in any variable of $f_1$ and $f_2$, and $d$ a bound for the absolute values of all the integer coefficients in $f_1$ and $f_2$, the worst case complexity of PRS_SR is proportional to

$$n^{2r+2}(r \log n + L(d))^2$$

(see Loos 1983).

### Exercises

1. Prove: If $I$ is a ufd then also $I[x]$ is a ufd.
2. Write a procedure in Maple (or your favorite computer algebra system) implementing the algorithm PRS_SR.
3. Compute the sequence of subresultants for the polynomials

$$f(x) = 3x^5 - 2x^4 - 18x^3 - 6x^2 + 15x + 9 ,$$
$$g(x) = x^4 - 3x^3 + x^2 - 2x - 3 .$$

   How does this sequence of subresultants compare to a polynomial remainder sequence for $f$ and $g$?
4. Let $K$ be an algebraically closed field, $n$ a positive integer, $H$ a hypersurface in $\mathbb{A}^n(K)$, the affine space of dimension $n$ over $K$. Let $f \in K[x_1, \ldots, x_n] \setminus K$ be a defining polynomial of $H$, i.e., $H = \{(a_1, \ldots, a_n) \mid f(a_1, \ldots, a_n) = 0\}$. Let $f = f_1^{m_1} \cdot \ldots \cdot f_r^{m_r}$ be the factorization of $f$ into irreducible factors. Let $I = I(H)$, i.e., $I$ is the ideal of polynomials in $K[x_1, \ldots, x_n]$ that vanish on $H$.
   Show that $I = \langle f_1 \cdot \ldots \cdot f_r \rangle$.
5. Let $K$ be a field, $f, g \in K[x, y]$ relatively prime.
   Show that there are only finitely many points $(a_1, a_2) \in \mathbb{A}^2(K)$ such that $f(a_1, a_2) = g(a_1, a_2) = 0$.

## 4.2 A modular gcd algorithm

For motivation let us once again look at the polynomials in Example 4.1.1,

$$f_1 = x^8 + x^6 - 3x^4 - 3x^3 + 8x^2 + 2x - 5 \ ,$$
$$f_2 = 3x^6 + 5x^4 - 4x^2 - 9x + 21 \ .$$

If $f_1$ and $f_2$ have a common factor $h$, then for some $q_1, q_2$ we have

$$f_1 = q_1 \cdot h, \qquad f_2 = q_2 \cdot h \ . \tag{4.2.1}$$

These relations stay valid if we take every coefficient in (4.2.1) modulo 5. But modulo 5 we can compute the gcd of $f_1$ and $f_2$ in a very fast way, since all the coefficients that will ever appear are bounded by 5. In fact the gcd of $f_1$ and $f_2$ modulo 5 is 1. By comparing the degrees on both sides of the equations in (4.2.1) we see that also over the integers $\gcd(f_1, f_2) = 1$. In this section we want to generalize this approach and derive a modular algorithm for computing the gcd of polynomials over the integers.

In any modular approach we need a bound for the number of moduli that we have to take, more precisely for the product of these moduli. In our case we need to know how big the coefficients of the gcd can be, given bounds for the coefficients of the inputs. Clearly the coefficients in the gcd can be bigger than the coefficients in the inputs, as can be seen from the following example:

$$\begin{aligned} a &= x^3 + x^2 - x - 1 = (x+1)^2(x-1) \ , \\ b &= x^4 + x^3 + x + 1 = (x+1)^2(x^2 - x + 1) \ , \\ \gcd(a, b) &= x^2 + 2x + 1 \quad\ \ = (x+1)^2 \ . \end{aligned}$$

The bound in Theorem 4.2.1 is derived from investigations in Landau (1905), Mignotte (1974, 1983).

**Theorem 4.2.1** (Landau–Mignotte bound). Let $a(x) = \sum_{i=0}^m a_i x^i$ and $b(x) = \sum_{i=0}^n b_i x^i$ be polynomials over $\mathbb{Z}$ ($a_m \neq 0 \neq b_n$) such that $b$ divides $a$. Then

$$\sum_{i=0}^n |b_i| \leq 2^n \left| \frac{b_n}{a_m} \right| \sqrt{\sum_{i=0}^m a_i^2} \ .$$

**Corollary.** Let $a(x) = \sum_{i=0}^m a_i x^i$ and $b(x) = \sum_{i=0}^n b_i x^i$ be polynomials over $\mathbb{Z}$ ($a_m \neq 0 \neq b_n$). Every coefficient of the gcd of $a$ and $b$ in $\mathbb{Z}[x]$ is bounded in absolute value by

$$2^{\min(m,n)} \cdot \gcd(a_m, b_n) \cdot \min\left( \frac{1}{|a_m|} \sqrt{\sum_{i=0}^m a_i^2}, \ \frac{1}{|b_n|} \sqrt{\sum_{i=0}^n b_i^2} \right) \ .$$

*Proof.* The gcd of $a$ and $b$ is a divisor of both $a$ and $b$ and its degree is bounded by the minimum of the degrees of $a$ and $b$. Furthermore the leading coefficient of the gcd divides $a_m$ and $b_n$ and therefore also $\gcd(a_m, b_n)$. □

The gcd of $a(x)$ mod $p$ and $b(x)$ mod $p$ may not be the modular image of the integer gcd of $a$ and $b$. An example for this is $a(x) = x - 3, b(x) = x + 2$. The gcd over $\mathbb{Z}$ is 1, but modulo 5 $a$ and $b$ are equal and their gcd is $x + 2$. But fortunately these situations are rare.

So what we want from a prime $p$ is the commutativity of the following diagram, where $\phi_p$ is the homomorphism from $\mathbb{Z}[x]$ to $\mathbb{Z}_p[x]$ defined as $\phi_p(f(x)) = f(x) \bmod p$.

$$\begin{array}{ccc} \mathbb{Z}[x] \times \mathbb{Z}[x] & \xrightarrow{\phi_p} & \mathbb{Z}_p[x] \times \mathbb{Z}_p[x] \\ \text{gcd in } \mathbb{Z}[x] \downarrow & & \downarrow \text{gcd in } \mathbb{Z}_p[x] \\ \mathbb{Z}[x] & \xrightarrow{\phi_p} & \mathbb{Z}_p[x] \end{array}$$

**Lemma 4.2.2.** Let $a, b \in \mathbb{Z}[x]^*$, $p$ a prime number not dividing the leading coefficients of both $a$ and $b$. Let $a_{(p)}$ and $b_{(p)}$ be the images of $a$ and $b$ modulo $p$, respectively. Let $c = \gcd(a, b)$ over $\mathbb{Z}$.
a. $\deg(\gcd(a_{(p)}, b_{(p)})) \geq \deg(\gcd(a, b))$.
b. If $p$ does not divide the resultant of $a/c$ and $b/c$, then $\gcd(a_{(p)}, b_{(p)}) = c \bmod p$.

*Proof.* a. $\gcd(a, b) \bmod p$ divides both $a_{(p)}$ and $b_{(p)}$, so it divides $\gcd(a_{(p)}, b_{(p)})$. Therefore $\deg(\gcd(a_{(p)}, b_{(p)})) \geq \deg(\gcd(a, b) \bmod p)$. But $p$ does not divide the leading coefficient of $\gcd(a, b)$, so $\deg(\gcd(a, b) \bmod p) = \deg(\gcd(a, b))$.

b. Let $c_{(p)} = c \bmod p$. $a/c$ and $b/c$ are relatively prime. $c_{(p)}$ is non-zero. So

$$\gcd(a_{(p)}, b_{(p)}) = c_{(p)} \cdot \gcd(a_{(p)}/c_{(p)}, b_{(p)}/c_{(p)}) .$$

If $\gcd(a_{(p)}, b_{(p)}) \neq c_{(p)}$, then the gcd of the right-hand side must be nontrivial. Therefore $\operatorname{res}(a_{(p)}/c_{(p)}, b_{(p)}/c_{(p)}) = 0$. The resultant, however, is a sum of products of coefficients, so $p$ has to divide $\operatorname{res}(a/c, b/c)$. □

Of course, the gcd of polynomials over $\mathbb{Z}_p$ is determined only up to multiplication by non-zero constants. So by "$\gcd(a_{(p)}, b_{(p)}) = c \bmod p$" we actually mean "$c \bmod p$ *is a* gcd of $a_{(p)}, b_{(p)}$."

From Lemma 4.2.2 we know that there are only finitely many primes $p$ which do not divide the leading coefficients of $a$ and $b$ but for which $\deg(\gcd(a_{(p)}, b_{(p)})) > \deg(\gcd(a, b))$. When these degrees are equal we call $p$ a *lucky* prime.

One possibility for computing the gcd of two integer polynomials $a$ and $b$ would be to determine the Landau–Mignotte bound $M$, choose a prime $p \geq 2M$

not dividing the leading coefficients of $a$ and $b$, compute the gcd $c_{(p)}$ of $a$ and $b$ modulo $p$, center the coefficients of $c_{(p)}$ around 0 (i.e., represent $\mathbb{Z}_p$ as $\{k \mid -p/2 < k \leq p/2\}$), interpret $c_{(p)}$ as an integer polynomial $c$, and test whether $c$ divides $a$ and $b$ in $\mathbb{Z}[x]$. If yes, we have found the gcd of $a$ and $b$, if no, $p$ was an unlucky prime and we choose a different prime. Since there are only finitely many unlucky primes, this algorithm terminates and produces the gcd. The drawback is that $p$ may be very big and coefficient arithmetic may be costly.

In the sequel we describe a modular algorithm that chooses several primes, computes the gcd modulo these primes, and finally combines these modular gcds by an application of the Chinese remainder algorithm. Since in $\mathbb{Z}_p[x]$ the gcd is defined only up to multiplication by constants, we are confronted with the so-called *leading coefficient problem*. The reason for this problem is that over the integers the gcd will, in general, have a leading coefficient different from 1, whereas over $\mathbb{Z}_p$ the leading coefficient can be chosen arbitrarily. So before we can apply the Chinese remainder algorithm we have to normalize the leading coefficient of $\gcd(a_{(p)}, b_{(p)})$. Let $a_m, b_n$ be the leading coefficients of $a$ and $b$, respectively. The leading coefficient of the gcd divides the gcd of $a_m$ and $b_n$. Thus, for primitive polynomials we may normalize the leading coefficient of $\gcd(a_{(p)}, b_{(p)})$ to $\gcd(a_m, b_n) \bmod p$ and in the end take the primitive part of the result. These considerations lead to the following modular gcd algorithm.

**Algorithm GCD_MOD**(in: $a, b$; out: $g$);
[$a, b \in \mathbb{Z}[x]^*$ primitive, $g = \gcd(a, b)$.
Integers modulo $m$ are represented as $\{k \mid -m/2 < k \leq m/2\}$.]
1. $d := \gcd(\mathrm{lc}(a), \mathrm{lc}(b))$;
   $M := 2 \cdot d \cdot$ (Landau–Mignotte bound for $a, b$);
   [in fact any other bound for the size of the coefficients can be used]
2. $p :=$ a new prime not dividing $d$;
   $c_{(p)} := \gcd(a_{(p)}, b_{(p)})$; [with $\mathrm{lc}(c_{(p)}) = 1$]
   $g_{(p)} := (d \bmod p) \cdot c_{(p)}$;
3. if $\deg(g_{(p)}) = 0$ then $\{g := 1$; return$\}$;
   $P := p$;
   $g := g_{(p)}$;
4. while $P \leq M$ do
      $\{p :=$ a new prime not dividing $d$;
      $c_{(p)} := \gcd(a_{(p)}, b_{(b)})$; [with $\mathrm{lc}(c_{(p)}) = 1$]
      $g_{(p)} := (d \bmod p) \cdot c_{(p)}$;
      if $\deg(g_{(p)}) < \deg(g)$ then goto (3);
      if $\deg(g_{(p)}) = \deg(g)$
      then $\{g :=$ CRA_2$(g, g_{(p)}, P, p)$;
         [actually CRA_2 is applied to the coefficients of $g$ and $g_{(p)}$]
         $P := P \cdot p\}$ $\}$;
5. $g := \mathrm{pp}(g)$;
   if $g|a$ and $g|b$ then return;
   goto (2).

Usually we do not need as many primes as the Landau–Mignotte bound tells us for determining the integer coefficients of the gcd in GCD_MOD. Whenever $g$ remains unchanged for a series of iterations through the "while"-loop, we might apply the test in step (5) and exit if the outcome is positive.

*Example 4.2.1.* We apply GCD_MOD for computing the gcd of

$$a = 2x^6 - 13x^5 + 20x^4 + 12x^3 - 20x^2 - 15x - 18 ,$$
$$b = 2x^6 + x^5 - 14x^4 - 11x^3 + 22x^2 + 28x + 8 .$$

$d = 2$. The bound in step (1) is

$$M = 2 \cdot 2 \cdot 2^6 \cdot 2 \cdot \min(\tfrac{1}{2}\sqrt{1666}, \tfrac{1}{2}\sqrt{1654}) \sim 10412 .$$

As the first prime we choose $p = 5$. $g_{(5)} = (2 \bmod 5)(x^3 + x^2 + x + 1)$. So $P = 5$ and $g = 2x^3 + 2x^2 + 2x + 2$.

Now we choose $p = 7$. We get $g_{(7)} = 2x^4 + 3x^3 + 2x + 3$. Since the degree of $g_{(7)}$ is higher than the degree of the current $g$, the prime 7 is discarded.

Now we choose $p = 11$. We get $g_{(11)} = 2x^3 + 5x^2 - 3$. By an application of CRA_2 to the coefficients of $g$ and $g_{(11)}$ modulo 5 and 11, respectively, we get $g = 2x^3 + 27x^2 + 22x - 3$. $P$ is set to 55.

Now we choose $p = 13$. We get $g_{(13)} = 2x^2 - 2x - 4$. All previous results are discarded, we go back to step (3), and we set $P = 13$, $g := 2x^2 - 2x - 4$.

Now we choose $p = 17$. We get $g_{(17)} = 2x^2 - 2x - 4$. By an application of CRA_2 to the coefficients of $g$ and $g_{(17)}$ modulo 13 and 17, respectively, we get $g = 2x^2 - 2x - 4$. $P$ is set to 221.

In general, we would have to continue choosing primes. But following the suggestion above, we apply the test in step (5) to our partial result and we see that $\mathrm{pp}(g)$ divides both $a$ and $b$. Thus, we get $\gcd(a, b) = x^2 - x - 2$.

The complexity of GCD_MOD is proportional to $m^3(\log m + L(d))^2$, where $m$ is the maximal degree of $a$ and $b$ and $d$ a bound for the absolute values of all the coefficients in $a$ and $b$ (see Loos 1983).

### Multivariate polynomials

We generalize the modular approach for univariate polynomials over $\mathbb{Z}$ to multivariate polynomials over $\mathbb{Z}$. So the inputs are elements of $\mathbb{Z}[x_1, \ldots, x_{n-1}][x_n]$, where the coefficients are in $\mathbb{Z}[x_1, \ldots, x_{n-1}]$ and the main variable is $x_n$. In this method we compute modulo irreducible polynomials $p(x)$ in $\mathbb{Z}[x_1, \ldots, x_{n-1}]$. In fact we use linear polynomials of the form $p(x) = x_{n-1} - r$ where $r \in \mathbb{Z}$. So reduction modulo $p(x)$ is simply evaluation at $r$.

For a polynomial $a \in \mathbb{Z}[x_1, \ldots, x_{n-2}][y][x]$ and $r \in \mathbb{Z}$ we let $a_{y-r}$ stand for $a \bmod y - r$. Obviously the proof of Lemma 4.2.2 can be generalized to this situation.

**Lemma 4.2.3.** Let $a, b \in \mathbb{Z}[x_1, \ldots, x_{n-2}][y][x]^*$ and $r \in \mathbb{Z}$ such that $y - r$ does not divide both $\mathrm{lc}_x(a)$ and $\mathrm{lc}_x(b)$. Let $c = \gcd(a, b)$.
a. $\deg_x(\gcd(a_{y-r}, b_{y-r})) \geq \deg_x(\gcd(a, b))$.
b. If $y - r \not| \operatorname{res}_x(a/c, b/c)$ then $\gcd(a_{y-r}, b_{y-r}) = c_{y-r}$.

The analogue to the Landau–Mignotte bound is even easier to derive: let $c$ be a factor of $a$ in $\mathbb{Z}[x_1, \ldots, x_{n-2}][y][x]$. Then $\deg_y(c) \leq \deg_y(a)$. So we get the algorithm GCD_MODm. For computing the gcd of $a, b \in \mathbb{Z}[x_1, \ldots, x_n]$, the algorithm is initially called as GCD_MODm$(a, b, n, n-1)$.

**Algorithm GCD_MODm**(in: $a, b, n, s$; out: $g$);
$[a, b \in \mathbb{Z}[x_1, \ldots, x_s][x_n]^*,\ 0 \leq s < n;\ g = \gcd(a, b).]$
0. if $s = 0$ then $\{g := \gcd(\mathrm{cont}(a), \mathrm{cont}(b))\mathrm{GCD\_MOD}(\mathrm{pp}(a), \mathrm{pp}(b))$; return$\}$;
1. $M := 1 + \min(\deg_{x_s}(a), \deg_{x_s}(b))$;
2. $r :=$ an integer s.t. $\deg_{x_n}(a_{x_s-r}) = \deg_{x_n}(a)$ or $\deg_{x_n}(b_{x_s-r}) = \deg_{x_n}(b)$;
   $g_{(r)} :=$ GCD_MODm$(a_{x_s-r}, b_{x_s-r}, n, s-1)$;
3. $m := 1$;
   $g := g_{(r)}$;
4. while $m \leq M$ do
   $\{r :=$ a new integer s.t. $\deg_{x_n}(a_{x_s-r}) = \deg_{x_n}(a)$ or $\deg_{x_n}(b_{x_s-r}) = \deg_{x_n}(b)$;
   $g_{(r)} :=$ GCD_MODm$(a_{x_s-r}, b_{x_s-r}, n, s-1)$;
   if $\deg_{x_n}(g_{(r)}) < \deg_{x_n}(g)$ then goto (3);
   if $\deg_{x_n}(g_{(r)}) = \deg(g)$
   then $\{$incorporate $g_{(r)}$ into $g$ by Newton interpolation (see Sect. 3.1);
   $m := m + 1\}$ $\}$;
5. if $g \in \mathbb{Z}[x_1, \ldots, x_s][x_n]$ and $g|a$ and $g|b$ then return;
   goto (2).

*Example 4.2.2.* We look at an example in $\mathbb{Z}[x, y]$. Let

$$a(x, y) = 2x^2y^3 - xy^3 + x^3y^2 + 2x^4y - x^3y - 6xy + 3y + x^5 - 3x^2 ,$$
$$b(x, y) = 2xy^3 - y^3 - x^2y^2 + xy^2 - x^3y + 4xy - 2y + 2x^2 .$$

We have

$$M = 1 + \min(\deg_x(a), \deg_x(b)) = 4 .$$

The algorithm proceeds as follows:
$r = 1$: $\gcd(a_{x-1}, b_{x-1}) = y + 1$.
$r = 2$: $\gcd(a_{x-2}, b_{x-2}) = 3y + 4$. Now we use Newton interpolation to obtain $g = (2x - 1)y + (3x - 2)$.
$r = 3$: $\gcd(a_{x-3}, b_{x-3}) = 5y + 9$. Now by Newton interpolation we obtain $g = (2x-1)y+x^2$ and this is the gcd (the algorithm would actually take another step).

The modular approach is the fastest currently known, both in theory and practice. The complexity of GCD_MODm is proportional to $m^{2n+1}(n \log m + L(d))^2$, where $n$ is the number of variables, $m$ is the maximal degree of $a$ and $b$, and $d$ a bound for the absolute values of all the integer coefficients in $a$ and $b$ (see Loos 1983).

### Exercises

1. Apply GCD_MOD for computing the gcd of $x^5 - x^4 - 3x^2 - 3x + 2$ and $x^4 - 2x^3 - 3x^2 + 4x + 4$ in $\mathbb{Z}[x]$. Use the primes $2, 3, 5, 7, \ldots$
2. Compute the gcd of the bivariate integer polynomials

$$f(x, y) = y^6 + xy^5 + x^3y - xy + x^4 - x^2 \;,$$
$$g(x, y) = xy^5 - 2y^5 + x^2y^4 - 2xy^4 + xy^2 + x^2y$$

   both by the subresultant algorithm and the modular algorithm.
3. What is the gcd $h$ of the polynomials $f, g$ in $\mathbb{Z}[x]$? Check whether the integer factors of the resultant of $f/h$ and $g/h$ are unlucky primes in the modular approach to gcd computation.

$$f = x^7 - 3x^5 - 2x^4 + 13x^3 - 15x^2 + 7x - 1 \;,$$
$$g = x^6 - 9x^5 + 18x^4 - 13x^3 + 2x^2 + 2x - 1 \;.$$

## 4.3 Computation of resultants

Since the 0th subresultant of two polynomials is equal to their resultant, we can use the algorithm PRS_SR for computing resultants. But, as in the case of gcd computations, we can also apply a modular method for computing multivariate polynomial resultants. Such an approach is described in Collins (1971).

Suppose we want to compute the resultant of the two polynomials $a, b \in \mathbb{Z}[x_1, \ldots, x_r]$ w.r.t the main variable $x_r$. We reduce $a$ and $b$ modulo various primes $p$ to $a_{(p)}$ and $b_{(p)}$, compute $\mathrm{res}_{x_r}(a_{(p)}, b_{(p)})$, and use the Chinese remainder algorithm for constructing $\mathrm{res}_{x_r}(a, b)$. For the subproblem of computing $\mathrm{res}_{x_r}(a_{(p)}, b_{(p)})$ over $\mathbb{Z}_p$ we use evaluation homomorphisms for the variables $x_1, \ldots, x_{r-1}$ and subsequent interpolation. Thus, the problem is ultimately reduced to a resultant computation in $\mathbb{Z}_p[x_r]$, which can easily be achieved by the subresultant prs algorithm.

**Lemma 4.3.1.** Let $I$, $J$ be integral domains, $\phi$ a homomorphism from $I$ into $J$. The homomorphism from $I[x]$ into $J[x]$ induced by $\phi$ will also be denoted $\phi$, i.e., $\phi(\sum_{i=0}^{m} c_i x^i) = \sum_{i=0}^{m} \phi(c_i) x^i$. Let $a(x), b(x)$ be polynomials in $I[x]$. If $\deg(\phi(a)) = \deg(a)$ and $\deg(\phi(b)) = \deg(b) - k$, then $\phi(\mathrm{res}_x(a, b)) = \phi(\mathrm{lc}(a))^k \mathrm{res}_x(\phi(a), \phi(b))$.

*Proof.* Let $M$ be the Sylvester matrix of $a$ and $b$, $M^*$ the Sylvester matrix of $a^* = \phi(a)$ and $b^* = \phi(b)$. If $k = 0$, then clearly $\phi(\mathrm{res}_x(a, b)) = \mathrm{res}_x(a^*, b^*)$.

If $k > 0$ then $M^*$ can be obtained from $\phi(M)$ by deleting its first $k$ rows and columns. Since the first $k$ columns of $\phi(M)$ contain $\phi(\mathrm{lc}(a))$ on the diagonal and are zero below the diagonal, $\phi(\mathrm{res}_x(a, b)) = \phi(\det(M)) = \det(\phi(M)) = \mathrm{lc}(a)^k \mathrm{res}_x(a^*, b^*)$. □

As in any application of the modular method, we need a bound on the number of required homomorphic images. The bound for the evaluation homomorphisms is obvious, namely if $\deg_{x_r}(a) = m_r$, $\deg_{x_r}(b) = n_r$, $\deg_{x_{r-1}}(a) = m_{r-1}$, $\deg_{x_{r-1}}(b) = n_{r-1}$, then from inspection of the Sylvester matrix we immediately see that $\deg_{x_{r-1}}(\mathrm{res}_{x_r}(a, b)) \leq m_r n_{r-1} + n_r m_{r-1}$. So if $x_{r-1}$ is evaluated at $m_r n_{r-1} + n_r m_{r-1} + 1$ points, the resultant can be reconstructed from the resultants of the evaluated polynomials. The method fails if the finite field $\mathbb{Z}_p$ does not contain enough evaluation points. In practice, however, $p$ will be much bigger than the required number of evaluation points, so this possibility of failure is not a practical one and we will ignore it.

For determining a bound for the integer coefficients in the resultant we use the following norm of multivariate polynomials which is inductively defined as

$$\mathrm{norm}(c(x_1)) = \|c(x_1)\|_1 = \sum_{i=0}^{m} |c_i| \quad \text{for } c(x_1) = \sum_{i=0}^{m} c_i x_1^i \in \mathbb{Z}[x_1] ,$$

$$\mathrm{norm}(c(x_1, \ldots, x_r)) = \sum_{i=0}^{m} \mathrm{norm}(c_i) \quad \text{for } c(x_1, \ldots, x_r) = \sum_{i=0}^{m} c_i x_r^i \in \mathbb{Z}[x_1, \ldots, x_r] .$$

For this definition of the norm we have $\mathrm{norm}(a + b) \leq \mathrm{norm}(a) + \mathrm{norm}(b)$, $\mathrm{norm}(a \cdot b) \leq \mathrm{norm}(a) \cdot \mathrm{norm}(b)$, and $|\alpha| \leq \mathrm{norm}(a)$ if $\alpha$ is any integer coefficient in $a$.

**Lemma 4.3.2.** Let $a(x_1, \ldots, x_r) = \sum_{i=0}^{m} a_i(x_1, \ldots, x_{r-1}) x_r^i$, $b(x_1, \ldots, x_r) = \sum_{i=0}^{n} b_i(x_1, \ldots, x_{r-1}) x_r^i$ be polynomials in $\mathbb{Z}[x_1, \ldots, x_r]$. Let $d = \max_{0 \leq i \leq m} \mathrm{norm}(a_i)$, $e = \max_{0 \leq i \leq n} \mathrm{norm}(b_i)$, $\alpha$ an integer coefficient in $\mathrm{res}_{x_r}(a, b)$. Then $|\alpha| \leq (m + n)! d^n e^m$.

*Proof.* Each non-zero term of the determinant of the Sylvester matrix, as the product of $n$ coefficients of $a$ and $m$ coefficients of $b$, has a norm of at most $d^n e^m$. Since there are at most $(m+n)!$ such terms, we have $\mathrm{norm}(\mathrm{res}_{x_r}(a, b)) \leq (m + n)! d^n e^m$, and hence $|\alpha| \leq (m + n)! d^n e^m$. □

These considerations lead to Collins's modular algorithm for computing the resultant of two multivariate polynomials over the integers.

**Algorithm RES_MOD**(in: $a, b$; out: $c$);
[$a, b \in \mathbb{Z}[x_1, \ldots, x_r]$, $r \geq 1$, $a$ and $b$ have positive degree in $x_r$; $c = \mathrm{res}_{x_r}(a, b)$.]

1. $m := \deg_{x_r}(a)$; $n := \deg_{x_r}(b)$;
   $d := \max_{0 \le i \le m} \mathrm{norm}(a_i)$; $e := \max_{0 \le i \le n} \mathrm{norm}(b_i)$;
   $P := 1$; $c := 0$; $B := 2(m+n)!d^n e^m$;
2. while $P \le B$ do
   {$p$ := a new prime such that $\deg_{x_r}(a) = \deg_{x_r}(a_{(p)})$ and $\deg_{x_r}(b) = \deg_{x_r}(b_{(p)})$;
   $c_{(p)}$ := RES_MODp($a_{(p)}, b_{(p)}$);
   $c$ := CRA_2($c, c_{(p)}, P, p$);
   [for $P = 1$ the output is simply $c_{(p)}$, otherwise CRA_2 is actually applied to the coefficients of $c$ and $c_{(p)}$]
   $P := P \cdot p$};
   return.

The subalgorithm RES_MODp computes multivariate resultants over $\mathbb{Z}_p$ by evaluation homomorphisms.

**Algorithm RES_MODp**(in: $a, b$; out: $c$);
[$a, b \in \mathbb{Z}_p[x_1, \ldots, x_r]$, $r \ge 1$, $a$ and $b$ have positive degree in $x_r$; $c = \mathrm{res}_{x_r}(a, b)$.]
0. if $r = 1$ then {$c$ := last element of PRS_SR($a, b$); return};
1. $m_r := \deg_{x_r}(a)$; $n_r := \deg_{x_r}(b)$; $m_{r-1} := \deg_{x_{r-1}}(a)$; $n_{r-1} := \deg_{x_{r-1}}(b)$;
   $B := m_r n_{r-1} + n_r m_{r-1} + 1$;
   $D(x_{r-1}) := 1$; $c(x_1, \ldots, x_{r-1}) := 0$; $\beta := -1$;
2. while $\deg(D) \le B$ do
   2.1. {$\beta := \beta + 1$; [if $\beta = p$ stop and report failure]
   if $\deg_{x_r}(a_{x_{r-1}=\beta}) < \deg_{x_r}(a)$ or $\deg_{x_r}(b_{x_{r-1}=\beta}) < \deg_{x_r}(a)$ then goto (2.1);
   $c_{(\beta)}(x_1, \ldots, x_{r-2})$ := RES_MODp($a_{x_{r-1}=\beta}, b_{x_{r-1}=\beta}$);
   $c := (c_{(\beta)}(x_1, \ldots, x_{r-2}) - c(x_1, \ldots, x_{r-2}, \beta))D(\beta)^{-1}D(x_{r-1}) + c(x_1, \ldots, x_{r-1})$; [so $c$ is the result of the Newton interpolation]
   $D(x_{r-1}) := (x_{r-1} - \beta)D(x_{r-1})$};
   return.

The complexity of RES_MOD is analyzed in Collins (1971) and it turns out to be dominated by $(m+1)^{2r+1}(\log d(m+1)) + (m+1)^{2r}(\log d(m+1))^2$, where $d$ is an upper bound for the norms of the inputs $a$ and $b$ and the degrees of $a$ and $b$ in any variable are not greater than $m$.

Solving systems of algebraic equations by resultants

**Theorem 4.3.3.** Let $K$ be an algebraically closed field, let

$$a(x_1, \ldots, x_r) = \sum_{i=0}^{m} a_i(x_1, \ldots, x_{r-1})x_r^i, \quad b(x_1, \ldots, x_r) = \sum_{i=0}^{n} b_i(x_1, \ldots, x_{r-1})x_r^i$$

be elements of $K[x_1, \ldots, x_r]$ of positive degrees $m$ and $n$ in $x_r$, and let $c(x_1,$

$\ldots, x_{r-1}) = \mathrm{res}_{x_r}(a, b)$. If $(\alpha_1, \ldots, \alpha_r) \in K^r$ is a common root of $a$ and $b$, then $c(\alpha_1, \ldots, \alpha_{r-1}) = 0$. Conversely, if $c(\alpha_1, \ldots, \alpha_{r-1}) = 0$, then one of the following holds:

a. $a_m(\alpha_1, \ldots, \alpha_{r-1}) = b_n(\alpha_1, \ldots, \alpha_{r-1}) = 0$,

b. for some $\alpha_r \in K$, $(\alpha_1, \ldots, \alpha_r)$ is a common root of $a$ and $b$.

*Proof.* $c = ua + vb$, for some $u, v \in K[x_1, \ldots, x_r]$. If $(\alpha_1, \ldots, \alpha_r)$ is a common root of $a$ and $b$, then the evaluation of both sides of this equation immediately yields $c(\alpha_1, \ldots, \alpha_{r-1}) = 0$.

Now assume $c(\alpha_1, \ldots, \alpha_{r-1}) = 0$. Suppose $a_m(\alpha_1, \ldots, \alpha_{r-1}) \neq 0$, so we are not in case (a). Let $\phi$ be the evaluation homomorphism $x_1 = \alpha_1, \ldots, x_{r-1} = \alpha_{r-1}$. Let $k = \deg(b) - \deg(\phi(b))$. By Lemma 4.3.1 we have $0 = c(\alpha_1, \ldots, \alpha_{r-1}) = \phi(c) = \phi(\mathrm{res}_{x_r}(a, b)) = \phi(a_m)^k \mathrm{res}_{x_r}(\phi(a), \phi(b))$. Since $\phi(a_m) \neq 0$, we have $\mathrm{res}_{x_r}(\phi(a), \phi(b)) = 0$. Since the leading term in $\phi(a)$ is non-zero, $\phi(a)$ and $\phi(b)$ must have a common non-constant factor, say $d(x_r)$ (see van der Waerden 1970: sect. 5.8). Let $\alpha_r$ be a root of $d$ in $K$. Then $(\alpha_1, \ldots, \alpha_r)$ is a common root of $a$ and $b$. Analogously we can show that (b) holds if $b_n(\alpha_1, \ldots, \alpha_{r-1}) \neq 0$. □

Theorem 4.3.3 suggests a method for determining the solutions of a system of algebraic, i.e., polynomial, equations over an algebraically closed field. Suppose, for example, that a system of three algebraic equations is given as

$$a_1(x, y, z) = a_2(x, y, z) = a_3(x, y, z) = 0 \ .$$

Let, e.g.,

$$b(x) = \mathrm{res}_z(\mathrm{res}_y(a_1, a_2), \mathrm{res}_y(a_1, a_3)) \ ,$$
$$c(y) = \mathrm{res}_z(\mathrm{res}_x(a_1, a_2), \mathrm{res}_x(a_1, a_3)) \ ,$$
$$d(z) = \mathrm{res}_y(\mathrm{res}_x(a_1, a_2), \mathrm{res}_x(a_1, a_3)) \ .$$

In fact, we might compute these resultants in any other order. By Theorem 4.3.3, all the roots $(\alpha_1, \alpha_2, \alpha_3)$ of the system satisfy $b(\alpha_1) = c(\alpha_2) = d(\alpha_3) = 0$. So if there are finitely many solutions, we can check for all of the candidates whether they actually solve the system.

Unfortunately, there might be solutions of $b$, $c$, or $d$, which cannot be extended to solutions of the original system, as we can see from the following example.

*Example 4.3.1.* Consider the system of algebraic equations

$$a_1(x, y, z) = 2xy + yz - 3z^2 = 0 \ ,$$
$$a_2(x, y, z) = x^2 - xy + y^2 - 1 = 0 \ ,$$
$$a_3(x, y, z) = yz + x^2 - 2z^2 = 0 \ .$$

We compute

$$\begin{aligned} b(x) &= \mathrm{res}_z(\mathrm{res}_y(a_1, a_3), \mathrm{res}_y(a_2, a_3)) \\ &= x^6(x-1)(x+1)(127x^4 - 167x^2 + 4) , \\ c(y) &= \mathrm{res}_z(\mathrm{res}_x(a_1, a_3), \mathrm{res}_x(a_2, a_3)) \\ &= (y-1)^3(y+1)^3(3y^2-1)(127y^4 - 216y^2 + 81) \cdot \\ &\quad \cdot (457y^4 - 486y^2 + 81) , \\ d(z) &= \mathrm{res}_y(\mathrm{res}_x(a_1, a_2), \mathrm{res}_x(a_1, a_3)) \\ &= 5184z^{10}(z-1)(z+1)(127z^4 - 91z^2 + 16) . \end{aligned}$$

All the solutions of the system, e.g., $(1, 1, 1)$, have coordinates which are roots of $b, c, d$. But there is no solution of the system having $y$-coordinate $1/\sqrt{3}$. So not every root of these resultants can be extended to a solution of the system.

## Exercises

1. Apply algorithm RES_MOD for computing the resultant of $a(x, y)$ and $b(x, y)$ w.r.t. $y$

$$\begin{aligned} a(x, y) &= xy^2 - x^3y - 2x^2y + xy + 2x^4 - 2x^2 , \\ b(x, y) &= 2x^2y^2 - 4x^3y + 4x^4 . \end{aligned}$$

2. Use resultant computations for solving the system of algebraic equations $f_1 = f_2 = f_3 = 0$ over $\mathbb{C}$

$$\begin{aligned} f_1(x, y, z) &= 2xy - yz + 2z , \\ f_2(x, y, z) &= x^2 + yz - 1 , \\ f_3(x, y, z) &= xz + yz - 2x . \end{aligned}$$

3. Solve over $\mathbb{C}$:

$$\begin{aligned} f_1(x, y, z) &= xz - xy^2 - 4x^2 - \tfrac{1}{4} = 0 , \\ f_2(x, y, z) &= y^2z + 2x + \tfrac{1}{2} = 0 , \\ f_3(x, y, z) &= x^2z + y^2 + \tfrac{1}{2}x = 0 . \end{aligned}$$

4. According to Theorem 4.3.3 a solution of $\mathrm{res}_{x_r}(a, b)$ can be extended to a common solution of $a$ and $b$ if $\mathrm{lc}_{x_r}(a)$ or $\mathrm{lc}_{x_r}(b)$ is a non-zero constant in $K$. This can be achieved by a suitable change of variables. Work out the details of an algorithm for solving systems of algebraic equations by resultants along these lines.

## 4.4 Squarefree factorization

By just computing gcds we can produce a so-called squarefree factorization of a polynomial, i.e., a partial solution to the problem of factoring polynomials which is to be treated in the next chapter. Throughout this section let $K$ be a computable field generated as $Q(I)$, where $I$ is a ufd. Whenever $I$ is a ufd, then also $I[x]$ is a ufd (see Sect. 4.1, Exercise 1).

*Definition 4.4.1.* A polynomial $a(x_1, \ldots, x_n)$ in $I[x_1, \ldots, x_n]$ is *squarefree* iff every nontrivial factor $b(x_1, \ldots, x_n)$ of $a$ (i.e., $b$ not similar to $a$ and not a constant) occurs with multiplicity exactly 1 in $a$.

By Gauss's lemma we know that for primitive polynomials the squarefree factorizations in $I[x]$ and $K[x]$ are the same. There is a simple criterion for deciding squarefreeness.

**Theorem 4.4.1.** Let $a(x)$ be a nonzero polynomial in $K[x]$, where $\mathrm{char}(K) = 0$ or $K = \mathbb{Z}_p$ for a prime $p$. Then $a(x)$ is squarefree if and only if $\gcd(a(x), a'(x)) = 1$. ($a'(x)$ is the derivative of $a(x)$.)

*Proof.* If $a(x)$ is not squarefree, i.e., for some non-constant $b(x)$ we have $a(x) = b(x)^2 \cdot c(x)$, then

$$a'(x) = 2b(x)b'(x)c(x) + b^2(x)c'(x) .$$

So $a(x)$ and $a'(x)$ have a non-trivial gcd.

On the other hand, if $a(x)$ is squarefree, i.e.,

$$a(x) = \prod_{i=1}^{n} a_i(x) ,$$

where the $a_i(x)$ are pairwise relatively prime irreducible polynomials, then

$$a'(x) = \sum_{i=1}^{n} \left( a_i'(x) \prod_{\substack{j=1 \\ j \neq i}}^{n} a_j(x) \right) .$$

Now it is easy to see that none of the irreducible factors $a_i(x)$ is a divisor of $a'(x)$. $a_i(x)$ divides all the summands of $a'(x)$ except the $i$-th. This finishes the proof for characteristic 0. In $\mathbb{Z}_p[x]$, $a_i'(x)$ cannot vanish, for otherwise we could write $a_i(x) = b(x^p) = b(x)^p$ for some $b(x)$, and this would violate our assumption of squarefreeness. Thus, $\gcd(a(x), a'(x)) = 1$. □

The problem of squarefree factorization for $a(x) \in K[x]$ consists of determining the squarefree pairwise relatively prime polynomials $b_1(x), \ldots, b_s(x)$, such that

$$a(x) = \prod_{i=1}^{s} b_i(x)^i . \qquad (4.4.1)$$

*Definition 4.4.2.* The representation of $a$ as in (4.4.1) is called the *squarefree factorization* of $a$.

In characteristic 0 (e.g., when $a(x) \in \mathbb{Z}[x]$), we can proceed as follows. We set $a_1(x) := a(x)$ and $a_2(x) := \gcd(a_1, a_1')$. Then

$$a_2(x) = \prod_{i=1}^{s} b_i(x)^{i-1} = \prod_{i=2}^{s} b_i(x)^{i-1}$$

and

$$c_1(x) := a_1(x)/a_2(x) = \prod_{i=1}^{s} b_i(x)$$

contains every squarefree factor exactly once. Now we set

$$a_3(x) := \gcd(a_2, a_2') = \prod_{i=3}^{s} b_i(x)^{i-2} ,$$

$$c_2(x) := a_2(x)/a_3(x) = \prod_{i=2}^{s} b_i(x) .$$

$c_2(x)$ contains every squarefree factor of muliplicity $\geq 2$ exactly once. So we have

$$b_1(x) = c_1(x)/c_2(x) .$$

Next we set

$$a_4(x) := \gcd(a_3, a_3') = \prod_{i=4}^{s} b_i(x)^{i-3} ,$$

$$c_3(x) := a_3(x)/a_4(x) = \prod_{i=3}^{s} b_i(x) .$$

So we have

$$b_2(x) = c_2(x)/c_3(x) .$$

Iterating this process until $c_{s+1}(x) = 1$, we ultimately get the desired squarefree factorization of $a(x)$. This process for computing a squarefree factorization is summarized in SQFR_FACTOR.

**Algorithm SQFR_FACTOR**(in: $a$; out: $F$);
[$a$ is a primitive polynomial in $\mathbb{Z}[x]$,
$F = [b_1(x), \ldots, b_s(x)]$ is the list of squarefree factors of $a$.]
1. $F :=$ [ ];
   $a_1 := a$;

$a_2 := \gcd(a_1, a_1');$
$c_1 := a_1/a_2;$
$a_3 := \gcd(a_2, a_2');$
$c_2 := a_2/a_3;$
$F := \mathrm{CONS}(c_1/c_2, F);$

2. while $c_2 \neq 1$ do
   $\{a_2 := a_3;\ a_3 := \gcd(a_3, a_3');$
   $c_1 := c_2;\ c_2 := a_2/a_3;$
   $F := \mathrm{CONS}(c_1/c_2, F)\};$

   $F := \mathrm{INV}(F)$; return.

If the polynomial $a(x)$ is in $\mathbb{Z}_p[x]$, the situation is slightly more complicated. First we determine

$$d(x) = \gcd(a(x), a'(x)) .$$

If $d(x) = 1$, then $a(x)$ is squarefree and we can set $a_1(x) = a(x)$ and stop. If $d(x) \neq 1$ and $d(x) \neq a(x)$, then $d(x)$ is a proper factor of $a(x)$ and we can carry out the process of squarefree factorization both for $d(x)$ and $a(x)/d(x)$. Finally, if $d(x) = a(x)$, then we must have $a'(x) = 0$, i.e., $a(x)$ must contain only terms whose exponents are a multiple of $p$. So we can write $a(x) = b(x^p) = b(x)^p$ for some $b(x)$, and the problem is reduced to the squarefree factorization of $b(x)$.

An algorithm for squarefree factorization in $\mathbb{Z}_p[x]$ along these lines is presented in Akritas (1989), namely PSQFFF.

Theorem 4.4.1 can be generalized to multivariate polynomials. For a proof we refer to the Exercises.

**Theorem 4.4.2.** Let $a(x_1, \ldots, x_n) \in K[x_1, \ldots, x_n]$ and $\mathrm{char}(K) = 0$. Then $a$ is squarefree if and only if $\gcd(a, \partial a/\partial x_1, \ldots, \partial a/\partial x_n) = 1$.

Squarefree factorization is only a first step in the complete factorization of a polynomial. However, it is relatively inexpensive and it is a prerequisite of many factorization algorithms.

### Exercises

1. Apply SQFR_FACTOR to the polynomial $a(x) = x^7 + x^6 - x^5 - x^4 - x^3 - x^2 + x + 1$ over the integers.
2. Prove Theorem 4.4.2.
3. Based on Theorem 4.4.1, derive an algorithm for squarefree factorization in $\mathbb{Z}_p[x]$, $p$ a prime.
4. Let $p(x) = 112x^4 + 58x^3 - 31x^2 + 107x - 66$. What are the squarefree factorizations of $p(x)$ modulo 3 and 11, respectively?

## 4.5 Squarefree partial fraction decomposition

*Definition 4.5.1.* Let $p(x)/q(x)$ be a *proper rational function* over the field $K$, i.e., $p, q \in K[x]$, $\gcd(p, q) = 1$, and $\deg(p) < \deg(q)$. Let $q = q_1 \cdot q_2^2 \cdots q_k^k$ be the squarefree factorization of $q$. Let $a_1(x), \ldots, a_k(x) \in K[x]$ be such that

$$\frac{p(x)}{q(x)} = \sum_{i=1}^{k} \frac{a_i(x)}{q_i(x)^i} \quad \text{with } \deg(a_i) < \deg(q_i^i) \text{ for } 1 \leq i \leq k . \tag{4.5.1}$$

Then the right-hand side of (4.5.1) is called the *incomplete squarefree partial fraction decomposition* (ispfd) of $p/q$.

Let $b_{ij}(x) \in K[x]$, $1 \leq j \leq i \leq k$, be such that

$$\frac{p(x)}{q(x)} = \sum_{i=1}^{k} \sum_{j=1}^{i} \frac{b_{ij}(x)}{q_i(x)^j} \quad \text{with } \deg(b_{ij}) < \deg(q_i) \text{ for } 1 \leq j \leq i \leq n . \tag{4.5.2}$$

Then the right-hand side of (4.5.2) is called the *(complete) squarefree partial fraction decomposition* (spfd) of $p/q$.

Both the incomplete and the complete squarefree partial fraction decomposition of a proper rational function are uniquely determined. For any proper rational function $p/q$ the ispfd can be computed by the following algorithm.

**Algorithm ISPFD**(in: $p, q$; out: $D$);
[$p/q$ is a proper rational function in $K(x)$,
$D = [[a_1, q_1], \ldots, [a_k, q_k]]$ is the ispfd of $p/q$, i.e., $p/q = \sum_{i=1}^{k}(a_i/q_i^i)$ with $\deg(a_i) < \deg(q_i^i)$ for $1 \leq i \leq k$.]
1. $[q_1, \ldots, q_k] :=$ SQFR_FACTOR$(q)$;
2. $c_0 := p$; $d_0 := q$; $i := 1$;
3. while $i < k$ do
   $\{d_i := d_{i-1}/q_i^i$;
   determine $c_i, a_i$ such that $\deg(c_i) < \deg(d_i)$, $\deg(a_i) < \deg(q_i^i)$, and $c_i \cdot q_i^i + a_i \cdot d_i = c_{i-1}\}$;
   $a_k := c_{k-1}$; return.

**Theorem 4.5.1.** The algorithm ISPFD is correct.

*Proof.* Immediately before execution of the body of the "while" statement for $i$, the relation

$$\frac{p}{q} = \frac{a_1}{q_1} + \ldots + \frac{a_{i-1}}{q_{i-1}^{i-1}} + \frac{c_{i-1}}{d_{i-1}}, \quad \text{where } d_{i-1} = q_i^i \ldots q_k^k , \tag{4.5.3}$$

holds, as can easily be seen by induction on $i$.

The polynomials $c_i$ and $a_i$ in step (3) can be computed by application of the corollary to Theorem 3.1.2. □

Once we have the incomplete spfd we can rather easily get the complete spfd by successive division. Namely if $a_i = s \cdot q_i + t$, then

$$\frac{a_i}{q_i^i} = \frac{s}{q_i^{i-1}} + \frac{t}{q_i^i} .$$

*Example 4.5.1.* Consider the proper rational function

$$\frac{p(x)}{q(x)} = \frac{4x^8 - 3x^7 + 25x^6 - 11x^5 + 18x^4 - 9x^3 + 8x^2 - 3x + 1}{3x^9 - 2x^8 + 7x^7 - 4x^6 + 5x^5 - 2x^4 + x^3} .$$

The squarefree factorization of $q(x)$ is

$$q(x) = (3x^2 - 2x + 1)(x^2 + 1)^2 x^3 .$$

Application of ISPFD yields the incomplete spfd

$$\frac{p(x)}{q(x)} = \frac{4x}{3x^2 - 2x + 1} + \frac{-x^3 + 2x + 2}{(x^2 + 1)^2} + \frac{x^2 - x + 1}{x^3} .$$

By successive division of the numerators by the corresponding $q_i$'s we finally get the complete spfd .

$$\frac{p(x)}{q(x)} = \frac{4x}{3x^2 - 2x + 1} + \frac{-x}{x^2 + 1} + \frac{3x + 2}{(x^2 + 1)^2} + \frac{1}{x} + \frac{-1}{x^2} + \frac{1}{x^3} .$$

### Exercises

1. Show the uniqueness of the incomplete and the complete spfd of proper rational functions.
2. Compute the complete spfd of $1/(x^4 - 2x^3 + 2x - 1)$.

## 4.6 Integration of rational functions

The problem we consider in this section is the integration of rational functions with rational coefficients, i.e., to compute

$$\int \frac{p(x)}{q(x)} \mathrm{d}x ,$$

where $p(x), q(x) \in \mathbb{Q}[x]$, $\gcd(p, q) = 1$, and $q(x)$ is monic. We exclude the trivial case $q = 1$.

From classical calculus we know that this integral can be expressed as

$$\int \frac{p(x)}{q(x)} \mathrm{d}x = \frac{g(x)}{q(x)} + c_1 \cdot \log(x - \alpha_1) + \ldots + c_n \cdot \log(x - \alpha_n) \ , \qquad (4.6.1)$$

where $g(x) \in \mathbb{Q}[x]$, $\alpha_1, \ldots, \alpha_n$ are the different roots of $q$ in $\mathbb{C}$, and $c_1, \ldots, c_n \in \mathbb{Q}(\alpha_1, \ldots, \alpha_n)$. This requires factorization of $q$ over $\mathbb{C}$ into its linear factors, decomposing $p/q$ into its complete partial fraction decomposition, and computation in the potentially extremely high degree algebraic extension $\mathbb{Q}(\alpha_1, \ldots, \alpha_n)$. Then the solution (4.6.1) is achieved by integration by parts and C. Hermite's reduction method.

However, as we will see in the sequel, complete factorization of the denominator can be avoided, resulting in a considerable decrease in computational complexity. Instead of factoring $q$ we will only use its squarefree factors.

First we compute the squarefree factorization of the denominator $q$, i.e.,

$$q = f_1 \cdot f_2^2 \cdot \ldots \cdot f_r^r \ ,$$

where the $f_i \in \mathbb{Q}[x]$ are squarefree, $f_r \neq 1$, $\gcd(f_i, f_j) = 1$ for $i \neq j$. Based on this squarefree factorization we compute the squarefree partial fraction decomposition of $p/q$, i.e.,

$$\begin{aligned} \frac{p}{q} &= g_0 + \sum_{i=1}^{r} \sum_{j=1}^{i} \frac{g_{ij}}{f_i^j} \\ &= g_0 + \frac{g_{11}}{f_1} + \frac{g_{21}}{f_2} + \frac{g_{22}}{f_2^2} + \ldots + \frac{g_{r1}}{f_r} + \ldots + \frac{g_{rr}}{f_r^r} \ , \end{aligned} \qquad (4.6.2)$$

where $g_0, g_{ij} \in \mathbb{Q}[x]$, $\deg(g_{ij}) < \deg(f_i)$, for all $1 \leq j \leq i \leq r$. Integrating $g_0$ is no problem, so let us consider the individual terms in (4.6.2).

Now let $g/f^n$ be one of the non-trivial terms in (4.6.2) with $n \geq 2$, i.e., $f$ is squarefree and $\deg(g) < \deg(f)$. We reduce the computation of

$$\int \frac{g(x)}{f(x)^n} \mathrm{d}x$$

to the computation of an integral of the form

$$\int \frac{h(x)}{f(x)^{n-1}} \mathrm{d}x \quad \text{where } \deg(h) < \deg(f) \ .$$

This is achieved by a reduction process due to C. Hermite.

Since $f$ is squarefree, we have $\gcd(f, f') = 1$. By the extended Euclidean algorithm E_EUCLID and the corollary to Theorem 3.1.2 compute $c, d \in \mathbb{Q}[x]$

such that

$$g = c \cdot f + d \cdot f' \quad \text{where } \deg(c), \deg(d) < \deg(f) .$$

By integration by parts we can now reduce

$$\begin{aligned}\int \frac{g}{f^n} &= \int \frac{c \cdot f + d \cdot f'}{f^n} = \int \frac{c}{f^{n-1}} + \int \frac{d \cdot f'}{f^n} \\ &= \int \frac{c}{f^{n-1}} - \frac{d}{(n-1) \cdot f^{n-1}} + \int \frac{d'}{(n-1) \cdot f^{n-1}} \\ &= -\frac{d}{(n-1) \cdot f^{n-1}} + \int \frac{\overbrace{c + d'/(n-1)}^{h}}{f^{n-1}} ,\end{aligned}$$

where $\deg(h) < \deg(f)$.

Now we collect all the rational partial results and the remaining integrals and put everything over a common denominator, so that we get polynomials $g(x), h(x) \in \mathbb{Q}[x]$ such that

$$\int \frac{p}{q} = g_0 + \frac{g}{\underbrace{f_2 \cdot f_3^2 \cdots f_r^{r-1}}_{\bar{q}}} + \int \frac{h}{\underbrace{f_1 \cdot \ \dots \ \cdot f_r}_{q^*}} , \tag{4.6.3}$$

where $\deg(g) < \deg(\bar{q})$ and $\deg(h) < \deg(q^*)$.

We could also determine $g$ and $h$ in (4.6.3) by first choosing undetermined coefficients for these polynomials, differentiating (4.6.3), and then solving the resulting linear system for the undetermined coefficients. However, the Hermite reduction process is usually faster. Let us prove that the decomposition in (4.6.3) is unique.

**Lemma 4.6.1.** Let $p, q, u, v \in \mathbb{Q}[x]$, $\gcd(p, q) = 1$, $\gcd(u, v) = 1$, and $p/q = (u/v)'$ (so $u/v$ is the integral of $p/q$). Let $w \in \mathbb{Q}[x]$ be a squarefree factor of $q$. Then $w$ divides $v$, and the multiplicity of $w$ in $q$ is strictly greater than the multiplicity of $w$ in $v$.

*Proof.* Clearly we can restrict ourselves to $w$ being irreducible (otherwise apply the lemma for all irreducible factors of $w$). Now, since

$$\left(\frac{u}{v}\right)' = \frac{u'v - uv'}{v^2} = \frac{p}{q} ,$$

$w$ must divide $v$. Assume now that $v = w^r \hat{w}$ with $\gcd(w, \hat{w}) = 1$. We show that $w^r$ does not divide $u'v - uv'$. Suppose it does. Since $w^r$ divides $u'v$ and

$\gcd(w, u) = 1$, $w^r$ would have to divide $v' = rw^{r-1}w'\hat{w} + w^r\hat{w}'$. Hence, $w$ would have to divide $w'\hat{w}$. But this is impossible since $w$ is irreducible. Therefore $w^{r+1}$ must divide the reduced denominator of $(u/v)'$. □

**Theorem 4.6.2.** The solution $g, h$ to Eq. (4.6.3) is unique.

*Proof.* Suppose there were two solutions. By subtraction we would get a solution for $p = 0$,

$$\int 0\, dx = \frac{g}{\bar{q}} + \int \frac{h}{q^*}\, dx \ .$$

So $(g/\bar{q})' = -h/q^*$. By Lemma 4.6.1, every factor in the denominator of $h/q^*$ must have multiplicity at least 2. This is impossible, since $q^*$ is squarefree. □

The integral $\int h/q^*$ can be computed in the following well-known way: Let $q^*(x) = (x - \alpha_1) \ldots (x - \alpha_n)$, where $\alpha_1, \ldots, \alpha_n$ are the distinct roots of $q^*$. Then

$$\begin{aligned} &\int \frac{h(x)}{q^*(x)}\, dx = \sum_{i=1}^{n} \int \frac{c_i}{x - \alpha_i}\, dx = \sum_{i=1}^{n} c_i \log(x - \alpha_i) \\ &\text{with } c_i = \frac{h(\alpha_i)}{q^{*\prime}(\alpha_i)}, \quad 1 \le i \le n \ . \end{aligned} \tag{4.6.4}$$

No part of the sum of logarithms in (4.6.4) can be a rational function, as we can see from the following theorem in Hardy (1916: p. 14).

**Theorem 4.6.3.** Let $\alpha_1, \ldots, \alpha_n$ be distinct elements of $\mathbb{C}$ and $c_1, \ldots, c_n \in \mathbb{C}$. If $\sum_{i=1}^{n} c_i \log(x - \alpha_i)$ is a rational function, then $c_i = 0$ for all $1 \le i \le n$.

*Example 4.6.1.* Let us integrate $x/(x^2 - 2)$ according to (4.6.4).

$$\begin{aligned} \int \frac{x}{x^2 - 2}\, dx &= \int \frac{1/2}{x - \sqrt{2}}\, dx + \int \frac{1/2}{x + \sqrt{2}}\, dx \\ &= \tfrac{1}{2}(\log(x - \sqrt{2}) + \log(x + \sqrt{2})) = \tfrac{1}{2} \log(x^2 - 2) \ . \end{aligned}$$

So obviously we do not always need the full splitting field of $q^*$ in order to express the integral of $h/q^*$. In fact, whenever we have two logarithms with the same constant coefficient, we can combine these logarithms.

The following theorem, which has been independently discovered by M. Rothstein (1976) and B. Trager (1976), answers the question of what is the smallest field in which we can express the integral of $h/q^*$.

**Theorem 4.6.4.** Let $p, q \in \mathbb{Q}[x]$ be relatively prime, $q$ monic and squarefree, and $\deg(p) < \deg(q)$. Let

$$\int \frac{p}{q} = \sum_{i=1}^{n} c_i \log v_i \ , \tag{4.6.5}$$

where the $c_i$ are distinct non-zero constants and the $v_i$ are monic squarefree pairwise relatively prime elements of $\bar{\mathbb{Q}}[x]$. Then the $c_i$ are the distinct roots of the polynomial

$$r(c) = \mathrm{res}_x(p - c \cdot q', q) \in \mathbb{Q}[c] \ ,$$

and

$$v_i = \gcd(p - c_i \cdot q', q) \quad \text{for } 1 \le i \le n \ .$$

*Proof.* Let $u_i = (\prod_{j=1}^{n} v_j)/v_i$, for $1 \le i \le n$. Then by differentiation of (4.6.5) we get

$$p \cdot \prod_{i=1}^{n} v_i = q \cdot \sum_{i=1}^{n} c_i v_i' u_i \ .$$

So $q | \prod_{i=1}^{n} v_i$ and on the other hand each $v_i | q v_i' u_i$, which implies that each $v_i | q$. Hence,

$$q = \prod_{i=1}^{n} v_i \quad \text{and} \quad p = \sum_{i=1}^{n} c_i v_i' u_i \ .$$

Consequently, for each $j$, $1 \le j \le n$, we have

$$\begin{aligned} v_j &= \gcd(0, v_j) = \gcd\Big(p - \sum_{i=1}^{n} c_i v_i' u_i, v_j\Big) \\ &= \gcd(p - c_j v_j' u_j, v_j) = \gcd\Big(p - c_j \sum_{i=1}^{n} v_i' u_i, v_j\Big) \\ &= \gcd(p - c_j q', v_j) \ , \end{aligned}$$

and for $l \ne j$ we have

$$\gcd(p - c_j q', v_l) = \gcd(p - c_j v_l' u_l, v_l) = \gcd((c_l - c_j) v_l' u_l, v_l) = 1 \ .$$

Thus we conclude that

$$v_i = \gcd(p - c_i q', q) \quad \text{for } 1 \le i \le n \ . \tag{4.6.6}$$

Equation (4.6.6) implies that $\mathrm{res}_x(p - c_i q', q) = 0$ for all $1 \le i \le n$. Conversely, if $c \in \bar{\mathbb{Q}}$ and $\mathrm{res}_x(p - cq', q) = 0$, then $\gcd(p - cq', q) = s(x) \in \bar{\mathbb{Q}}[x]$ with $\deg(s) > 0$. Thus, any irreducible factor $t(x)$ of $s(x)$ divides $p - cq' = \sum_{i=1}^{n} c_i v_i' u_i - c \sum_{i=1}^{n} v_i' u_i$. Since $t$ divides one and only one $v_j$, we get

$t|(c_j - c)v_j' u_j$, which implies that $c_j - c = 0$. Thus, the $c_j$ are exactly the distinct roots of $r(c)$. □

*Example 4.6.1* (continued). We apply Theorem 4.6.4. $r(c) = \text{res}_x(p - cq', q) = \text{res}_x(x - c(2x), x^2 - 2) = -2(2c-1)^2$. There is only one root of $r(c)$, namely $c_1 = 1/2$. We get the argument of the corresponding logarithm as $v_1 = \gcd(x - \frac{1}{2}(2x), x^2 - 2) = x^2 - 2$. So

$$\int \frac{x}{x^2 - 2}\,dx = \tfrac{1}{2}\log(x^2 - 2)\ .$$

*Example 4.6.2*. Let us consider integrating the rational function

$$\frac{p(x)}{q(x)} = \frac{4x^8 - 3x^7 + 25x^6 - 11x^5 + 18x^4 - 9x^3 + 8x^2 - 3x + 1}{3x^9 - 2x^8 + 7x^7 - 4x^6 + 5x^5 - 2x^4 + x^3}\ .$$

The squarefree factorization of $q(x)$ is

$$q(x) = (3x^2 - 2x + 1)(x^2 + 1)^2 x^3\ ,$$

so the squarefree partial fraction decomposition of $p/q$ is

$$\frac{p(x)}{q(x)} = \frac{4x}{3x^2 - 2x + 1} + \frac{-x}{x^2 + 1} + \frac{3x + 2}{(x^2 + 1)^2} + \frac{1}{x} + \frac{-1}{x^2} + \frac{1}{x^3}\ .$$

Now let us consider the third term of this decomposition, i.e., we determine

$$\int \frac{3x + 2}{(x^2 + 1)^2}\,dx\ .$$

By the extended Euclidean algorithm we can write

$$3x + 2 = 2 \cdot (x^2 + 1) + (-x + \tfrac{3}{2}) \cdot (2x)\ .$$

Integration by parts yields

$$\begin{aligned}
\int \frac{3x + 2}{(x^2 + 1)^2}\,dx &= \int \frac{2}{x^2 + 1}\,dx + \int \frac{(-x + \frac{3}{2}) \cdot (2x)}{(x^2 + 1)^2}\,dx \\
&= \int \frac{2}{x^2 + 1}\,dx + \frac{(-x + \frac{3}{2}) \cdot (-1)}{x^2 + 1} - \int \frac{1}{x^2 + 1}\,dx \\
&= \frac{x - \frac{3}{2}}{x^2 + 1} + \int \frac{1}{x^2 + 1}\,dx\ .
\end{aligned}$$

The remaining integral is purely logarithmic, namely

$$\int \frac{1}{x^2+1}\,\mathrm{d}x = \frac{i}{2}\cdot\log(1-ix) - \frac{i}{2}\cdot\log(1+ix) = \arctan(x)\ .$$

### Exercises

1. Prove the statement (4.6.4).
2. Integrate $8x/(x^4-2)$ both by the classical formula given in (4.6.4) and according to Theorem 4.6.4. What is the smallest extension of $\mathbb{Q}$ over which the integral can be expressed?
3. Let $c_i$ be one of the constants appearing in Theorem 4.6.4. Do the conjugates of $c_i$ also appear? If so, what do the corresponding $v_i$ look like?
4. Use a computer algebra system to finish the computation of Example 4.6.2.

## 4.7 Bibliographic notes

Like for many other topics of computer algebra, D. E. Knuth (1981) provides an excellent exposition of problems and analyses in gcd computation. The modular method is presented in Brown (1971). See also Moses and Yun (1973) and Char et al. (1989). For polynomial gcds with algebraic number coefficients we refer to Langemyr and McCallum (1989), Smedley (1989), and Encarnación (1994). In Kaltofen (1988) gcds are computed for polynomials represented by straight-line programs. Possible implementations resulting from Hermite's method of integration of rational functions are discussed in Tobey (1967). Horowitz (1971) contains a detailed analysis of algorithms for squarefree partial fraction decomposition. Recently integration of rational functions has been reexamined in Lazard and Rioboo (1990).

# 5 Factorization of polynomials

## 5.1 Factorization over finite fields

Similar to what we have done for the computation of gcds of polynomials, we will reduce the computation of the factors of an integral polynomial to the computation of the factors of the polynomial modulo a prime number. So we have to investigate this problem first, i.e., we consider the problem of factoring a polynomial $a(x) \in \mathbb{Z}_p[x]$, $p$ a prime number. W.l.o.g. we may assume that $\mathrm{lc}(a) = 1$.

In the sequel we describe E. R. Berlekamp's (1968) algorithm for factoring squarefree univariate polynomials in $\mathbb{Z}_p[x]$. Throughout this section let $a(x)$ be a squarefree polynomial of degree $n$ in $\mathbb{Z}_p[x]$, $p$ a prime number, having the following factorization into irreducible factors

$$a(x) = \prod_{i=1}^{r} a_i(x) .$$

By Theorem 3.1.11, for every choice of $s_1, \ldots, s_r \in \mathbb{Z}_p$ there exists a uniquely determined polynomial $v(x) \in \mathbb{Z}_p[x]$ such that

$$\begin{aligned} &v(x) \equiv s_i \bmod a_i(x) \quad \text{for } 1 \le i \le r , \quad \text{and} \\ &\deg(v) < \deg(a_1) + \ldots + \deg(a_r) = n . \end{aligned} \tag{5.1.1}$$

In (5.1.1) it is essential that $a$ is squarefree, i.e., the $a_i$'s are relatively prime.

**Lemma 5.1.1.** For every $a_i, a_j$, $i \neq j$, there exist $s_1, \ldots, s_r \in \mathbb{Z}_p$ such that the corresponding solution $v(x)$ of (5.1.1) generates a factorization $b \cdot c$ of $a$ with $a_i | b$ and $a_j | c$.

*Proof.* If $r = 1$ there is nothing to prove. So assume $r \geq 2$. Choose $s_i \neq s_j$ and the other $s_k$'s arbitrary. Let $v$ be the corresponding solution of (5.1.1). Then

$$a_i(x) \mid \gcd(a(x), v(x) - s_i) \quad \text{and} \quad a_j(x) \nmid \gcd(a(x), v(x) - s_i) . \qquad \square$$

So we could solve the factorization problem over $\mathbb{Z}_p$, if we could get a complete overview of the solutions $v(x)$ of (5.1.1) for all the choices of $s_1, \ldots, s_r \in \mathbb{Z}_p$. Fortunately this can be achieved by linear algebra methods.

If $v(x)$ satisfies (5.1.1), then

$$v(x)^p \equiv s_i^p = s_i \equiv v(x) \bmod a_i(x) \quad \text{for } 1 \leq i \leq r .$$

So we have

$$v(x)^p \equiv v(x) \bmod a(x) \quad \text{and } \deg(v) < n . \tag{5.1.2}$$

Every solution of (5.1.1) for some $s_1, \ldots, s_r$ solves (5.1.2).

But what about the converse of this implication? Is every solution of (5.1.2) also a solution of (5.1.1) for some $s_1, \ldots, s_r$? From the fact that $GF(p)$ is the splitting field of $x^p - x$, we get that

$$v(x)^p - v(x) = (v(x) - 0)(v(x) - 1) \ldots (v(x) - (p - 1)) .$$

So if $v(x)$ satisfies (5.1.2), then $a(x)$ divides $v(x)^p - v(x)$ and therefore every irreducible factor $a_i(x)$ must divide one of the factors $v(x) - s$ of $v(x)^p - v(x)$. Thus, every solution of (5.1.2) is also a solution of (5.1.1) for some $s_1, \ldots, s_r$. In particular, there are exactly $p^r$ solutions of (5.1.2).

By Fermat's little theorem and Theorem 2.5.2 the solutions of (5.1.2) constitute a vector space over $\mathbb{Z}_p$. So we can get a complete overview of the solutions of (5.1.2), if we can compute a basis for this vector space.

Let the $(n \times n)$-matrix $Q(a)$ over $\mathbb{Z}_p$,

$$Q(a) = Q = \begin{pmatrix} q_{0,0} & \cdots & q_{0,n-1} \\ \vdots & & \vdots \\ q_{n-1,0} & \cdots & q_{n-1,n-1} \end{pmatrix} ,$$

be defined by

$$x^{pk} \equiv q_{k,n-1}x^{n-1} + \ldots + q_{k,1}x + q_{k,0} \bmod a(x) \quad \text{for } 0 \leq k \leq n - 1 .$$

That is, the entries in the $k$-th row of $Q$ are the coefficients of $\operatorname{rem}(x^{pk}, a(x))$. Using the representation of $v(x) = v_{n-1}x^{n-1} + \ldots + v_0$ as the vector $(v_0, \ldots, v_{n-1})$, we have

$$v \cdot Q = v \quad \Longleftrightarrow$$

$$v(x) = \sum_{j=0}^{n-1} v_j x^j = \sum_{j=0}^{n-1} \sum_{k=0}^{n-1} v_k \cdot q_{k,j} x^j \equiv \sum_{k=0}^{n-1} v_k x^{pk} = v(x^p) = v(x)^p \bmod a(x) .$$

We summarize all these results in the following theorem.

**Theorem 5.1.2.** With the notation used above, a polynomial $v(x) = v_{n-1}x^{n-1} + \ldots + v_1 x + v_0$ in $\mathbb{Z}_p[x]$ solves (5.1.2) if and only if the vector $(v_0, \ldots, v_{n-1})$ is in the null-space of the matrix $Q - I$ ($I$ the identity matrix of dimension $n$), i.e., $v \cdot (Q - I) = (0, \ldots, 0)$.

Now we are ready to formulate Berlekamp's algorithm for factoring squarefree univariate polynomials in $\mathbb{Z}_p[x]$.

**Algorithm FACTOR_B**(in: $a$, $p$; out: $F$);
[$p$ is a prime number, $a$ is a squarefree polynomial in $\mathbb{Z}_p[x]$,
$F$ is the list of prime factors of $a$.]

1. form the $(n \times n)$-matrix $Q$ over $\mathbb{Z}_p$, where the $k$-th line $(q_{k,0}, \ldots, q_{k,n-1})$ of $Q$ satisfies
   $\mathrm{rem}(x^{pk}, a(x)) = q_{k,n-1}x^{n-1} + \ldots + q_{k,0}$, for $0 \leq k \leq n-1$;
2. by column operations transform the matrix $Q - I$ into (e.g., lower-right) triangular form;
   from the triangular form read off the rank $n - r$ of the matrix $Q - I$;
   [There are exactly $r$ linearly independent solutions $v^{[1]}, \ldots, v^{[r]}$ of $v \cdot (Q - I) = 0$.
   Let $v^{[1]}$ be the trivial solution $(1, 0, \ldots, 0)$.
   So (after interpretation of vectors as polynomials) there are $p^r$ solutions $t_1 \cdot v^{[1]} + \ldots + t_r \cdot v^{[r]}$ of (5.1.2), and therefore $r$ irreducible factors of $a(x)$.]
3. if $r = 1$, then $a(x)$ is irreducible and we set $F := [a]$;
   otherwise, compute $\gcd(a(x), v^{[2]}(x) - s)$ for $s \in \mathbb{Z}_p$ and put the factors of $a$ found in this way into the list $F$;
   as long as $F$ contains fewer than $r$ factors, choose the next $v^{[k]}(x)$, $k = 3, \ldots, r$, and compute $\gcd(f(x), v^{[k]}(x) - s)$ for $f$ in $F$;
   add the factors found in this way to $F$;
   [ultimately, $F$ will contain all the factors of $a(x)$]
   return.

*Example 5.1.1.* Let us use FACTOR_B for factoring the polynomial

$$a(x) = x^5 + x^3 + 2x^2 + x + 2$$

in $\mathbb{Z}_3[x]$. First we have to check for squarefreeness. $a'(x) = 2x^4 + x + 1$, so $\gcd(a, a') = 1$ in $\mathbb{Z}_3[x]$ and therefore $a(x)$ is squarefree.

The rows of the $(5 \times 5)$-matrix $Q$ are the coefficients of $x^0, x^3, x^6, x^9, x^{12}$ modulo $a(x)$. So

$$Q = \begin{pmatrix} 1 & 0 & 0 & 0 & 0 \\ 0 & 0 & 0 & 1 & 0 \\ 0 & 1 & 2 & 1 & 2 \\ 0 & 1 & 1 & 2 & 2 \\ 2 & 0 & 2 & 1 & 1 \end{pmatrix}.$$

$Q - I$ can be transformed into the triangular form

$$\begin{pmatrix} 0 & 0 & 0 & 0 & 0 \\ 0 & 0 & 0 & 1 & 0 \\ 0 & 0 & 1 & 1 & 2 \\ 0 & 0 & 1 & 1 & 2 \\ 1 & 0 & 0 & 0 & 0 \end{pmatrix}.$$

We read off $r = 2$, i.e., there are 2 irreducible factors of $a(x)$. The null-space of $Q - I$ is spanned by

$$v^{[1]} = (1, 0, 0, 0, 0) \quad \text{and} \quad v^{[2]} = (0, 0, 2, 1, 0) \ .$$

Now we get the factors by appropriate gcd computations:

$$\gcd(a(x), v^{[2]}(x) + 2) = x^2 + x + 2 \ ,$$
$$\gcd(a(x), v^{[2]}(x) + 1) = x^3 + 2x^2 + 1 \ .$$

The basic operations in FACTOR_B are the setting up and solution of a system of linear equations and the gcd computations for determining the actual factors. The complexity of FACTOR_B is proportional to $n^3 + prn^2$, where $n$ is the degree of the polynomial (compare Knuth 1981: sect. 4.6.2).

### Exercises

1. Why does the polynomial input $a(x)$ to the Berlekamp algorithm have to be squarefree? What happens with Lemma 5.1.1 if $a(x)$ is not squarefree? Produce an example of a non-squarefree polynomial which is not factored by the Berlekamp algorithm.
2. Apply the process of squarefree factorization and Berlekamp's algorithm for factoring $a(x) = x^7 + 4x^6 + 2x^5 + 4x^3 + 3x^2 + 4x + 2$ modulo 5.
3. How many factors does $u(x) = x^4 + 1$ have in $\mathbb{Z}_p[x]$, $p$ a prime? (Hint: Consider the cases $p = 2$, $p = 8k + 1$, $p = 8k + 3$, $p = 8k + 5$, $p = 8k + 7$ separately.) How many factors does $u(x)$ have in $\mathbb{Z}[x]$?
4. Write a Maple procedure for implementing FACTOR_B.

## 5.2 Factorization over the integers

Before developing algorithms for actually producing a factorization of a reducible polynomial, we might want to decide whether a given polynomial is in fact irreducible. A powerful criterion for irreducibility is due to Eisenstein, a proof can be found, for instance, in van der Waerden (1970).

**Theorem 5.2.1** (Eisenstein's irreducibility criterion). Let $R$ be a ufd and $f(x) = a_n x^n + a_{n-1} x^{n-1} + \cdots + a_1 x + a_0$ a primitive polynomial of positive degree $n$ in $R[x]$. If there is an irreducible element $p$ of $R$ such that

$$p \nmid a_n,\ p | a_i \text{ for all } i < n, \text{ and } p^2 \nmid a_0 \ ,$$

or

$$p \nmid a_0,\ p | a_i \text{ for all } i > 0, \text{ and } p^2 \nmid a_n \ ,$$

then $f(x)$ is irreducible in $R[x]$.

Univariate polynomials

According to the corollary to Gauss's lemma (Theorem 4.1.1) factorizations of univariate integral polynomials are essentially the same in $\mathbb{Z}[x]$ and $\mathbb{Q}[x]$. For reasons of efficiency we concentrate on the case of integral polynomials. The factorization of integers is a much harder problem than the factorization of polynomials. For this reason we do not intend to factor the content of integral polynomials. Throughout this section we assume that the polynomial to be factored is a primitive non-constant polynomial.

The problem of factoring a primitive univariate integral polynomial $a(x)$ consists in finding pairwise relatively prime irreducible polynomials $a_i(x)$ and positive integers $m_i$ such that

$$a(x) = \prod_{i=1}^{r} a_i(x)^{m_i} .$$

As for polynomials over finite fields we will first compute a squarefree factorization of $a(x)$. By application of SQFR_FACTOR our factorization problem is reduced to the problem of factoring a primitive squarefree polynomial. So from now on let us assume that $a(x)$ is primitive and squarefree.

As in the case of polynomial gcds we would like to use the fast factorization algorithm modulo a prime $p$. However, the approach of choosing several primes and combining the results by the Chinese remainder algorithm does not work for factorization. We do not know which of the factors modulo the different primes correspond to each other. So we choose a different approach. The problem of factorization over $\mathbb{Z}$ is reduced to factorization modulo $p$ and a subsequent lifting of the result to a factorization modulo $p^k$ (compare Sect. 3.2). If $k$ is high enough, the integer factors can be constructed.

**Theorem 5.2.2** (Hensel lemma). Let $p$ be a prime number and $a(x), a_1(x), \ldots, a_r(x) \in \mathbb{Z}[x]$. Let $(a_1 \bmod p), \ldots, (a_r \bmod p)$ be pairwise relatively prime in $\mathbb{Z}_p[x]$ and $a(x) \equiv a_1(x) \cdot \ldots \cdot a_r(x) \bmod p$. Then for every natural number $k$ there are polynomials $a_1^{(k)}(x), \ldots, a_r^{(k)}(x) \in \mathbb{Z}[x]$ such that

$$a(x) \equiv a_1^{(k)}(x) \cdot \ldots \cdot a_r^{(k)}(x) \bmod p^k$$

and

$$a_i^{(k)}(x) \equiv a_i(x) \bmod p \quad \text{for } 1 \leq i \leq r .$$

*Proof.* Let $R = \mathbb{Z}[x]$, $I = \langle p \rangle$ and

$$f_1 = x_1 \cdot \ldots \cdot x_r - a(x) \in R[x_1, \ldots, x_r] .$$

Then

$$f_1(a_1, \ldots, a_r) \equiv 0 \bmod I .$$

The Jacobian matrix of $f_1$ evaluated at $a_1, \ldots, a_r$ is

$$U = (u_{11} \ldots u_{1r}), \quad \text{where}$$

$$u_{1j} = \frac{\partial f_1}{\partial x_j}(a_1, \ldots, a_r) = \prod_{\substack{i=1,\\ j \neq j}}^{r} a_i \quad \text{for } 1 \leq j \leq r \ .$$

Since the elements of $U$ are relatively prime modulo $p$, there is a matrix $V$ in $R^{r \times 1}$ such that $UV \equiv 1 \bmod p$. Thus, all the requirements of the lifting theorem (Theorem 3.2.2) are satisfied and the statement is proved. □

As we have seen in Sect. 3.2, the result of the lifting process is by no means unique. So we want to choose a particular path in the lifting process which will allow us to reconstruct the factors over $\mathbb{Z}$ from the factors modulo $p$.

**Lemma 5.2.3.** Let $a(x) \in \mathbb{Z}[x]$ be primitive and squarefree. Let $p$ be a prime number not dividing $\mathrm{lc}(a)$. Let $a_1(x), \ldots, a_r(x) \in Z_p[x]$ be pairwise relatively prime such that $a \equiv a_1 \cdot \ldots \cdot a_r \bmod p$ and $\mathrm{lc}(a_1) = \mathrm{lc}(a) \bmod p$, $\mathrm{lc}(a_2) = \ldots = \mathrm{lc}(a_r) = 1$. Then for every natural number $k$ there are polynomials $a_1^{(k)}(x), \ldots, a_r^{(k)}(x) \in \mathbb{Z}_{p^k}[x]$ with $\mathrm{lc}(a_1^{(k)}) = \mathrm{lc}(a) \bmod p^k$, $\mathrm{lc}(a_2^{(k)}) = \ldots = \mathrm{lc}(a_r^{(k)}) = 1$ such that

$$a(x) \equiv a_1^{(k)}(x) \cdot \ldots \cdot a_r^{(k)}(x) \bmod p^k$$

and

$$a_i^{(k)}(x) \equiv a_i(x) \bmod p \quad \text{for } 1 \leq i \leq r \ .$$

*Proof.* We proceed by induction on $k$. For $k = 1$ we can obviously choose $a_i^{(1)} = a_i$ and all the requirements are satisfied.

So now assume that the $a_i^{(k)}$ satisfy the requirements. That is, for some $\hat{d}(x) \in \mathbb{Z}_p[x]$ we have

$$a - \prod_{i=1}^{r} a_i^{(k)} \equiv p^k \hat{d} \bmod p^{k+1} \ .$$

We replace the leading coefficient of $a_1^{(k)}$ by ($\mathrm{lc}(a) \bmod p^{k+1}$). Then for some $d(x) \in \mathbb{Z}_p[x]$ we have

$$a - \prod_{i=1}^{r} a_i^{(k)} \equiv p^k d \bmod p^{k+1} \ ,$$

where $\deg(d) < \deg(a)$. We will determine $b_i(x) \in \mathbb{Z}_p[x]$ with $\deg(b_i) < \deg(a_i)$ such that

$$a_i^{(k+1)} = a_i^{(k)} + p^k b_i \ .$$

Using this ansatz, we get

$$a - \prod_{i=1}^{r} a_i^{(k+1)} \equiv \underbrace{a - \prod_{i=1}^{r} a_i^{(k)}}_{p^k d} - p^k \Big( \sum_{i=1}^{r} b_i \underbrace{\prod_{j=1, j \neq i}^{r} a_j}_{=:\tilde{a}_i} \Big) \bmod p^{k+1} .$$

So the $a_i^{(k+1)}$'s will constitute a factorization modulo $p^{k+1}$ if and only if

$$d \equiv \sum_{i=1}^{r} b_i \cdot \tilde{a}_i \bmod p .$$

A solution is guaranteed by an appropriate generalization of Theorem 3.1.2 and the corollary. See Sect. 3.1, Exercise 11 and algorithm LIN_COMB below. □

The following algorithm LIN_COMB (based on Wang 1979) will be used as a subalgorithm in the lifting process.

**Algorithm LIN_COMB**(in: $[a_1, \ldots, a_r]$; out: $[b_1, \ldots, b_r]$);
[$a_i \in K[x]$ ($K$ a field) pairwise relatively prime;
$b_i \in K[x]$, $\deg(b_i) < \deg(a_i)$ and $1 = \sum_{i=1}^{r} b_i \tilde{a}_i$, where $\tilde{a}_i = \prod_{j=1, j \neq i}^{r} a_j$]
1. $d := 1$; $i := 0$;
   for $j = 2$ to $r$ do $a_j^* := \prod_{k=j}^{r} a_k$;
2. while $i < r - 1$ do
   {$i := i + 1$;
   compute $u, v$ such that $d = ua_i + va_{i+1}^*$,
   $\deg(u) < \deg(a_{i+1}^*)$, $\deg(v) < \deg(a_i)$
   [corollary to Theorem 3.1.2];
   $b_i := v$; $d := u$};
3. $b_r := d$;
   return.

We summarize these algorithmic ideas in LIFT_FACTORS.

**Algorithm LIFT_FACTORS**(in: $a$, $[a_1, \ldots, a_r]$, $p$, $K$; out: $F$);
[$a$ is a primitive squarefree polynomial in $\mathbb{Z}[x]$, $p$ is a prime number not dividing $\mathrm{lc}(a)$ and s.t. ($a$ mod $p$) is squarefree in $\mathbb{Z}_p[x]$,
$a_1, \ldots, a_r \in Z_p[x]$ pairwise relatively prime, $\mathrm{lc}(a_1) = \mathrm{lc}(a) \bmod p$, $\mathrm{lc}(a_2) = \ldots = \mathrm{lc}(a_r) = 1$, and $a \equiv a_1 \cdot \ldots \cdot a_r \bmod p$, $K \in \mathbb{N}$;
$F = [\bar{a}_1, \ldots, \bar{a}_r]$, $\bar{a}_i \in \mathbb{Z}_{p^K}[x]$, such that $a \equiv \bar{a}_1 \cdot \ldots \cdot \bar{a}_r \bmod p^K$, $\mathrm{lc}(\bar{a}_1) = \mathrm{lc}(a) \bmod p^K$, $\mathrm{lc}(\bar{a}_2) = \ldots = \mathrm{lc}(\bar{a}_r) = 1$, and $\bar{a}_i \equiv a_i \bmod p$.]
1. by an application of LIN_COMB to $[a_1, \ldots, a_r]$ compute $v_i \in \mathbb{Z}_p[x]$ s.t. $\deg(v_i) < \deg(a_i)$ and $1 \equiv \sum_{i=1}^{r} v_i \tilde{a}_i \bmod p$, where $\tilde{a}_i = \prod_{j=1, j \neq i}^{r} a_j$;
   (Sect. 3.1, Exercise 11)
2. for $i = 1$ to $r$ do $\bar{a}_i := a_i$;
   $k := 1$;

3. while $k < K$ do
   {replace $\mathrm{lc}(\bar{a}_1)$ by $(\mathrm{lc}(a) \bmod p^{k+1})$;
   $\tilde{d} := (a - \prod_{i=1}^{r} \bar{a}_i \bmod p^{k+1})$;
   $d := \tilde{d}/p^k$;
   for $i = 1$ to $r$ do
     $\{b_i := \mathrm{rem}(dv_i, a_i)$;
     $\bar{a}_i := \bar{a}_i + p^k b_i\}$;
   $k := k + 1\}$;
4. $F := [\bar{a}_1, \ldots, \bar{a}_r]$;
   return.

As for the general lifting algorithm LIFT there is also a quadratic lifting scheme for LIFT_FACTORS. The interested reader is referred to Wang (1979).

Now we put all the subalgorithms together and we get the Berlekamp–Hensel algorithm FACTOR_BH for factoring primitive univariate squarefree polynomials over the integers.

**Algorithm FACTOR_BH**(in: $a$; out: $F$);
[$a$ is a primitive squarefree polynomial in $\mathbb{Z}[x]$; $F = [a_1, \ldots, a_r]$, where $a_1, \ldots, a_r$ are primitive irreducible polynomials in $\mathbb{Z}[x]$ such that $a = a_1 \cdot \ldots \cdot a_r$.]
1. choose a prime number $p$ such that $p \nmid \mathrm{lc}(a)$ and $a$ is squarefree modulo $p$ (i.e., $p$ does not divide the discriminant of $a$);
2. $[u_1, \ldots, u_s] :=$ FACTOR_B$(a, p)$;
   normalize the $u_i$'s such that $\mathrm{lc}(u_1) = \mathrm{lc}(a) \bmod p$ and $\mathrm{lc}(u_2) = \ldots = \mathrm{lc}(u_s) = 1$;
3. determine a natural number $B$ which bounds the absolute value of any coefficient in a factor of $a$ over the integers (for instance, use the Landau–Mignotte bound, Theorem 4.2.1);
   $K := \min\{k \in \mathbb{N} \mid p^k \geq 2|\mathrm{lc}(a)|B\}$;
4. $[v_1, \ldots, v_s] :=$ LIFT_FACTORS$(a, [u_1, \ldots, u_s], p, K)$;
5. [combine factors]
   $\bar{a} := a$;
   $C := \{2, \ldots, s\}$; [$v_1$ will be included in the last factor]
   $i := 0$;
   $m := 0$;
   while $m < |C|$ do
     $\{m := m + 1$;
     for all $\{i_1, \ldots, i_m\} \subseteq C$ do
       { [integers modulo $p^K$ are centered around 0, i.e., the representation of $\mathbb{Z}_{p^K}$ is $\{q \mid -p^K/2 < q \leq p^K/2\}$]
       $\tilde{b} := (\mathrm{lc}(\bar{a}) \cdot v_{i_1} \cdot \ldots \cdot v_{i_m} \bmod p^K)$, interpreted as a polynomial over the integers;
       $b := \mathrm{pp}(\tilde{b})$;
       if $b | \bar{a}$
       then
         $\{i := i + 1$;

$a_i := b;$
$\bar{a} := \bar{a}/b;$
$C := C \setminus \{i_1, \ldots, i_m\}$
$\} \} \};$
$i := i + 1;$
$a_i := \bar{a};$
6. $F := [a_1, \ldots, a_i];$
return.

Step (5) is necessary, because irreducible factors over the integers might factor further modulo a prime $p$. In fact, there are irreducible polynomials over the integers which factor modulo every prime number. An example of this is $x^4 + 1$ (see Sect. 5.1, Exercise 3).

The complexity of FACTOR_BH would be polynomial in the size of the input except for step (5). Since in step (5), in the worst case, we have to consider all possible combinations of factors modulo $p$, this might lead to a combinatorial explosion, rendering the algorithm FACTOR_BH exponential in the size of the input. Nevertheless, in practical examples the combinations of factors does not present an insurmountable problem. Basically all the major computer algebra systems employ some variant of FACTOR_BH as the standard factoring algorithm for polynomials over the integers.

*Example 5.2.1.* We want to factor the primitive squarefree integral polynomial

$$a(x) = 6x^7 + 7x^6 + 4x^5 + x^4 + 6x^3 + 7x^2 + 4x + 1 .$$

We use FACTOR_BH in the process. A suitable prime is 5, $a(x)$ stays squarefree modulo 5.

By an application of the Berlekamp algorithm FACTOR_B, $a(x)$ is factored modulo 5 into

$$a(x) \equiv \underbrace{(x-2)}_{u_1} \cdot \underbrace{(x^2-2)}_{u_2} \cdot \underbrace{(x^2+2)}_{u_3} \cdot \underbrace{(x^2-x+2)}_{u_4} \mod 5 .$$

By an application of LIFT_FACTORS we lift this factorization to a factorization modulo 25, getting

$$a(x) \equiv \underbrace{(6x+3)}_{v_1} \cdot \underbrace{(x^2-7)}_{v_2} \cdot \underbrace{(x^2+7)}_{v_3} \cdot \underbrace{(x^2+9x-8)}_{v_4} \mod 25 .$$

The Landau–Mignotte bound for $a$ is rather big. Let us assume that by some additional insight we know that $K = 2$ is good enough for constructing the integral factors. Now we have to try combinations of factors modulo 25 to get the factors over the integers. So we set $\bar{a} := a$ and $C := \{2, 3, 4\}$. Testing the factors $v_2, v_3, v_4$ we see that only $v_4$ yields a factor over the integers:

$$a_1(x) := \mathrm{pp}(\mathrm{lc}(\bar{a}) \cdot v_4 \bmod 25) = 3x^2 + 2x + 1 .$$

So now $\bar{a} := \bar{a}/a_1 = 2x^5 + x^4 + 2x + 1$. The combination of $v_2$ and $v_3$ yields the factor

$$a_2(x) := \mathrm{pp}(\mathrm{lc}(\bar{a}) \cdot v_2 \cdot v_3 \bmod 25) = x^4 + 1 .$$

We set $\bar{a} := \bar{a}/a_2 = 2x + 1$. Now $C$ has become empty, and the last factor is

$$a_3(x) := \bar{a}(x) = 2x + 1 .$$

FACTOR_BH returns $F = [a_1, a_2, a_3]$, i.e., the factorization

$$a(x) = (3x^2 + 2x + 1) \cdot (x^4 + 1) \cdot (2x + 1) .$$

Multivariate polynomials

L. Kronecker (1882) describes a method for reducing the factorization of a multivariate polynomial over a unique factorization domain $I$ to the factorization of a univariate polynomial over $I$. This reduction is achieved by a mapping of the form

$$\begin{aligned} S_d\colon\ I[x_1, \ldots, x_n] &\longrightarrow I[y] \\ h(x_1, \ldots, x_n) &\longmapsto h(y, y^d, \ldots, y^{d^{n-1}}) , \end{aligned}$$

for $d \in \mathbb{N}$. Clearly $S_d$ is a homomorphism. $S_d$ can be inverted for those polynomials $h$, for which the maximal degree in any of the variables is less than $d$. So by $S_d^{-1}$ let us denote the additive mapping from $I[y]_{/\langle y^{d^n}\rangle}$ to $I[x_1, \ldots, x_n]$ satisfying

$$S_d^{-1}(c\, y^\alpha) = c\, x_1^{\alpha_1} \ldots x_n^{\alpha_n} ,$$

where $\alpha_1 + \alpha_2 d + \ldots + \alpha_n d^{n-1}$ is the representation of $\alpha$ in the positional number system with radix $d$.

**Lemma 5.2.4.** Let $f \in I[x_1, \ldots, x_n]$, $d > \max_{1 \le i \le n} \deg_{x_i}(f)$, $g$ a factor of $f$. Then there are irreducible factors $g_1, \ldots, g_s$ of $S_d(f)$ such that $g = S_d^{-1}(\prod_{j=1}^s g_j)$.

*Proof.* No factor of $f$ can have a degree higher than the respective degree of $f$ in any variable. Let $f = g \cdot h$. Then

$$S_d(f) = S_d(g) \cdot S_d(h) = \left(\prod_{j=1}^{s} g_j\right) \cdot S_d(h)$$

for some irreducible $g_1, \ldots, g_s \in I[y]$. So

$$g = S_d^{-1}(S_d(g)) = S_d^{-1}\left(\prod_{j=1}^{s} g_j\right) . \qquad \square$$

This lemma immediately leads to Kronecker's factorization algorithm for multivariate polynomials.

**Algorithm FACTOR_K**(in: $f$; out: $F$);
[$f \in I[x_1, \ldots, x_n]$;
$F = [f_1, \ldots, f_s]$, where $f_1, \ldots, f_s$ are the irreducible factors of $f$.]
1. [compute degree bound]
   $d := (\max_{1 \le i \le n} \deg_{x_i}(f)) + 1$;
2. [reduce to univariate case]
   factor $S_d(f)$ into irreducibles $g_1, \ldots, g_s$ in $I[y]$;
3. [recover multivariate factors by combination]
   $F := [\ ]$; $C := \{1, \ldots, s\}$; $i := 0$;
   while $C \neq \emptyset$ do
     {$i := i + 1$;
     for all $g_{j_1}, \ldots, g_{j_i}$ such that $j_k \in C$ do
       {$g := S_d^{-1}(g_{j_1} \cdot \ldots \cdot g_{j_i})$;
       if $g | f$
       then {$F :=$ CONS($g$, $F$); $C := C \setminus \{j_1, \ldots, j_i\}$; $f := f/g$}
       }
     };
   return.

The complexity of FACTOR_K depends on how fast polynomials in $I[y]$ can be factored. In any case, the factor combination in step (3) makes the algorithm exponential in the degree of the input $f$.

When we are dealing with multivariate polynomials over the integers, we need not use the quite time consuming Kronecker algorithm, but we can instead use evaluation homomorphisms to construct a lifting approach.

By application of Theorem 4.4.2 the problem can be reduced to the factorization of squarefree polynomials. By gcd computations we extract the primitive part w.r.t. the main variable $x_n$. So now let

$$f(x_1, \ldots, x_{n-1}, x_n)$$

be a primitive squarefree polynomial in $\mathbb{Z}[x_1, \ldots, x_n]$. We choose an evaluation point $(a_1, \ldots, a_{n-1}) \in \mathbb{Z}^{n-1}$ which preserves the degree and squarefreeness, factor the univariate polynomial

$$f(a_1, \ldots, a_{n-1}, x_n) ,$$

and finally lift this factorization modulo the prime ideal

$$\mathcal{P} = \langle x_1 - a_1, \ldots, x_{n-1} - a_{n-1} \rangle$$

to a factorization moludo $\mathcal{P}^t$ for high enough $t$. For a complete formulation we refer to Wang and Rothschild (1975), Wang (1978), and Musser (1976).

*Example 5.2.2* (from Kaltofen 1983). Let us factor the squarefree integral polynomial

$$f(x_1, x_2, x) = x^3 + ((x_1 + 2)x_2 + 2x_1 + 1)x^2 + \\ + ((x_1 + 2)x_2^2 + (x_1^2 + 2x_1 + 1)x_2 + 2x_1^2 + x_1)x + \\ + (x_1 + 1)x_2^3 + (x_1 + 1)x_2^2 + (x_1^3 + x_1^2)x_2 + x_1^3 + x_1^2 .$$

Step 1. First we choose an evaluation point that preserves the degree and squarefreeness and has as many zero components as possible.
$(x_1, x_2) = (0, 0)$: $f(0, 0, x) = x^3 + x^2$ is not squarefree, but
$(x_1, x_2) = (1, 0)$: $f(1, 0, x) = x^3 + 3x^2 + 3x + 2$ is squarefree.

By the change of variables $x_1 = w + 1$, $x_2 = z$ we move the evaluation point to the origin,

$$f(w + 1, z, x) = x^3 + 3x^2 + 3x + 2 + \\ + w^3 + (2x + 4)w^2 + (2x^2 + 5x + 5)w + \\ + (w + 2)z^3 + ((x + 1)w + (3x + 2))z^2 + \\ + (w^3 + (x + 4)w^2 + (x^2 + 4x + 5)w + (3x^2 + 4x + 2))z .$$

By $f_{ij}(x)$ we denote the coefficient of $w^j z^i$ in $f$.

Step 2. Factor $f_{00}$ (i.e., $f$ evaluated at $(w, z) = (0, 0)$) in $\mathbb{Z}[x]$. We get

$$x^3 + 3x^2 + 3x + 2 = \underbrace{(x + 2)}_{g_{00}} \underbrace{(x^2 + x + 1)}_{h_{00}} .$$

Step 3. Compute degree bounds for $w$ and $z$ in factors of

$$f(w + 1, z, x) = g(w, z, x)h(w, z, x) ,$$

i.e., $\deg_w(g), \deg_w(h) \leq 3$, and $\deg_z(g), \deg_z(h) \leq 3$.

Step 4. Lift $g_{00}$ and $h_{00}$ to highest degrees in $w$ and $z$. We use the ansatz

$$g(w, z, x) = g_{00}(x) + g_{01}(x)w + g_{02}(x)w^2 + g_{03}(x)w^3 + \\ + (g_{10}(x) + g_{11}(x)w + g_{12}(x)w^2 + g_{13}(x)w^3)z + \\ + (g_{20}(x) + g_{21}(x)w + g_{22}(x)w^2 + g_{23}(x)w^3)z^2 + \\ + (g_{30}(x) + g_{31}(x)w + g_{32}(x)w^2 + g_{33}(x)w^3)z^3 ,$$

and analogously for $h(w, z, x)$. First we lift to a factorization of $f(w+1, 0, x)$:

a formal multiplication of $g$ and $h$ leads to the equations

$$\begin{aligned} f_{01} &= g_{01}h_{00} + g_{00}h_{01} , \\ f_{02} &= g_{00}h_{02} + g_{01}h_{01} + g_{02}h_{00} , \\ f_{03} &= g_{00}h_{03} + g_{01}h_{02} + g_{02}h_{01} + g_{03}h_{00} . \end{aligned}$$

These equations can be solved by a modification of the extended Euclidean algorithm, yielding

$$f(w+1,0,x) = ((x+2) + 1 \cdot w)\,((x^2+x+1) + (x+2)w + w^2) .$$

Now we lift to a factorization of $f(w+1, z, x)$: again by the extended Euclidean algorithm we successively solve

$$\begin{aligned} f_{10} &= g_{00}h_{10} + g_{10}h_{00} , \\ f_{11} - g_{01}h_{10} - g_{10}h_{01} &= g_{00}h_{11} + g_{11}h_{00} , \\ f_{20} - g_{10}h_{10} &= g_{00}h_{20} + g_{20}h_{00} . \end{aligned}$$

All the other equations have 0 as their left-hand sides.

We get the factor candidates

$$\begin{aligned} f(w+1, z, x) = {} & ((x+2) + w + (2+w)z) \cdot \\ & \cdot ((x^2+x+1) + (x+2)w + w^2 + xz + z^2) , \end{aligned}$$

which are the actual factors. By resubstituting $w = x_1 - 1, z = x_2$ we get the factorization

$$f(x_1, x_2, x) = (x + x_1x_2 + x_1 + x_2 + 1) \cdot (x^2 + (x_1 + x_2)x + x_1^2 + x_2^2) .$$

## Exercises

1. Let $a(x) = 5x^3 + 9x^2 - 146x - 120 \in \mathbb{Z}[x]$. Lift the factorization

$$a(x) \equiv (2x+1)(x+1)x \bmod 3$$

to a factorization modulo 27. Is the result a factorization over the integers?

2. Modify LIFT_FACTORS to produce a quadratic lifting algorithm.
3. Apply FACTOR_BH for factoring the integral polynomial

$$a(x) = 2x^6 - 6x^5 - 101x^4 + 302x^3 + 148x^2 - 392x - 49.$$

As the prime use 5. All the coefficients of factors of $a$ are bounded in absolute value by 12.

## 5.3 A polynomial-time factorization algorithm over the integers

In previous sections we have dealt with the Berlekamp–Hensel algorithm for factoring univariate and multivariate polynomials over $\mathbb{Z}$. While this algorithm yields a factorization in reasonable time in most cases, it suffers from an exponential worst case complexity. For a long time it was an open problem in computer algebra whether the factorization of a polynomial $f$ could be achieved in time polynomial in the size of $f$. A. K. Lenstra et al. (1982) introduced an algorithm which is able to factor univariate polynomials over $\mathbb{Z}$ in polynomial time in the size of the input. Simultaneously E. Kaltofen (1982, 1985a) showed that the problem of factoring multivariate polynomials can be reduced to the problem of factoring univariate polynomials by an algorithm that takes time polynomial in the size of the input. These two results taken together provide a polynomial-time factorization algorithm for multivariate polynomials over $\mathbb{Z}$.

In this section we describe the approach of A. K. Lenstra et al. Their factorization algorithm relates the factors of a polynomial $f$ to a certain lattice in $\mathbb{R}^m$ and determines a reduced basis for this lattice. From this basis the factors of $f$ can be determined.

Lattices

From a vector space basis $b_1, \ldots, b_n$ for $\mathbb{R}^n$ an orthogonal basis $b_1^*, \ldots, b_n^*$ can be computed by the Gram–Schmidt orthogonalization process. Let $< \cdot, \cdot >$ denote the inner product of two vectors in $\mathbb{R}^n$ and let $\| \cdot \|$ denote the Euclidean length of a vector in $\mathbb{R}^n$; so $\|a\|^2 = \; < a, a >$. In the Gram–Schmidt orthogonalization process the vectors $b_i^*$ and the real numbers $\mu_{ij}$, $1 \leq j < i \leq n$, are inductively defined by the formulas

$$\begin{aligned} b_i^* &= b_i - \sum_{j=1}^{i-1} \mu_{ij} b_j^* \;, \\ \mu_{ij} &= \; < b_i, b_j^* > / \|b_j^*\|^2 \;. \end{aligned} \tag{5.3.1}$$

*Definition 5.3.1.* Let $n$ be a positive integer and $b_1, \ldots, b_n \in \mathbb{R}^n$ linearly independent vectors over $\mathbb{R}$. The set

$$L = \sum_{i=1}^{n} \mathbb{Z} b_i = \left\{ \sum_{i=1}^{n} a_i b_i \;\middle|\; a_1, \ldots, a_n \in \mathbb{Z} \right\}$$

is called the *lattice spanned by* $b_1, \ldots, b_n$. We say that $b_1, \ldots, b_n$ form a *basis* of the lattice $L$. An arbitrary subset $M$ of $\mathbb{R}^n$ is a *lattice* iff there are $b_1, \ldots, b_n$ such that $M$ is the lattice spanned by $b_1, \ldots, b_n$. $n$ is called the *rank* of the lattice. If $L$ is the lattice spanned by $b_1, \ldots, b_n$ in $\mathbb{R}^n$, then the *determinant* $\det(L)$ of $L$ is defined as

$$\det(L) = |\det(b_1, \ldots, b_n)| \;,$$

where the $b_i$'s are written as column vectors. The determinant is independent of the particular basis of the lattice (see, for instance, Cassels 1971). A basis $b_1, \ldots, b_n$ for a lattice $L$ satisfying

$$|\mu_{ij}| \leq 1/2 \quad \text{for } 1 \leq j < i \leq n$$

and

$$\|b_i^* + \mu_{i\,i-1} b_{i-1}^*\|^2 \geq \tfrac{3}{4}\|b_{i-1}^*\|^2 \quad \text{for } 1 < i \leq n ,$$

where $b_i^*$ and $\mu_{ij}$ are defined as in (5.3.1), is a *reduced* basis.

Observe (Exercise 2) that if $b_1, \ldots, b_n$ are a reduced basis for a lattice in $\mathbb{R}^n$, then

$$\|b_i^*\|^2 \geq \tfrac{1}{2}\|b_{i-1}^*\|^2 \quad \text{for } 1 < i \leq n .$$

**Theorem 5.3.1.** Let $b_1, \ldots, b_n$ be a reduced basis for a lattice $L$ in $\mathbb{R}^n$ and let $b_1^*, \ldots, b_n^*$ be the orthogonal basis produced by the Gram–Schmidt orthogonalization process. Then we have

a. $\|b_j\|^2 \leq 2^{i-1} \cdot \|b_i^*\|^2$ for $1 \leq j \leq i \leq n$,
b. $\|b_1\|^2 \leq 2^{n-1} \cdot \|x\|^2$ for every $x \in L \setminus \{0\}$,
c. if $x_1, \ldots, x_t \in L$ are linearly independent, then $\max\{\|b_1\|^2, \ldots, \|b_t\|^2\} \leq 2^{n-1} \cdot \max\{\|x_1\|^2, \ldots, \|x_t\|^2\}$.

*Proof.* A proof of this proposition can be found in Lenstra et al. (1982). □

The algorithm BASIS_REDUCTION transforms an arbitrary basis for a lattice $L$ into a reduced basis, as described in Lenstra et al. (1982).

**Algorithm BASIS_REDUCTION**(in: $a_1, \ldots, a_n$; out: $b_1, \ldots, b_n$);
[$a_1, \ldots, a_n$ is a basis for a lattice $L$ in $\mathbb{R}^n$; $b_1, \ldots, b_n$ is a reduced basis for $L$]

**Subalgorithm REDUCE**(in: $k, l$);
if $|\mu_{kl}| > 1/2$
then {$r :=$ integer nearest to $\mu_{kl}$;
  $b_k := b_k - r \cdot b_l$;
  for $j = 1$ to $l - 1$ do $\mu_{kj} := \mu_{kj} - r \cdot \mu_{lj}$;
  $\mu_{kl} := \mu_{kl} - r$};
return.

**Subalgorithm UPDATE**(in: $k$);
$\mu := \mu_{k\,k-1}$; $B := B_k + \mu^2 B_{k-1}$; $\mu_{k\,k-1} := \mu B_{k-1}/B$;
$B_k := B_{k-1} B_k / B$; $B_{k-1} := B$;
$(b_{k-1}, b_k) := (b_k, b_{k-1})$;
for $j = 1$ to $k - 2$ do $(\mu_{k-1\,j}, \mu_{kj}) := (\mu_{kj}, \mu_{k-1\,j})$;

```
for i = k + 1 to n do
    (μ_{i k-1}, μ_{ik}) := (μ_{i k-1} μ_{k k-1} + μ_{ik}(1 - μ μ_{k k-1}),  μ_{i k-1} - μ μ_{ik});
return.

1. [Gram–Schmidt orthogonalization]
   for i = 1 to n do
      {b_i := a_i; b_i^* := a_i;
      for j = 1 to i - 1 do
         {μ_{ij} := <b_i, b_j^*>/B_j;
         b_i^* := b_i^* - μ_{ij} b_j^*};
      B_i := ||b_i^*||^2};
2. [reduction]
   k := 2;
   while k ≤ n do
      {l := k - 1;
      REDUCE(k, l);
      if B_k < (3/4 - μ_{k k-1}^2) B_{k-1}
   2.1. then {UPDATE(k);
             if k > 2 then k := k - 1}
   2.2. else {for l = k - 2 downto 1 do REDUCE(k, l);
             k := k + 1}}
   return.
```

*Example 5.3.1.* We apply the algorithm BASIS_REDUCTION to the basis $a_1 = (1, 2, 1)$, $a_2 = (0, 1, 1)$, $a_3 = (1, 0, 1)$ in $\mathbb{R}^3$. In step (1) the basis $b_1, b_2, b_3$ is initialized to $a_1, a_2, a_3$ and the Gram–Schmidt orthogonalization process yields the orthogonal basis $b_1^* = (1, 2, 1)$, $b_2^* = (-\frac{1}{2}, 0, \frac{1}{2})$, $b_3^* = (\frac{2}{3}, -\frac{2}{3}, \frac{2}{3})$, the transformation coefficients $\mu_{21} = \frac{1}{2}$, $\mu_{31} = \frac{1}{3}$, $\mu_{32} = 0$, and the (squares of the) norms $B_1 = 6$, $B_2 = \frac{1}{2}$, $B_3 = \frac{4}{3}$.

In step (2) we set $k = 2$ and go to the top of the "while" loop. $l$ is set to 1. REDUCE(2, 1) does nothing. $B_2 < (\frac{3}{4} - \mu_{21}^2)B_1$, so the "then" branch is executed. UPDATE(2) interchanges $b_1$ and $b_2$, and reassigns $B_1 := 2$, $B_2 := \frac{3}{2}$, $\mu_{21} := \frac{3}{2}$, $\mu_{31} := \frac{1}{2}$, $\mu_{32} := \frac{1}{3}$. We return to the top of the "while" loop, $k = 2$. $l$ is set to 1. REDUCE(2, 1) now results in the reassignments $b_2 := b_2 - b_1 = (1, 1, 0)$, $\mu_{21} := \frac{1}{2}$. $B_2 \geq (\frac{3}{4} - \mu_{21}^2)B_1$, so the "else" branch is executed, yielding $k = 3$. We return to the top of the "while" loop, $k = 3$. $l$ is set to 2. REDUCE(3, 2) does nothing. $B_3 \geq (\frac{3}{4} - \mu_{32}^2)B_2$, so the "else" branch is executed. REDUCE(3, 1) does nothing, and $k$ is set to 4. We return to the top of the "while" loop. Now $k > 3$, so the algorithm terminates with the reduced basis $b_1 = (0, 1, 1), b_2 = (1, 1, 0), b_3 = (1, 0, 1)$.

**Theorem 5.3.2.** The algorithm BASIS_REDUCTION is correct.

*Proof.* a. Partial correctness: Initially the basis $b_1, \ldots, b_n$ is set to $a_1, \ldots, a_n$, a basis for the lattice $L$. In step (1) orthogonal vectors $b_1^*, \ldots, b_n^*$ and corresponding coefficients $\mu_{ij}$ are computed, so that (5.3.1) holds. During the algorithm the $b_i$'s are changed several times, but they always form a basis for the lattice $L$ and the $\mu_{ij}$'s, $b_i^*$'s are changed accordingly so that (5.3.1) remains valid. Actually, after step (1) it is not necessary to store the values of the $b_i^*$'s, but it suffices to keep track of the numbers $B_i = \|b_i^*\|^2$.

We show that whenever the "while" loop in step (2) is entered, the following invariant holds:

$$|\mu_{ij}| \leq \tfrac{1}{2} \quad \text{for } 1 \leq j < i < k$$

and

$$\|b_i^* + \mu_{i\,i-1} b_{i-1}^*\|^2 \geq \tfrac{3}{4}\|b_{i-1}^*\|^2 \quad \text{for } 1 < i < k\,. \tag{5.3.2}$$

So when the algorithm terminates, i.e., $k = n + 1$, we have that $b_1, \ldots, b_n$ is a reduced basis for the lattice $L$ generated by the input basis $a_1, \ldots, a_n$.

Invariant (5.3.2) obviously holds when the "while" loop is entered for the first time, since for $k = 2$ the condition (5.3.2) is empty. $k$ will always be in the range $2 \leq k \leq n + 1$ during the execution of the algorithm.

Suppose that we come to the top of the "while" loop, $k \leq n$, and the invariant (5.3.2) holds. First the variables are changed by the subalgorithm REDUCE such that $|\mu_{k\,k-1}| \leq \frac{1}{2}$. Next we check whether $\|b_k^* + \mu_{k\,k-1} b_{k-1}^*\|^2 < \frac{3}{4}\|b_{k-1}^*\|^2$. (Notice that $\|b_k^* + \mu_{k\,k-1} b_{k-1}^*\|^2 = \|b_k^*\|^2 + \mu_{k\,k-1}^2\|b_{k-1}^*\|^2$, since the vectors $b_{k-1}^*$ and $b_k^*$ are orthogonal.) If this is the case, $b_k$ and $b_{k-1}$ are interchanged by UPDATE and all the other $b_i$'s are left unchanged. The $B_i$'s and $\mu_{ij}$'s are updated accordingly. Then $k$ is replaced by $k - 1$. Now (5.3.2) holds. If the condition does not hold, we first achieve $|\mu_{kj}| \leq \frac{1}{2}$ for $1 \leq j \leq k - 1$. This is done by the subalgorithm REDUCE. Then we replace $k$ by $k + 1$. Now (5.3.2) holds.

The details in the correctness proofs of the subalgorithms are left to the reader and can partially be found in Lenstra et al. (1982).

b. Termination: Let

$$d_i = \left|(< b_j, b_l >)_{1 \leq j,l \leq i}\right| \quad \text{for } 1 \leq i \leq n\,,$$

$d_0 = 1$. By Exercise 3, $d_i = \prod_{j=1}^{i} \|b_j^*\|^2 \in \mathbb{R}^+$. Let $D = \prod_{i=1}^{n-1} d_i$. In step (2.1) $B_{k-1} = \|b_{k-1}^*\|^2$ is replaced by $B_k + \mu_{k\,k-1}^2 B_{k-1}$, which is less than $\frac{3}{4}B_{k-1}$. So $B_{k-1}$ is reduced by a factor of $\frac{3}{4}$. But the numbers $d_i$ are bounded from below by a positive real bound that depends only on the lattice $L$ (see Lenstra et al. 1982). So there is also a positive real bound for $D$ and hence an upper bound for the number of times that (2.1) is executed. The number of times that (2.2) is executed can be at most $n - 1$ more than the number of times that (2.1) is executed. So the algorithm BASIS_REDUCTION terminates. □

In Lenstra et al. (1982) a detailed complexity analysis of BASIS_REDUCTION is given. We just quote the result, their proposition 1.26.

**Theorem 5.3.3.** Let $L \subset \mathbb{Z}^n$ be a lattice with basis $a_1, \ldots, a_n$, and let $B \in \mathbb{R}$, $B \geq 2$, be such that $\|a_i\|^2 \leq B$ for $1 \leq i \leq n$. Then the number of arithmetic operations needed by BASIS_REDUCTION for the input $a_1, \ldots, a_n$ is $\mathcal{O}(n^4 \log B)$, and the integers on which these operations are performed each have length $\mathcal{O}(n \log B)$.

So if we use classical multiplication, the complexity of BASIS_REDUCTION is $\mathcal{O}(n^6 (\log B)^3)$, which can be reduced to $\mathcal{O}(n^{5+\epsilon} (\log B)^{2+\epsilon})$, for every $\epsilon > 0$, by fast multiplication techniques.

## Factors and lattices

Throughout this section let $p$ be a prime number, $k$ a positive integer, $f$ a primitive, squarefree polynomial of degree $n$, $n > 0$, in $\mathbb{Z}[x]$, and $h$ a monic polynomial of degree $l$, $0 < l \leq n$, in $\mathbb{Z}_{p^k}[x]$, such that

$$h \text{ divides } (f \bmod p^k) \text{ in } \mathbb{Z}_{p^k}[x] \,, \tag{5.3.3}$$

$$(h \bmod p) \text{ is irreducible in } \mathbb{Z}_p[x] \,, \tag{5.3.4}$$

$$(f \bmod p) \text{ is squarefree in } \mathbb{Z}_p[x] \,. \tag{5.3.5}$$

**Theorem 5.3.4.** a. There is a uniquely determined irreducible factor $h_0$ of $f$ in $\mathbb{Z}[x]$, up to sign, such that $(h \bmod p)$ divides $(h_0 \bmod p)$.

b. Further, if $g$ divides $f$ in $\mathbb{Z}[x]$, then the following are equivalent:

i. $(h \bmod p)$ divides $(g \bmod p)$ in $\mathbb{Z}_p[x]$,

ii. $h$ divides $(g \bmod p^k)$ in $\mathbb{Z}_{p^k}[x]$,

iii. $h_0$ divides $g$ in $\mathbb{Z}[x]$.

*Proof.* a. Let $f = \prod_{i=0}^{s} h_i$ be the factorization of $f$ in $\mathbb{Z}[x]$. So $(h \bmod p)$ divides one of the factors $(h_i \bmod p)$, say $(h_0 \bmod p)$. The uniqueness of $h_0$ follows from (5.3.5).

b. Obviously (ii) $\Longrightarrow$ (i) and (iii) $\Longrightarrow$ (i).

Now we prove (i) $\Longrightarrow$ (iii): assume (i). Because of (5.3.5) $(h \bmod p)$ does not divide $(f/g \bmod p)$ in $\mathbb{Z}_p[x]$. So also $(h_0 \bmod p)$ does not divide $(f/g \bmod p)$ in $\mathbb{Z}_p[x]$ and furthermore $h_0 \nmid f/g \in \mathbb{Z}[x]$. Therefore, $h_0$ must be a factor of $g$ in $\mathbb{Z}[x]$.

Finally we prove (i) $\Longrightarrow$ (ii): $(h \bmod p)$ and $(f/g \bmod p)$ are relatively prime in $\mathbb{Z}_p[x]$, so for certain $r, s \in \mathbb{Z}_p[x]$ we have

$$r \cdot h + s \cdot (f/g) \equiv 1 \pmod{p} \,.$$

By the lifting theorem (Theorem 3.2.2) we get $r', s' \in \mathbb{Z}_{p^k}[x]$ such that

$$r' \cdot h + s' \cdot (f/g) \equiv 1 \pmod{p^k} ,$$

or

$$r' \cdot (g \bmod p^k) \cdot h + s' \cdot f \equiv g \pmod{p^k} .$$

But $h$ divides the left-hand side of this congruence, so we have that $h$ divides $g$ modulo $p^k$. □

**Corollary.** $h$ divides $(h_0 \bmod p^k)$ in $\mathbb{Z}_{p^k}[x]$.

*Proof.* Follows from (ii) for $g = h_0$. □

In the sequel we denote by $h_0$ the polynomial in Theorem 5.3.4. Observe that $h_0$ is a primitive polynomial since $f$ is primitive.

The set of polynomials in $\mathbb{R}[x]$ of degree not greater than $m$, for $m \in \mathbb{N}$, is a vector space over $\mathbb{R}$ isomorphic to $\mathbb{R}^{m+1}$. The isomorphism is given by viewing a polynomial as its coefficient vector, i.e., by identifying

$$\sum_{i=0}^{m} a_i x^i \quad \text{with } (a_0, \ldots, a_m) .$$

The length $\|f\| = \sqrt{\sum_{i=0}^{m} a_i^2}$ of a polynomial $f = \sum_{i=0}^{m} a_i x^i$ equals the Euclidean length $\|(a_0, \ldots, a_m)\|$ of the corresponding vector.

In the following we let $m$ be an integer greater or equal to $l$. We let $L_{m,h}$ be the set of polynomials of degree not greater than $m$ in $\mathbb{Z}[x]$ that are divisible by $h$ modulo $p^k$, i.e.,

$$L_{m,h} = \{g \in \mathbb{Z}[x] \mid \deg(g) \le m \text{ and } h|(g \bmod p^k) \text{ in } \mathbb{Z}_{p^k}[x]\} .$$

By the above isomorphism $L_{m,h}$ is a lattice in $\mathbb{R}^{m+1}$ and it is spanned by the basis

$$\{p^k x^i \mid 0 \le i < l\} \cup \{h x^j \mid 0 \le j \le m - l\} .$$

**Theorem 5.3.5.** Let $b \in L_{m,h}$ satisfy $p^{kl} > \|f\|^m \cdot \|b\|^n$. Then $h_0$ divides $b$ in $\mathbb{Z}[x]$, and in particular $\gcd(f, b) \ne 1$.

*Proof.* We may assume $b \ne 0$. Let $s = \deg(b)$, $g = \gcd(f, b)$ in $\mathbb{Z}[x]$, and $t = \deg(g)$. Observe that $0 \le t \le s \le m$. In order to show that $h_0$ divides $b$, it suffices to show that $h_0$ divides $g$, which by Theorem 5.3.4 (b) is equivalent to

$$(h \bmod p) \text{ divides } (g \bmod p) \text{ in } \mathbb{Z}_p[x] . \tag{5.3.6}$$

Assume that (5.3.6) does not hold. Then ($h$ mod $p$) divides ($f/g$ mod $p$). Consider the set of polynomials

$$M = \{\lambda f + \mu b \mid \lambda, \mu \in \mathbb{Z}[x], \ \deg(\lambda) < s - t, \ \deg(\mu) < n - t\} .$$

Let

$$M' = \left\{ \sum_{i=t}^{n+s-t-1} a_i x^i \;\middle|\; \sum_{i=0}^{n+s-t-1} a_i x^i \in M \right\} ,$$

i.e., $M'$ is the projection of $M$ onto its last coordinates.

As shown in Lenstra et al. (1982), the projections of

$$\{x^i f \mid 0 \le i < s - t\} \cup \{x^j b \mid 0 \le j < n - t\}$$

on $M'$ are linearly independent. They also span $M'$ as a lattice in $\mathbb{R}^{n+s-2t}$, so $M'$ is a lattice of rank $n + s - 2t$. From Hadamard's inequality we obtain

$$\det(M') \le \|f\|^{s-t} \cdot \|b\|^{n-t} \le \|f\|^m \cdot \|b\|^n < p^{kl} . \tag{5.3.7}$$

Let $b_t, b_{t+1}, \ldots, b_{n+s-t-1}$ be a basis of $M'$ with $\deg(b_j) = j$ (Exercise 4). Observe that $t + l - 1 \le n + s - t - 1$, since $g$ divides $b$ and ($h$ mod $p$) divides ($f/g$ mod $p$). The leading coefficients of $b_t, b_{t+1}, \ldots, b_{t+l-1}$ are divisible by $p^k$ (Exercise 5). So

$$\det(M') = \left| \prod_{i=t}^{n+s-t-1} \mathrm{lc}(b_i) \right| \ge p^{kl} ,$$

a contradiction to (5.3.7).

Therefore, (5.3.6) must hold, which completes the proof. □

**Lemma 5.3.6.** Let $q(x) = b_0 + b_1 x + \cdots + b_l x^l \in \mathbb{Z}[x]$ be a divisor of $p(x) \in \mathbb{Z}[x]$.

a. $|b_i| \le \binom{l}{i} \|p\|$ for $0 \le i \le l$.

b. $\|q\| \le \binom{2l}{l}^{1/2} \|p\|$.

*Proof.* a. Theorem 2 in Mignotte (1974).

b. By (a) we get

$$\|q\| \le \sqrt{\sum_{i=0}^{l} \binom{l}{i}^2} \cdot \|p\| .$$

Vandermonde's equation

$$\sum_{k=0}^{n} \binom{r}{k} \binom{s}{n-k} = \binom{r+s}{n}$$

applied to $s = r = n = l$ yields

$$\|q\| \leq \binom{2l}{l}^{1/2} \cdot \|p\| . \qquad \square$$

**Theorem 5.3.7.** Let $b_1, \ldots, b_{m+1}$ be a reduced basis for $L_{m,h}$, and let

$$p^{kl} > 2^{mn/2} \cdot \binom{2m}{m}^{n/2} \cdot \|f\|^{m+n} . \tag{5.3.8}$$

a. $\deg(h_0) \leq m$ if and only if $\|b_1\| < \sqrt[n]{p^{kl}/\|f\|^m}$.
b. Assume that there exists an index $j \in \{1, \ldots, m+1\}$ for which

$$\|b_j\| < \sqrt[n]{p^{kl}/\|f\|^m} . \tag{5.3.9}$$

Let $t$ be the largest such $j$. Then $\deg(h_0) = m + 1 - t$, $h_0 = \gcd(b_1, \ldots, b_t)$ and (5.3.9) holds for all $j$ with $1 \leq j \leq t$.

*Proof.* a. "$\Longleftarrow$": If $\|b_1\|$ is bounded in this way, then by Theorem 5.3.5 $h_0$ divides $b_1$, and since $\deg(b_1) \leq m$ we get $\deg(h_0) \leq m$.

"$\Longrightarrow$": If $\deg(h_0) \leq m$ then $h_0 \in L_{m,h}$. So by Theorem 5.3.1 (b) and Lemma 5.3.6 (b)

$$\|b_1\| \leq 2^{m/2} \cdot \|h_0\| \leq 2^{m/2} \cdot \binom{2m}{m}^{1/2} \cdot \|f\| .$$

Using (5.3.8) we get the desired bound for $\|b_1\|$.

b. Let

$$J = \{j \mid 1 \leq j \leq m+1 \text{ and } j \text{ satisfies (5.3.9)}\} .$$

By Theorem 5.3.5 for every $j \in J$ the polynomial $h_0$ divides $b_j$. So $h_0$ divides $h_1$ for

$$h_1 = \gcd(\{b_j \mid j \in J\}) .$$

Each $b_j$, $j \in J$, is divisible by $h_1$ and has degree not greater than $m$, so it belongs to the lattice

$$\mathbb{Z} \cdot h_1 + \mathbb{Z} \cdot h_1 \cdot x + \ldots + \mathbb{Z} \cdot h_1 \cdot x^{m-\deg(h_1)}$$

of rank $m + 1 - \deg(h_1)$. Moreover, the $b_j$'s are linearly independent, so

$$|J| \leq m + 1 - \deg(h_1) . \tag{5.3.10}$$

As in (a) we show that

$$\|h_0 \cdot x^i\| = \|h_0\| \leq \binom{2m}{m}^{1/2} \cdot \|f\| \quad \text{for all } i \geq 0 \; .$$

For $i \in \{0, 1, \ldots, m - \deg(h_0)\}$ we have $h_0 \cdot x^i \in L_{m,h}$. So from Theorem 5.3.1 (c) we obtain

$$\|b_j\| \leq 2^{m/2} \cdot \binom{2m}{m}^{1/2} \cdot \|f\|$$

for $1 \leq j \leq m + 1 - \deg(h_0)$. So by (5.3.8)

$$\{1, \ldots, m + 1 - \deg(h_0)\} \subseteq J \; . \tag{5.3.11}$$

But $h_0$ divides $h_1$, so from (5.3.10) and (5.3.11) we obtain that

$$\begin{aligned} &\deg(h_0) = \deg(h_1) = m + 1 - t \; , \\ &J = \{1, \ldots, t\} \; , \\ &h_1 = a \cdot h_0 \quad \text{for some } a \in \mathbb{Z} \; . \end{aligned}$$

Furthermore, we get $\deg(h_0) \leq m$ by (a), so $h_0 \in L_{m,h}$.

$h_0$ is primitive, so for proving that $h_0$ is equal to $h_1$, up to sign, it suffices to show that $h_1$ also is primitive. Let $j$ be an arbitrary element of $J$. $h_0$ divides $\mathrm{pp}(b_j)$. Since $h_0 \in L_{m,h}$, also $\mathrm{pp}(b_j) \in L_{m,h}$. But $b_j$ belongs to a basis for $L_{m,h}$. So $b_j$ must be primitive, and hence also the factor $h_1$ of $b_j$ must be primitive. So $h_0 = \pm h_1$. □

The factorization algorithm

Before we describe the Lenstra–Lenstra–Lovász factorization algorithm, we start with two subalgorithms.

**Algorithm LLL_SUBALG1**(in: $f, p, k, h, m$; out: $h_0$);
[$f \in \mathbb{Z}[x]$ a primitive, squarefree polynomial,
$p$ a prime number not dividing $\mathrm{lc}(f)$ and such that ($f$ mod $p$) is squarefree,
$k$ a positive integer,
$h$ a polynomial in $\mathbb{Z}_{p^k}[x]$ such that $\mathrm{lc}(h) = 1$, $h$ divides ($f$ mod $p^k$), ($h$ mod $p$) is irreducible in $\mathbb{Z}_p[x]$,
$m$ an integer greater or equal to $\deg(h)$, such that $p^{k\deg(h)} > 2^{mn/2} \cdot \binom{2m}{m}^{n/2} \cdot \|f\|^{m+n}$;
$h_0$ is the irreducible factor of $f$ for which ($h$ mod $p$) divides ($h_0$ mod $p$), if this factor has degree $\leq m$, $h_0 =$ error otherwise.]
1. $n := \deg(f)$; $l := \deg(h)$;
2. $(b_1, \ldots, b_{m+1}) :=$ BASIS_REDUCTION($p^k x^0, \ldots, p^k x^{l-1}, hx^0, \ldots, hx^{m-l}$);

3. if $\|b_1\| \geq \sqrt[n]{p^{kl}/\|f\|^m}$
   then $h_0 :=$ error
   else $\{t :=$ largest integer such that $\|b_t\| < \sqrt[n]{p^{kl}/\|f\|^m}$;
     $h_0 := \gcd(b_1, \ldots, b_t)\}$;
   return.

**Theorem 5.3.8.** a. Algorithm LLL_SUBALG1 is correct.
b. The number of arithmetic operations needed by algorithm LLL_SUBALG1 is $\mathcal{O}(m^4 k \log p)$, and the integers on which these operations are performed each have length $\mathcal{O}(m k \log p)$.

*Proof.* a. $b_1, \ldots, b_{m+1}$ form a reduced basis for the lattice $L_{m,h}$. If $\|b_1\| \geq \sqrt[n]{p^{kl}/\|f\|^m}$, then by Theorem 5.3.7 (a) $\deg(h_0) > m$, so "error" is the correct answer. Otherwise by Theorem 5.3.7 (b) $h_0 = \gcd(b_1, \ldots, b_t)$, where $t$ is the greatest index such that $\|b_t\| < \sqrt[n]{p^{kl}/\|f\|^m}$.

b. Every vector $a$ in the initial basis for $L_{m,h}$ is bounded by $\|a\|^2 \leq 1 + l \cdot p^{2k} =: B$. From $l \leq n$ and the input condition for $m$ we see that $m$ is dominated by $k \log p$. Since $\log l < l \leq m$ we get $\log B$ is dominated by $k \log p$. Application of Theorem 5.3.3 yields the desired bounds for step (2).

In step (3) we need to compute the greatest common divisor of $b_1, \ldots, b_t$. Every coefficient $c$ in $b_j$, $1 \leq j \leq t$, is bounded by $\sqrt[n]{p^{kl}/\|f\|^m}$, so $\log c < k \log p$. If we use the subresultant gcd algorithm, every coefficient in the computation of $\gcd(b_1, b_2)$ is bounded by $m k \log p + m \log m \sim m k \log p$ (see Knuth 1981: sect. 4.6.1 (26)). By the Landau–Mignotte bound the coefficients in the gcd are of size $\mathcal{O}(2^m \|b_1\|)$, so their length is dominated by $m + \log B \sim \log B$. The same bounds hold for all the successive gcd computations. One gcd computation takes $\mathcal{O}(m^2)$ arithmetic operations. We need at most $m$ gcd computations, so the number of arithmetic operations for all of them is dominated by $m^3$. □

**Algorithm LLL_SUBALG2**(in: $f, p, h$; out: $h_0$);
[$f \in \mathbb{Z}[x]$ a primitive, squarefree polynomial,
$p$ a prime number not dividing $\mathrm{lc}(f)$ and such that ($f$ mod $p$) is squarefree,
$h$ an irreducible polynomial in $\mathbb{Z}_p[x]$ such that $\mathrm{lc}(h) = 1$ and $h$ divides ($f$ mod $p$);
$h_0$ is the irreducible factor of $f$ for which $h$ divides ($h_0$ mod $p$).]
1. $n := \deg(f)$; $l := \deg(h)$;
   if $l = n$ then $\{h_0 := f$; return$\}$;
   [Now $l < n$.]
2. $k :=$ least positive integer such that $p^{kl} > 2^{(n-1)n/2} \cdot \binom{2(n-1)}{n-1}^{n/2} \cdot \|f\|^{2n-1}$;
   $[h'', h'] :=$ LIFT_FACTORS($f$, [($f/h$ mod $p$), $h$], $p$, $k$);
   [$h' \in \mathbb{Z}_{p^k}[x]$ is congruent to $h$ modulo $p$, $\mathrm{lc}(h) = 1$ and $h'$ divides ($f$ mod $p^k$).]
3. $u :=$ greatest integer such that $l \leq (n-1)/2^u$;
   while $u > 0$ do

$\{m := \lfloor (n-1)/2^u \rfloor;$
$h_0 :=$ LLL_SUBALG1$(f, p, k, h', m)$;
if $h_0 \neq$ error then return;
$u := u - 1\}$;

4. $h_0 := f$;
return.

**Theorem 5.3.9.** a. Algorithm LLL_SUBALG2 is correct.

b. Let $m_0$ be the degree of the result $h_0$. Then the number of arithmetic operations needed by algorithm LLL_SUBALG2 is $\mathcal{O}(m_0(n^5 + n^4 \log \|f\| + n^3 \log p))$, and the integers on which these operations are performed each have length $\mathcal{O}(n^3 + n^2 \log \|f\| + n \log p)$.

*Proof.* a. If $l = n$ then $f$ is irreducible in $\mathbb{Z}_p[x]$, so it is irreducible over the integers. After executing step (2) we have

$$f \equiv h' \cdot h'' \pmod{p^k} ,$$

where $(h' \bmod p) = h$ and $\mathrm{lc}(h') = 1$. Now let $m$ be such that $l \leq m \leq n - 1$. $m$ satisfies the input condition of LLL_SUBALG1. So if the irreducible factor $h_0$ of $f$ corresponding to $h'$ has degree not greater than $m$, then LLL_SUBALG1 will compute it. If LLL_SUBALG1 returns "error" for all values of $m$, then there is no proper factor corresponding to $h'$, so the result is $h_0 = f$.

b. Since $k$ is the least positive integer satisfying the condition in step (2), we have

$$p^{k-1} \leq p^{(k-1)l} \leq 2^{(n-1)n/2} \cdot \binom{2(n-1)}{n-1}^{n/2} \cdot \|f\|^{2n-1} .$$

Using the fact that

$$\log \binom{2(n-1)}{n-1}^{n/2} \leq \log \left(2^{2(n-1)}\right)^{n/2} = (n-1)n \log 2 ,$$

we see that

$$k \log p = (k-1) \log p + \log p \preceq n^2 + n \log \|f\| + \log p .$$

Let $m_1$ be the largest value of $m$ considered in LLL_SUBALG2. Since we start with small values of $m$ and stop as soon as $m$ is greater than the degree of $h_0$, it follows that $m_1 < 2m_0$. All the other values of $m$ are of the form $\lfloor \frac{m_1}{2} \rfloor$, $\lfloor \frac{m_1}{4} \rfloor$, ..., $\lfloor \frac{m_1}{2^u} \rfloor$. So if we sum over all these values of $m$ we get

$$\sum_{\substack{m \text{ considered} \\ \text{in LLL\_SUBALG2}}} m \le \frac{m_1}{2^u} + \ldots + \frac{m_1}{2} + m_1 = m_1 \cdot \frac{\left(\frac{1}{2}\right)^{u+1} - 1}{\left(\frac{1}{2}\right) - 1} \le 2m_1 < 4m_0 \ .$$

Therefore, $\sum m^4 \le (\sum m)^4 = \mathcal{O}(m_0^4)$. Applying Theorem 5.3.8 (b) we deduce that the number of arithmetic operations needed for executing step (3) is dominated by

$$\begin{aligned} m_0^4 k \log p &\preceq m_0^4(n^2 + n \log \|f\| + \log p) \\ &\preceq m_0(n^5 + n^4 \log \|f\| + n^3 \log p) \end{aligned}$$

and that the integers on which these operations are performed each have length dominated by

$$m_0(n^2 + n \log \|f\| + \log p) \preceq n^3 + n^2 \log \|f\| + n \log p \ .$$

The same bounds hold for the Hensel lifting in step (2). □

Now we are ready to combine all these subalgorithms and get the factorization algorithm FACTOR_LLL.

**Algorithm FACTOR_LLL**(in: $f$; out: $F$);
[$f \in \mathbb{Z}[x]$ a primitive, squarefree polynomial of positive degree;
$F = [f_1, \ldots, f_s]$, the $f_i$'s are the distinct irreducible factors of $f$ in $\mathbb{Z}[x]$.]
1. $n := \deg(f)$;
   $R := \mathrm{res}(f, f')$; [$R \neq 0$, since $f$ is squarefree]
2. $p :=$ smallest prime not dividing $R$;
   [so ($f$ mod $p$) has degree $n$ and is squarefree in $\mathbb{Z}_p[x]$]
   *pfactors* := FACTOR_B($f$, $p$);
3. $F := [\ ]$;
   $\bar{f} := f$;
   while $\deg(\bar{f}) > 0$ do
     {$h :=$ FIRST(*pfactors*);
     $h := h/\mathrm{lc}(h)$;
     $h_0 :=$ LLL_SUBALG2($\bar{f}$, $p$, $h$);
     [$h_0$ is the irreducible factor of $\bar{f}$ for which $h$ divides ($h_0$ mod $p$)]
     $F :=$ CONS($h_0$, $F$);
     $\bar{f} := \bar{f}/h_0$;
     for $g \in$ *pfactors* do
       if $g$ divides ($h_0$ mod $p$)
       then remove $g$ from *pfactors*;
4. return.

**Theorem 5.3.10.** a. Algorithm FACTOR_LLL is correct.

b. The number of arithmetic operations needed by algorithm FACTOR_LLL is $\mathcal{O}(n^6+n^5 \log \|f\|)$, and the integers on which these operations are performed each have length $\mathcal{O}(n^3 + n^2 \log \|f\|)$.

*Proof.* a. Since $p \nmid \mathrm{res}(f, f')$, $\deg(f \bmod p) = n$ and $f$ is squarefree modulo $p$. So the Berlekamp factorization algorithm can be applied to $f$ and $p$, and it computes the list of irreducible factors of $f$ modulo $p$. The correctness of FACTOR_LLL now follows immediately from the correctness of LLL_SUBALG2.

b. $p$ is the least prime not dividing $\mathrm{res}(f, f')$. So by Hardy and Wright (1979: sect. 22.2), there is a positive bound $A$ such that $p = 2$ or

$$e^{Ap} < \prod_{q<p, q \text{ prime}} q \leq |\mathrm{res}(f, f')| \;.$$

By Hadamard's inequality

$$|\mathrm{res}(f, f')| \leq n^n \cdot \|f\|^{2n-1} \;.$$

Therefore,

$$p < (n \log n + (2n - 1) \log \|f\|)/A$$

or $p = 2$. Therefore the terms involving $\log p$ in Theorem 5.3.9 are absorbed by the other terms.

The algorithm FACTOR_B needs $\mathcal{O}(n^3 + prn^2)$ arithmetic operations, where $r$ is the actual number of factors. Substituting the bound for $p$ and using $r \leq n$, we get that the number of arithmetic operations for the application of FACTOR_B is dominated by $n^4 \log n + n^4 \log \|f\|$.

The number of arithmetic operations needed in the execution of LLL_SUBALG2 in step (3) is dominated by $m_0(n^5 + n^4 \log \|\bar{f}\|)$, where $m_0$ is the degree of the factor $h_0$ corresponding to $h$. Lemma 5.3.6 implies that $\log \|\bar{f}\| \preceq n + \log \|f\|$. All the degrees of the irreducible factors of $f$ add up to $n$, so the number of arithmetic operations needed in step (3) is $\preceq n^6 + n^5 \log \|f\|$.

By Theorem 5.3.9 the integers considered in FACTOR_LLL are of length $\mathcal{O}(n^3 + n^2 \log \|f\|)$. □

**Corollary.** If classical algorithms for the arithmetic operations are used then the complexity of FACTOR_LLL is $\mathcal{O}(n^{12} + n^9 (\log \|f\|)^3)$. If fast algorithms (see Sect. 2.1) are used then the complexity of FACTOR_LLL is $\mathcal{O}(n^{9+\epsilon} + n^{7+\epsilon} (\log \|f\|)^{2+\epsilon})$.

*Example 5.3.2.* We want to demonstrate how the algorithm FACTOR_LLL factors the polynomial

$$f = x^4 - 3x + 1$$

over the integers. As the prime in step (2) we choose $p = 5$. Modulo 5 the polynomial factors into

$$f \equiv (x+1) \cdot (x^3 - x^2 + x + 1) \pmod{5} .$$

In step (3) we have to find the factors over the integers corresponding to these factors modulo 5.

So, for instance, we call LLL_SUBALG2 with $f$, $p = 5$, $h = x^3 - x^2 + x + 1$. The least positive integer $k$ such that $5^{3k} > 2^6\binom{6}{3}^2\|f\|^7$ is $k = 4$. By an application of LIFT_FACTORS we lift the factorization modulo 5 to a factorization modulo $5^4$

$$f \equiv (x - 139) \cdot \underbrace{(x^3 + 139x^2 - 54x - 9)}_{h'} \pmod{5^4} .$$

In step (3) $u = 0$ and LLL_SUBALG2 immediately returns the result $h_0 = f$. So we have detected that $f$ is irreducible over the integers.

Just for demonstration purposes we also apply LLL_SUBALG2 to the arguments $f$, $p = 5$, $h = x + 1$. In this case $k = 12$, and we have to lift the factorization to a factorization modulo $5^{12}$

$$f \equiv \underbrace{(x + 46966736)}_{h'} \cdot$$
$$\cdot (x^3 - 46966736x^2 + 22915571x - 42196259) \pmod{5^{12}} .$$

In step (3) $u = 1$, and LLL_SUBALG1 is called with the arguments $f$, $p = 5$, $k = 12$, $h = h'$, $m = 1$. BASIS_REDUCTION($5^{12}, x + 46966736$) yields the reduced basis $b_1 = (-10212, -6217)$, $b_2 = (6781, -19779)$. At this point $\|b_1\| > \sqrt[4]{5^{12}/\|f\|}$, so $h_0$ is set to "error." Next LLL_SUBALG1 is called with $f, p, k$ as before and $m = 3$. Again the result is "error." So LLL_SUBALG2 returns $h_0 = f$ as the only factor of $f$ over the integers.

*Example 5.3.3.* Let us also consider an example, where FACTOR_LLL really detects a factor of the given polynomial. Let

$$f = x^4 + x^3 + 2x^2 + x + 1 .$$

As the prime $p$ we choose 41. Modulo 41 we get the factorization

$$f \equiv \underbrace{(x - 9)}_{h} \cdot (x + 9) \cdot (x^2 + x + 1) \pmod{41} .$$

In LLL_SUBALG2 this factorization is lifted to a factorization modulo $41^5$,

$$f \equiv \underbrace{(x + 46464143)}_{h'} \cdot$$
$$\cdot (x^3 - 46464142x^2 - 46464142x - 46464143) \pmod{41^5} .$$

Let us only consider the value $m = 3$ as the bound for the degree of the irreducible factor. Application of BASIS_REDUCTION to the basis

$$41^5,\ h',\ h'x,\ h'x^2$$

yields the reduced basis

$$\begin{gathered}(1, 0, 1, 0, 0)\ ,\\ (-5237, -2476, 5238, 0, 0)\ ,\\ (1238, -10475, -1238, 0, 0)\ ,\\ (-1107107, -522435, 1103118, 1, 0)\ ,\\ (51440777964301, 24274494548205, -51255432497874, 0, 1)\ .\end{gathered}$$

Only the first element in the reduced basis satisfies the condition in step (3) of LLL_SUBALG1, so we get

$$h_0 = (1, 0, 1, 0, 0) = x^2 + 1\ .$$

After removing the two factors that divide $h_0$ modulo 41, we are left with only one factor, and the factorization

$$f = (x^2 + 1) \cdot (x^2 + x + 1)$$

has been computed.

## Exercises

1. Show that $\det(L)$, $L$ a lattice in $\mathbb{R}^n$, is independent of the basis.
2. Prove: If $b_1, \ldots, b_n$ are a reduced basis for a lattice in $\mathbb{R}^n$, then $\|b_i^*\|^2 \geq \frac{1}{2}\|b_{i-1}^*\|^2$ for $1 < i \leq n$.
3. Let $b_1, \ldots, b_n$ be a basis for a lattice in $\mathbb{R}^n$, $b_1^*, \ldots, b_n^*$ the orthogonal basis produced by the Gram–Schmidt process, and $d_i = |(< b_j, b_l >)_{1\leq j,l\leq i}|$ for $1 \leq i \leq n$. Then $d_i = \prod_{j=1}^{i} \|b_j^*\|^2$ for $1 \leq i \leq n$.
4. Let $b_1, \ldots, b_n \in \mathbb{Z}^n$ be a basis for a lattice $L$ in $\mathbb{R}^n$. Then $L$ has a basis $c_1, \ldots, c_n \in \mathbb{Z}^n$ such that $C = (c_1, \ldots, c_n)$ (the $c_i$'s written as columns) is an upper triangular matrix.
5. By the notation of Theorem 5.3.5 show that every polynomial $q \in M$ with $\deg(q) < t + l$ is divisible by $p^k$.

6. Compute a reduced basis for the lattice $L_{1,h}$, $h = x + 46966736$ in Example 5.3.2.
7. Apply the algorithm FACTOR_LLL for computing the irreducible factors of the integral polynomial $x^4 - 1$.

## 5.4 Factorization over algebraic extension fields

We describe an algorithm that has been presented in van der Waerden (1970) and slightly improved by B. Trager (1976). For further reading we refer to Wang (1976).

Let $K$ be a computable field of characteristic 0 such that there is an algorithm for factoring polynomials in $K[x]$. Let $\alpha$ be algebraic over $K$ with minimal polynomial $p(y)$ of degree $n$. Throughout this section we call $K$ the *ground field* and $K(\alpha)$ the *extension field.* Often we will write a polynomial $f(x) \in K(\alpha)[x]$ as $f(x, \alpha)$ to indicate the occurrence of $\alpha$ in the coefficients. Let $\alpha = \alpha_1, \alpha_2, \ldots, \alpha_n$ be the roots of $p(y)$ in a splitting field of $p$ over $K$. By $\phi_j$, $1 \leq j \leq n$, we denote the canonical field isomorphism that takes $\alpha$ into $\alpha_j$, i.e.,

$$\begin{aligned} \phi_j\colon\ K(\alpha) &\longrightarrow K(\alpha_j) \\ \alpha &\longmapsto \alpha_j \\ a &\longmapsto a \quad \text{for all } a \in K\ . \end{aligned}$$

$\phi_j$ can be extended to $\phi_j\colon K(\alpha)[x] \longrightarrow K(\alpha_j)[x]$ by letting it act on the coefficients.

We will reduce the problem of factorization in $K(\alpha)[x]$ to factorization in $K[x]$. This reduction will be achieved by associating a $g \in K[x]$ with the given $f \in K(\alpha)[x]$ such that the factors of $f$ are in a computable 1–1 correspondence with the factors of $g$, i.e.,

$$\begin{aligned} f \in K(\alpha)[x] &\longleftrightarrow g \in K[x] \\ \text{factors of } f &\overset{1-1}{\longleftrightarrow} \text{factors of } g\ . \end{aligned}$$

A candidate for such a function is the *norm*, which maps an element in the extension field to the product of all its conjugates over $K$. This product is an element of $K$.

$$\begin{aligned} \mathrm{norm}_{[K(\alpha)/K]}\colon\ K(\alpha) &\longrightarrow K \\ \beta &\longmapsto \prod_{\beta' \sim \beta} \beta'\ , \end{aligned}$$

where $\beta' \sim \beta$ means that $\beta'$ is conjugate to $\beta$ relative to $K(\alpha)$ over $K$. That is, if $\beta = q(\alpha)$ is the normal representation of $\beta$ in $K(\alpha)$ (compare Sect. 2.4), then

$$\mathrm{norm}_{[K(\alpha)/K]}(\beta) = \prod_{i=1}^{n} q(\alpha_i)\ .$$

If the field extension is clear from the context, we write just norm$(\cdot)$ instead of $\text{norm}_{[K(\alpha)/K]}(\cdot)$. Since the norm is symmetric in the $\alpha_i$'s, by the fundamental theorem on symmetric functions it can be expressed in terms of the coefficients of $p$ and thus lies in $K$. The norm can be generalized from $K(\alpha)$ to $K(\alpha)[x]$ by defining the norm of a polynomial $h(x, \alpha)$ to be $\prod_{i=1}^{n} h(x, \alpha_i)$, which can be computed as

$$\text{norm}(h(x, \alpha)) = \text{res}_y(h(x, y), p(y)) .$$

Clearly the norm can be generalized to multivariate polynomials. One important property of the norm is multiplicativity, i.e.,

$$\text{norm}(f \cdot g) = \text{norm}(f) \cdot \text{norm}(g) . \tag{5.4.1}$$

**Theorem 5.4.1.** If $f(x, \alpha)$ is irreducible over $K(\alpha)$, then $\text{norm}(f) = h(x)^j$ for some irreducible $h \in K[x]$ and some $j \in \mathbb{N}$.

*Proof.* Assume $\text{norm}(f) = g(x)h(x)$ and $g, h$ are relatively prime. For $1 \leq i \leq n$ let $f_i(x) = f(x, \alpha_i)$. Clearly $f = f_1$ divides $\text{norm}(f) = \prod f_i$. So, since $f$ is irreducible, $f|g$ or $f|h$. W.l.o.g. let us assume that $f|h$, i.e., $h(x) = f_1(x, \alpha) \cdot \tilde{h}(x, \alpha)$. Then $h(x) = \phi_j(h) = \phi_j(f_1)\phi_j(\tilde{h}) = f_j\tilde{h}(x, \alpha_j)$. Therefore, $f_j|h$ for $1 \leq j \leq n$. Since $g$ and $h$ are relatively prime, this implies that $\gcd(f_j, g) = 1$ for $1 \leq j \leq n$. Thus, $\gcd(\text{norm}(f), g) = 1$, i.e., $g = 1$. □

The previous theorem yields a method for finding minimal polynomials for elements $\beta \in K(\alpha)$. Let $\beta = q(\alpha)$, $b(x) = \text{norm}(x - \beta) = \text{norm}(x - q(\alpha))$. $x - \beta | b(x)$, so $b(\beta) = 0$. Therefore the minimal polynomial $p_\beta(x)$ has to be one of the irreducible factors of $b(x)$. By Theorem 5.4.1, $b(x) = p_\beta(x)^j$ for some $j \in \mathbb{N}$. So $p_\beta(x)$ can be determined by squarefree factorization of $b(x)$.

$K(\alpha)[x]$ is a Euclidean domain, so by successive application of the Euclidean algorithm the problem of factoring in $K(\alpha)[x]$ can be reduced to the problem of factoring squarefree polynomials in $K(\alpha)[x]$ (see Sect. 4.4). From now on let us assume that $f(x, \alpha) \in K(\alpha)[x]$ is squarefree.

**Theorem 5.4.2.** Let $f(x, \alpha) \in K(\alpha)[x]$ be such that $F(x) = \text{norm}(f)$ is squarefree. Let $F(x) = \prod_{i=1}^{r} G_i(x)$ be the irreducible factorization of $F(x)$. Then $\prod_{i=1}^{r} g_i(x, \alpha)$, where $g_i(x, \alpha) = \gcd(f, G_i)$ over $K(\alpha)$, is the irreducible factorization of $f(x, \alpha)$ over $K(\alpha)$.

*Proof.* The statement follows from
a. every $g_i$ divides $f$,
b. every irreducible factor of $f$ divides one of the $g_i$'s,
c. the $g_i$'s are relatively prime, and
d. every $g_i$ is irreducible.

Ad (a): This is obvious from $g_i = \gcd(f, G_i)$.

Ad (b): Let $v(x, \alpha)$ be an irreducible factor of $f$ over $K(\alpha)$. By Theorem 5.4.1, $\text{norm}(v) = w(x)^k$ for some irreducible $w(x) \in K[x]$. $v|f$ implies $\text{norm}(v)|\text{norm}(f)$. Since $\text{norm}(f)$ is squarefree, $\text{norm}(v)$ is irreducible and must be one of the $G_i$'s. So $v|g_i(x, \alpha)$.

Ad (c): Suppose the irreducible factor $v$ of $f$ divides both $g_i$ and $g_j$ for $i \neq j$. Then the irreducible polynomial $\text{norm}(v)$ divides both $\text{norm}(G_i) = G_i^n$ and $\text{norm}(G_j) = G_j^n$. This would mean that $G_i$ and $G_j$ have a common factor.

Ad (d): Clearly every $g_i$ is squarefree. Assume that $v_1(x, \alpha)$ and $v_2(x, \alpha)$ are distinct irreducible factors of $f$ and that both of them divide $g_i = \gcd(f, G_i)$. $v_1|G_i$ implies $\text{norm}(v_1)|\text{norm}(G_i) = G_i(x)^n$. Because of the squarefreeness of $\text{norm}(f)$, we must have $\text{norm}(v_1) = G_i$. Similarly we get $\text{norm}(v_2) = G_i$. But $(v_1 \cdot v_2)|f$ implies $\text{norm}(v_1 \cdot v_2) = G_i(x)^2|\text{norm}(f)$, in contradiction to the squarefreeness of $\text{norm}(f)$. □

So we can solve our factorization problem over $K(\alpha)$, if we can show that we can restrict our problem to the situation in which $\text{norm}(f)$ is squarefree. The following lemmata and theorem will guarantee exactly that.

**Lemma 5.4.3.** If $f(x)$ is a squarefree polynomial in $K[x]$, then there are only finitely many $s \in K$ for which $\text{norm}(f(x - s\alpha))$ is not squarefree.

*Proof.* Let $\beta_1, \ldots, \beta_m$ be the distinct roots of $f$. Then the roots of $f(x - s\alpha_j)$ are $\beta_i + s\alpha_j$, $1 \leq i \leq m$. Thus, the roots of $G(x) = \text{norm}(f(x - s\alpha_j)) = \prod_{k=1}^n f(x - s\alpha_k)$ are $\beta_i + s\alpha_k$ for $1 \leq i \leq m$, $1 \leq k \leq n$. $G$ can have a multiple root only if

$$s = \frac{\beta_j - \beta_i}{\alpha_k - \alpha_l},$$

where $k \neq l$. There are only finitely many such values. □

**Lemma 5.4.4.** If $f(x, \alpha)$ is a squarefree polynomial in $K(\alpha)[x]$, then there exists a squarefree polynomial $g(x) \in K[x]$ such that $f|g$.

*Proof.* Let $G(x) = \text{norm}(f(x, \alpha)) = \prod g_i(x)^i$ be the squarefree factorization of the norm of $f$. Since $f$ is squarefree, $f|g := \prod g_i(x)$. □

**Theorem 5.4.5.** For any squarefree polynomial $f(x, \alpha) \in K(\alpha)[x]$ there are only finitely many $s \in K$ for which $\text{norm}(f(x - s\alpha))$ is not squarefree.

*Proof.* Let $g(x)$ be as in Lemma 5.4.4. By Lemma 5.4.3 there are only finitely many $s \in K$ for which $\text{norm}(g(x - s\alpha))$ is not squarefree. But $f|g$ implies $\text{norm}(f(x - s\alpha))|\text{norm}(g(x - s\alpha))$. If $\text{norm}(f(x - s\alpha))$ is not squarefree, then neither is $\text{norm}(g(x - s\alpha))$. □

**Algorithm SQFR_NORM**(in: $f$; out: $g, s, N$);
[$f \in K(\alpha)[x]$ squarefree; $s \in \mathbb{N}$, $g(x) = f(x - s\alpha)$,
$N(x) = \text{norm}(g(x, \alpha))$ is squarefree.]
1. $s := 0$; $g(x, \alpha) := f(x, \alpha)$;
2. $N(x) := \text{res}_y(g(x, y), p(y))$;
3. while $\deg(\gcd(N(x), N'(x))) \neq 0$ do
   $\{s := s + 1$;
   $g(x, \alpha) := g(x - \alpha, \alpha)$;
   $N(x) := \text{res}_y(g(x, y), p(y))\}$;
   return.

So over a field of characteristic 0 we can always find a transformation of the form $f(x - s\alpha)$, $s \in \mathbb{N}$, such that $\text{norm}(f(x - s\alpha))$ is squarefree. These considerations give rise to an algorithm for computing a linear change of variable which transforms $f$ to a polynomial with squarefree norm.

Now we are ready to present an algorithm for factoring polynomials over the extension field.

**Algorithm FACTOR_ALG**(in: $f$; out: $F$);
[$f \in K(\alpha)[x]$ squarefree; $F = [f_1, \ldots, f_r]$, where $f_1, \ldots, f_r$ are the irreducible factors of $f$ over $K(\alpha)$.]
1. $[g, s, N] := \text{SQFR\_NORM}(f)$;
2. $L :=$ list of irreducible factors of $N(x)$ over $K$;
3. if $\text{LENGTH}(L) = 1$ then return($[f]$);
4. $F := [\,]$;
   for each $H(x)$ in $L$ do
   $\{h(x, \alpha) := \gcd(H(x), g(x, \alpha))$;
   $g(x, \alpha) := g(x, \alpha)/h(x, \alpha)$;
   $F := \text{CONS}(h(x + s\alpha, \alpha), F)\}$;
   return.

*Example 5.4.1.* We apply the factorization algorithm FACTOR_ALG to the domain $\mathbb{Q}(\sqrt[3]{2})[x]$, i.e., $K = \mathbb{Q}$, $\alpha$ a root of $p(y) = y^3 - 2$. Let us factor the polynomial

$$f(x, \alpha) = x^4 + \alpha x^3 - 2x - 2\alpha \ .$$

$f(x, \alpha)$ is squarefree. First we have to transform $f$ to a polynomial $g$ with squarefree norm. The norm of $f$ itself is

$$\text{norm}(f) = \text{res}_y(f(x, y), p(y)) = -(x^3 - 2)^3(x^3 + 2) \ ,$$

i.e., it is not squarefree. The transformation $x \mapsto x - \alpha$ does not work, but $x \mapsto x - 2\alpha$ does:

$$g(x, \alpha) := f(x - 2\alpha, \alpha) = x^4 - 7\alpha x^3 + 18\alpha^2 x^2 - 42x + 18\alpha \ ,$$

$$N(x) = \text{norm}(g) = x^{12} - 56x^9 + 216x^6 - 6048x^3 + 11664 \ ,$$

and $N(x)$ is squarefree. The factorization of $N(x)$ is

$$N(x) = (x^3 - 2)(x^3 - 54)(x^6 + 108) .$$

Computing the gcd of all the factors of $N(x)$ with $g(x, \alpha)$ gives us the factorization of $g(x, \alpha)$:

$$g(x, \alpha) = (x - \alpha)(x - 3\alpha)(x^2 - 3\alpha x + 3\alpha^2) ,$$

which can be transformed by $x \mapsto x + 2\alpha$ to the factorization

$$f(x, \alpha) = (x + \alpha)(x - \alpha)(x^2 + \alpha x + \alpha^2) .$$

Computation of primite elements for multiple field extensions

Over a field $K$ of characteristic 0 every algebraic extension field $K(\alpha)$ is separable, i.e., $\alpha$ is a root of multiplicity 1 of its minimal polynomial. So every multiple algebraic extension

$$K \subset K(\alpha_1) \subset \ldots \subset K(\alpha_1, \ldots, \alpha_n)$$

can be expressed as a simple algebraic extension, i.e.,

$$K(\alpha_1, \ldots, \alpha_n) = K(\gamma)$$

for some $\gamma$ algebraic over $K$. Such a $\gamma$ is called a *primitive element* for the field extension. We will describe how to compute such primitive elements. Clearly it suffices to find primitive elements for double field extensions

$$K \subset K(\alpha) \subset K(\alpha, \beta) ,$$

where $p(\alpha) = 0$ for some irreducible $p(x) \in K[x]$ and $q(\beta, \alpha) = 0$ for some irreducible $q(x, \alpha) \in K(\alpha)[x]$. Let $n = \deg(p)$ and $m = \deg(q)$.

**Theorem 5.4.6.** If $N(x) = \mathrm{norm}_{[K(\alpha)/K]}(q(x, \alpha))$ is squarefree, then $K(\alpha, \beta) = K(\beta)$, and $N(x)$ is the minimal polynomial for $\beta$ over $K$.

*Proof.* Let $\alpha_1, \ldots, \alpha_n$ be the roots of $p(x)$, and $\beta_{i1}, \ldots, \beta_{im}$ the roots of $q(x, \alpha_i)$. $\mathrm{norm}_{[K(\alpha)/K]}(q) = \prod_{i=1}^{n} q(x, \alpha_i)$, so if this norm is squarefree, then all the $\beta_{ij}$ must be different. So for every $\beta$ in $\{\beta_{ij} \mid 1 \le i \le n, 1 \le j \le m\}$ there is a uniquely determined $\alpha$ in $\{\alpha_1, \ldots, \alpha_n\}$ such that $q(\beta, \alpha) = 0$. Thus, $\gcd(q(\beta, x), p(x))$ must be linear,

$$\gcd(q(\beta, x), p(x)) = x - r(\beta) \quad \text{for some } r(y) \in K[y] ,$$

and therefore $\alpha = r(\beta)$. So $K(\alpha, \beta) = K(\beta)$.

$\beta$ is a root of $N(x) = \text{norm}_{[K(\alpha)/K]}(q)$. By Theorem 5.4.1, and the squarefreeness of $N(x)$, $N(x)$ must be the minimal polynomial for $\beta$ over $K$. □

**Algorithm PRIMITIVE_ELEMENT**(in: $p, q$; out: $N, A, B$);
[$p$ and $q$ are the minimal polynomials for $\alpha$ and $\beta$, respectively, as above; $N(x)$ is the minimal polynomial of $\gamma$ over $K$ such that $K(\alpha, \beta) = K(\gamma)$, $A$ and $B$ are the normal representations of $\alpha$ and $\beta$ in $K(\gamma)$, respectively.]
1. $[g, s, N] :=$ SQFR_NORM$(q(x, \alpha))$;
2. $A :=$ solution of the linear equation $\gcd(g(\gamma, x), p(x)) = 0$ in $K(\gamma)$, where $N(\gamma) = 0$;
3. $B := \gamma - sA$;
   return.

*Example 5.4.2.* Let us compute a primitive element $\gamma$ for the multiple extension $\mathbb{Q}(\sqrt{2}, \sqrt{3})$, i.e., for $\mathbb{Q}(\alpha, \beta)$, where $\alpha$ is a root of $p(x) = x^2 - 2$ and $\beta$ is a root of $q(x, \alpha) = q(x) = x^2 - 3$.

The norm of $q$ is not squarefree, in fact $\text{norm}_{[\mathbb{Q}(\sqrt{2})/\mathbb{Q}]}(q) = (x^2 - 3)^2$. So we need a linear transformation of the form $x \mapsto x - s\alpha$, and in fact $s = 1$ works.

$$g(x, \alpha) := q(x - \alpha, \alpha) = x^2 - 2\alpha x - 1 \ ,$$
$$\begin{aligned} N(x) &= \text{norm}_{[\mathbb{Q}(\sqrt{2})/\mathbb{Q}]}(g(x, \alpha)) \\ &= (x^2 - 2\alpha x - 1)(x^2 + 2\alpha x - 1) = x^4 - 10x^2 + 1 \ . \end{aligned}$$

$N(x)$ is irreducible. Let $\gamma$ be a root of $N(x)$. So $\gamma = \beta + \alpha$. We get the representation of $\alpha$ in $K(\gamma)$ as the solution of the linear equation

$$\begin{aligned} \gcd(g(\gamma, x), p(x)) &= \gcd(-2\gamma x + (\gamma^2 - 1), x^2 - 2) \\ &= x + \tfrac{1}{2}(-\gamma^3 + 9\gamma) = 0 \ , \end{aligned}$$

i.e., $\alpha = A(\gamma) = \frac{1}{2}(\gamma^3 - 9\gamma)$. Finally $\beta = B(\gamma) = \gamma - A(\gamma) = -\frac{1}{2}(\gamma^3 - 11\gamma)$.

## Exercises

1. Prove: Let $p(y)$ be the minimal polynomial for $\alpha$ over the field $K$. Then $\text{norm}_{[K(\alpha)/K]}(h(x, \alpha))$ and $\text{res}_y(h(x, y), p(y))$ agree up to a non-zero multiplicative constant.
2. Factor $f(x) = x^5 + \alpha^2 x^4 + (\alpha + 1)x^3 + (\alpha^2 + \alpha - 1)x^2 + \alpha^2 x + \alpha^2$ over $\mathbb{Q}(\alpha)$, where $\alpha^3 - \alpha + 1 = 0$.

### 5.5 Factorization over an algebraically closed field

Whereas we have already considered the problem of factoring univariate polynomials over a given algebraic extension, i.e., $f(x) \in K(\alpha)[x]$, and factoring multivariate polynomials over $K(\alpha)$ can be achieved by methods similar to the ones for integral polynomials (see Sect. 5.2), we now want to give an algorithmic approach to finding factors of bivariate polynomials over an algebraically closed field. Observe that factoring univariate polynomials over an algebraically closed field really amounts to factorization in finite extensions of the field of definition.

Let $K$, the field of definition, be a computable field of characteristic 0 and let $\bar{K}$ be the algebraic closure of $K$. Then the problem is
given: $f(x, y) \in K[x, y]$,
find: an irreducible factor $g(x, y)$ of $f(x, y)$ in $\bar{K}[x, y]$.

*Definition 5.5.1.* If $f \in K[x, y]$ has no non-trivial factor in $\bar{K}[x, y]$, then $f$ is called *absolutely irreducible*. A factorization over $\bar{K}$ is called an *absolute factorization*.

Clearly every factorization of $f(x, y) \in K[x, y]$ can be expressed in some algebraic extension $K(\gamma)$ of the field of definition $K$. Our problem now is to find the appropriate $\gamma$ and then factor $f$ in $K(\gamma)[x, y]$.

B. Trager (1984) and E. Kaltofen (1985c) describe two different ways of finding such factors. Duval (1991) gives a geometrical method for factoring over an algebraically closed field. In Bajaj et al. (1993) a factorization algorithm for $\mathbb{C}[x, y]$ is described. The first polynomial time algorithm for factoring over $\mathbb{C}$ seems to have been given by Chistov and Grigoryev (1983).

We will limit ourselves to describing Kaltofen's algorithm. W.l.o.g. we may assume that the input polynomial $f(x, y)$ is squarefree. We will view $f$ as a polynomial in the main variable $y$. The basic idea is the following. We consider the algebraic curve $\mathcal{C}$ in the affine plane $\bar{K} \times \bar{K} = \mathbb{A}^2(\bar{K})$ defined by the polynomial equation $f(x, y) = 0$. The factors of $f$ over $\bar{K}$ correspond exactly to the irreducible components of $\mathcal{C}$ in $\mathbb{A}^2(\bar{K})$. Now choose $a \in K$ such that $f(a, y)$ is squarefree and of degree $n = \deg_y(f)$. This means that the line $\mathcal{L} : x = a$ is not a tangent or asymptote of $\mathcal{C}$ and that none of the intersections of $\mathcal{C}$ and $\mathcal{L}$ are singularities of $\mathcal{C}$. Now if $\beta$ is a root of $f(a, y)$, then $\mathcal{P} = (a, \beta)$ is a simple point on $\mathcal{C}$. So there is a uniquely determined irreducible component of $\mathcal{C}$ passing through $\mathcal{P}$. Our goal is to compute the polynomial $g(x, y) \in \bar{K}[x, y]$ defining this component. $g$ will be an irreducible factor of $f$. See Fig. 8.

Before we start the actual factorization process, we normalize $f$ in the following way:

1. Clearly if $\text{cont}_y(f)$ (as a polynomial in $x$) is non-constant then this content is a factor of $f$. So let us assume that $\text{cont}_y(f) = 1$.
2. Check whether $f(x, y)$ is squarefree, i.e., $\gcd(f, \frac{\partial f}{\partial y}) = 1$. Otherwise determine a squarefree factor and proceed with this factor.

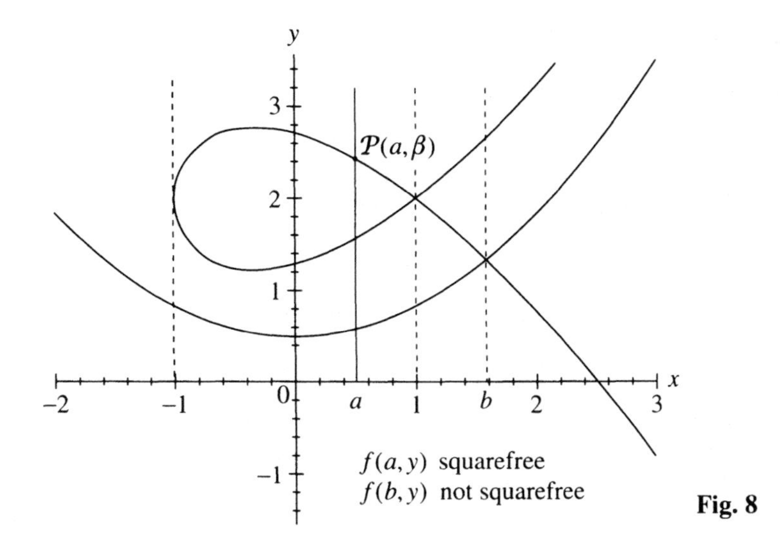


**Fig. 8**

3. Make $f$ monic in $y$ by replacing $f$ by the monic polynomial

$$\hat{f}(x, y) = \mathrm{lc}_y(f)^{\deg_y(f)-1} \cdot f\bigl(x, \frac{y}{\mathrm{lc}_y(f)}\bigr) .$$

   The factor structure is preserved by this transformation. In fact, if $\hat{g}(x, y)$ is a factor of $\hat{f}(x, y)$ then $\mathrm{pp}_y(\hat{g}(x, \mathrm{lc}_y(f)y))$ is a factor of $f$.
4. Find an integer $a$ in the range $|a| \leq \deg_x(f)\deg_y(f)$ such that $f(a, y)$ is squarefree. Replace $f$ by $f(x + a, y)$, i.e., change the coordinate system such that now $f(0, y)$ is squarefree (no singularities and vertical tangents or asymptotes at $x = 0$). Such an integer must exist because of Lemma 5.5.1.

**Lemma 5.5.1.** Let $f(x, y) \in K[x, y]$ be monic of degree $n$ in $y$ and squarefree. Then there exists $a \in \mathbb{Z}$ with $|a| \leq \deg_x(f) \cdot n$ such that $f(a, y)$ is squarefree.

*Proof.* Let $m = \deg_x(f)$. Since $f$ is squarefree, its discriminant

$$D(x) = \mathrm{res}_y\bigl(f, \frac{\partial f}{\partial y}\bigr)$$

is different from 0. $\deg_x(D) \leq (2n - 1)m$. So $D(x)$ cannot vanish on all the integer values $-m \cdot n, \ldots, 0, \ldots, m \cdot n$. □

Now we are prepared to describe the algorithm for finding a factor of $f$.

**Algorithm FIND_ALG_FACTOR**(in: $f$; out: $p, g$);
[$f(x, y) \in K[x, y]$ monic in $y$, $f(0, y)$ squarefree;
$p \in K[z]$ minimal polynomial for $\beta$ over $K$, $g(x, y) \in K(\beta)[x, y]$, and $g$ an irreducible factor of $f$ in $K(\beta)[x, y]$.]

1. [find point $\mathcal{P} = (0, \beta)$ on the curve $\mathcal{C}$ defined by $f$]
   $p(z) :=$ irreducible factor of $f(0, z)$ of least degree in $K[z]$;
   [let $\beta$ be a root of $p(z)$]
2. [compute a power series approximation of the branch of $\mathcal{C}$ through $\mathcal{P}$ in $K(\beta)[[y]]$ ]
   $m := \deg_x(f)$; $n := \deg_y(f)$; $k := (2n-1)\cdot m$;
   determine (by linear algebra over $K(\beta)$) $a_1, \ldots, a_k \in K(\beta)$ such that
   $f(x, \beta + a_1x + \ldots + a_kx^k) \equiv 0 \bmod x^{k+1}$;
   $Y := \beta + a_1x + \ldots + a_kx^k$;
3. [find the minimal polynomial for $Y$ in $K(\beta)[x, y]$ ]
   for $i = 1$ to $n-1$ do
   {try to solve the equation
   $$Y^i + u_{i-1}(x)Y^{i-1} + \ldots + u_1(x)Y + u_0(x) \equiv 0 \bmod x^{k+1} \qquad (5.5.1)$$
   for polynomials $u_j(x) \in K(\beta)[x]$ with $\deg(u_j) \le m$
   (this leads to a linear system over $K(\beta)$ for the coefficients of the $u_j$'s);
   if a solution to (5.5.1) exists
   then {$g := y^i + u_{i-1}y^{i-1} + \ldots + u_1y + u_0$;
   return$(p, g)$}};
4. [no factor has been found]
   return$(z, f)$.

We do not present a correctness proof of the algorithm FIND_ALG_FACTOR but rather refer to theorem 2 in Kaltofen (1985a) and theorem 2 in Kaltofen (1985c). In particular, the correctness of FIND_ALG_FACTOR implies the following theorem.

**Theorem 5.5.2.** Let $f(x, y) \in K[x, y]$ be monic in $y$ such that $f(0, y)$ is squarefree. Let $\beta$ be algebraic over $K$ such that $f(0, \beta) = 0$. Then $f$ is absolutely irreducible if and only if $f$ is irreducible in $K(\beta)[x, y]$.

In fact, J. R. Sendra (pers. comm.) has observed that if the curve $\mathcal{C}$ defined by the polynomial $f(x, y)$ over the field $K$ has enough points in the affine plane over $K$, then irreducibility of $f$ over $K$ already means absolute irreducibility of $f$.

**Theorem 5.5.3.** Let $f(x, y) \in K[x, y]$ be irreducible over $K$. If the curve defined by $f$ has infinitely many points $P_i = (p_i, q_i), i \in \mathbb{N}$, in $K^2$, then $f$ is absolutely irreducible.

*Proof.* For the same reason as in the proof of Lemma 5.5.1, possibly after an affine change of coordinates, $f(p_i, y)$ will be squarefree for some $i \in \mathbb{N}$. So, after another change of coordinates moving $P_i$ to $(0, 0)$ we will have $f(0, y)$ squarefree. Now the irreducible factor $p(z)$ constructed in step (1) of FIND_ALG_FACTOR is linear, in fact $p(z) = z$. The statement now follows from Theorem 5.5.2. □

*Example 5.5.1.* Let us find a factor of

$$h(x, y) = x^2 + y^2$$

in $\mathbb{C}[x, y]$. $h$ is primitive w.r.t. $y$, squarefree and monic. $h(0, y) = y^2$ is not squarefree, but $h(1, y) = y^2 + 1$ is. So let us replace $h$ by

$$f(x, y) = h(x + 1, y) = y^2 + x^2 + 2x + 1 \ .$$

Now we are prepared for an application of the algorithm FIND_ALG_FACTOR.

In step (1) $p(z) = z^2 + 1$, e.g., $\beta = \sqrt{-1} = i$.

In step (2) we set $m = 2, n = 2, k = 6$, and we determine $a_1, \ldots, a_6 \in \mathbb{Q}(i)$ such that we have

$$f(x, i + a_1 x + \cdots + a_6 x^6) \equiv 0 \bmod x^7 \ .$$

This leads to the linear system

$$\begin{aligned}
&i \cdot a_1 + 1 = 0 \ , \\
&2i \cdot a_2 + a_1^2 + 1 = 0 \ , \\
&i \cdot a_3 + a_1 a_2 = 0 \ , \\
&2i \cdot a_4 + a_1 a_3 + a_2^2 = 0 \ , \\
&i \cdot a_5 + a_1 a_4 + a_2 a_3 = 0 \ , \\
&2i \cdot a_6 + 2a_1 a_5 + 2a_2 a_4 + a_3^2 = 0 \ ,
\end{aligned}$$

which has the solution $(a_1, \ldots, a_6) = (i, 0, 0, 0, 0, 0)$. So $Y = i + i \cdot x$.

In step (3) we look for the polynomial of smallest degree having $Y$ as a root. In fact, in our example we only have to try polynomials of degree 1. The solution of

$$Y + u_0(x) = (i + i \cdot x) + u_0(x) \equiv 0 \bmod x^7$$

is $u_0(x) = -i \cdot x - i$, and this gives rise to the factor

$$g(x, y) = y + u_0(x) = y - ix - i$$

of $f$ in $\mathbb{Q}(i)[x, y] \subset \mathbb{C}[x, y]$. Inverting the linear change we get the factor

$$g(x - 1, y) = y - ix$$

of $h(x, y) = (y - ix)(y + ix)$.

The algorithm FIND_ALG_FACTOR lends itself to recursive application, leading to an algorithm for absolute factorization. However, if we start with a polynomial over $\mathbb{Z}$ (or $\mathbb{Q}$), we will soon introduce high degree algebraic extensions, which make the factorization costly.

### Exercises

1. Transform $f(x, y) = x^2y^2 - 4xy^2 + 4y^2 + x^2$ into a polynomial suited as input to FIND_ALG_FACTOR.
2. Determine the absolute factorization of the integral polynomial $f(x, y) = y^4 + x^4y^2 + 2x^2y + 1$.

## 5.6 Bibliographic notes

Alternatives to the Berlekamp algorithm for factoring polynomials over finite fields are described in Cantor and Zassenhaus (1981), Shoup (1991), Niederreiter and Göttfert (1993). Methods for testing irreducibility are discussed in Kaltofen (1985c, 1987). For factorization over algebraic number fields we refer also to Landau (1985). Absolute factorization is considered in Dvornicich and Traverso (1987), Heintz and Sieveking (1981). The lattice reduction algorithm is reconsidered in Pohst (1987), see also Pohst and Zassenhaus (1989). For a different approach to polynomial-time factorization, see Yokoyama et al. (1994). For further results on Hensel factorization, see Zassenhaus (1975, 1978, 1985). A good overview of problems and algorithms in factorization of polynomials can be found in Kaltofen (1983, 1986, 1992) and Sharpe (1987).

# 6 Decomposition of polynomials

## 6.1 A polynomial-time algorithm for decomposition

The first polynomial time algorithms for decomposition of polynomials have been presented by Gutierrez et al. (1988) and practically at the same time by D. Kozen and S. Landau (1989). We follow the approach of Kozen and Landau.

The problem to be considered is to decide whether a given polynomial $f(x)$ can be written as the functional composition of other polynomials, i.e., whether

$$f(x) = g(h(x)) = (g \circ h)(x)$$

for some polynomials $g, h$. We assume all the polynomials to have coefficients in a computable field $K$.

Decompositions of polynomials are interesting, e.g., for the solution of polynomial equations. If $f = g \circ h$, we can find the roots of $f$ by solving for the roots $\alpha$ of $g$ and then for the roots of $h - \alpha$.

In factoring polynomials we have to identify factorizations that differ only by a constant. A similar restriction has to be enforced in decomposing polynomials, since for every $a \in K^*$ the linear polynomials

$$l_a^{(1)}(x) = ax + b \quad \text{and} \quad l_a^{(2)}(x) = \frac{1}{a}(x - b)$$

are inverses of each other under composition. Thus, every polynomial admits a *trivial decomposition* of the form

$$f(x) = l_a^{(1)} \circ l_a^{(2)} \circ f(x) .$$

*Definition 6.1.1.* Let $f(x) \in K[x]$ be monic and of degree $> 1$. A *(functional) decomposition* of $f$ is a sequence $g_1, \ldots, g_k$ of polynomials in $K[x]$ such that

$$f = g_1 \circ g_2 \circ \ldots \circ g_k, \quad \text{i.e.,} \quad f(x) = g_1(g_2(\ldots g_k(x)\ldots)) .$$

The $g_i$ are called the *components* of the decomposition of $f$. If all decompositions of $f$ are trivial, i.e., all but one of the components are linear, then $f$ is called *indecomposable*. A *complete decomposition* is one in which all components are of degree $> 1$ and indecomposable.

Obviously it is sufficient to decompose monic polynomials. If $f(x)$ admits a decomposition $f = g \circ h$ and $c = \text{lc}(h)$, then also $f = g' \circ h'$, where $g'(x) = g(c \cdot x)$ and $h'(x) = h(x)/c$. So it suffices to search for decompositions into monic components. For a similar reason we can assume that $h(0) = 0$, i.e., the constant coefficient of $h$ is 0.

Complete decompositions are not unique, even if we disregard trivial decompositions. Further ambiguities in decomposition are

$$x^n \circ x^m = x^m \circ x^n \quad \text{for } m, n \in \mathbb{N}$$

and (6.1.1)

$$T_n \circ T_m = T_m \circ T_n \quad \text{for } m, n \in \mathbb{N},$$

where $T_n(x)$ is the $n$-th Chebyshev polynomial ($T_0(x) = 1$, $T_1(x) = x$, $T_n(x) = 2xT_{n-1}(x) - T_{n-2}(x)$ for $n > 1$). However, J. F. Ritt (1922) has shown that these are the only ambiguities.

**Theorem 6.1.1.** A monic polynomial $f(x) \in K[x]$, $\deg(f) > 1$, has a unique complete decomposition up to trivial decompositions and the ambiguities (6.1.1), provided $\text{char}(K) = 0$ or $\text{char}(K) > \deg(f)$.

Now let $f(x) \in K[x]$, $\deg(f) = n$, be the monic polynomial that we want to decompose. Let $n = r \cdot s$ be a non-trivial factorization of the degree of $f$. Then we want to decide whether $f$ can be decomposed into polynomials $g$ and $h$ of degrees $r$ and $s$, respectively, and if so compute such a decomposition.

$$\begin{aligned} f &= x^{rs} + a_{rs-1}x^{rs-1} + \ldots + a_0 , \\ g &= x^r + b_{r-1}x^{r-1} + \cdots + b_0 , \\ h &= x^s + c_{s-1}x^{s-1} + \cdots + c_1 x . \end{aligned} \tag{6.1.2}$$

Let $\beta_1, \ldots, \beta_r$ be the (not necessarily different) roots of $g$ in an algebraic extension of $K$. So

$$g(x) = \prod_{i=1}^{r} (x - \beta_i) ,$$

and therefore

$$f(x) = g(h(x)) = \prod_{i=1}^{r} (h(x) - \beta_i) .$$

**Lemma 6.1.2.** Let $f_1, f_2, g \in K[x]$ be monic. If $f_1$ and $f_2$ agree on their first $k$ coefficients, then so do $f_1 g$ and $f_2 g$.

*Proof.* Exercise 1. □

**Lemma 6.1.3.** Let $f, h$ be as in (6.1.2). $h^r(x)$ and $f(x)$ agree on their first $s$ coefficients.

*Proof.* $h(x)$ and $h(x) - \beta_1$ agree on their first $s$ coefficients. We write this as $h(x) \sim_s h(x) - \beta_1$. So by Lemma 6.1.2 also

$$h^2 \sim_s h(h - \beta_1) \sim_s (h - \beta_2)(h - \beta_1) .$$

Proceeding in this way, we finally arrive at

$$h^r \sim_s \prod_{i=1}^{r} (h - \beta_i) = f . \qquad \square$$

Now let $q_k(x)$ be the initial segment of length $k + 1$ of $h$, i.e.,

$$q_k(x) = x^s + c_{s-1}x^{s-1} + \ldots + c_{s-k}x^{s-k} , \tag{6.1.3}$$

for $0 \leq k \leq s$. Obviously $q_0 = x^s, q_s = q_{s-1} = h$ and $q_k = q_{k-1} + c_{s-k}x^{s-k}$ for $1 \leq k \leq s$.

**Lemma 6.1.4.** Let $h, q_k$ be as in (6.1.2) and (6.1.3). $h^r(x)$ and $q_k^r(x)$ agree on their first $k + 1$ coefficients.

*Proof.* $h \sim_{k+1} q_k$. The statement follows by $r - 1$ applications of Lemma 6.1.2. $\square$

Lemma 6.1.4 provides a recursive method for determining the coefficients of $h$. Suppose the first coefficients $c_{s-1}, \ldots, c_{s-k+1}$ of $h$ are known and we want to determine $c_{s-k}$, $1 \leq k \leq s - 1$. The $(k + 1)$-st coefficient of $q_k^r$ is the coefficient of $x^{rs-k}$, and by Lemmas 6.1.3 and 6.1.4 it is equal to $a_{rs-k}$, the $(k + 1)$-st coefficient of $f$. Furthermore,

$$\begin{aligned} q_k^r(x) &= (q_{k-1}(x) + c_{s-k}x^{s-k})^r \\ &= q_{k-1}^r(x) + r \cdot c_{s-k}x^{s-k} \cdot q_{k-1}^{r-1}(x) + \mathcal{O}(x^{rs-2k}) . \end{aligned}$$

Thus, $a_{rs-k} = d_k + r \cdot c_{s-k}$, where $d_k$ is the coefficient of $x^{rs-k}$ in $q_{k-1}^r$ and $r \cdot c_{s-k}$ is the coefficient of $x^{rs-k}$ in $r \cdot c_{s-k} \cdot x^{s-k} \cdot q_{k-1}^{r-1}$. Now we can compute the next coefficient in $h$ as

$$c_{s-k} = \frac{a_{rs-k} - d_k}{r} .$$

Once the coefficients of $h$ are determined, the coefficients of $g$ have to satisfy a system of linear equations. If this system is solvable, we get a candidate for a decomposition of $f$. Otherwise, $f$ cannot be decomposed into polynomials of degree $r$ and $s$, respectively. We still have to check the candidate, since the decomposition problem involves $rs$ equations, but there are only $r+s$ unknowns. We summarize this derivation in the following algorithm.

**Algorithm DECOMPOSE**(in: $f, s$; out: $[g, h]$ or "no decomposition");
$[f(x) \in K[x]$, $s | \deg(f)$; $g, h \in K[x]$ such that $\deg(h) = s$ and $f = g \circ h$ if such a decomposition exists, otherwise the message "no decomposition" is returned.]

1. $n := \deg(f)$; $r := n/s$;
2. for $i := 0$ to $r$ do $q_o^i := x^{is}$;
3. [determine candidate for $h$]
   for $k := 1$ to $s - 1$ do
      $\{d_k := \text{coeff}(q_{k-1}^r, n-k)$;
      $c_{s-k} := \frac{1}{r}(\text{coeff}(f, n-k) - d_k)$;
      calculate $c_{s-k}^j$ for $0 \leq j \leq r$;
      for $j := 0$ to $r$ do
         $q_k^j := \sum_{i=0}^{j} \binom{j}{i} c_{s-k}^i \cdot x^{i(s-k)} \cdot q_{k-1}^{j-i}$ };
   $h := q_{s-1}$;
4. [determine candidate for $g$]
   $A :=$ the $(r+1)\times(r+1)$-matrix, where $A_{ij} = \text{coeff}(h^j, is) = \text{coeff}(q_{s-1}^j, is)$;
   $a := (a_0, a_s, \ldots, a_{rs})^T$, where $a_i = \text{coeff}(f, i)$;
   if $A \cdot b = a$ is unsolvable
   then return("no decomposition")
   else $\{(b_0, \ldots, b_r)^T :=$ solution of $A \cdot b = a$;
      $g := b_r x^r + \cdots + b_0\}$;
5. [consistency check]
   if $f = g \circ h$
   then return($[g, h]$)
   else return("no decomposition").

**Theorem 6.1.5.** The number of arithmetic operations in DECOMPOSE is $\mathcal{O}(n^2 r)$, where $n = \deg(f)$ and $r = n/s$.

*Proof.* The calculation of $h$ in step (3) takes $\mathcal{O}(n^2 r)$ operations. The matrix $A$ in step (4) is triangular with all diagonal elements 1. So we can compute the candidate for $g$ by $\mathcal{O}(r^2)$ operations. □

In fact, as shown in Kozen and Landau (1989), using interpolation this complexity bound can be improved to $n^2$, provided that $K$ contains at least $n + 1$ elements.

*Example 6.1.1.* Let us try to decompose the polynomial

$$f(x) = x^6 + 6x^4 + x^3 + 9x^2 + 3x - 5$$

in $\mathbb{Q}[x]$ into components $g$ and $h$ of degrees 2 and 3, respectively. So $n = 6$ and $r = 2, s = 3$.

In step (2) we determine

$$q_0^0(x) = 1, \quad q_0^1(x) = x^3, \quad q_0^2(x) = x^6 .$$

Now we successively determine longer initial segments of $h$ in step (3).

$$\begin{aligned} k = 1:\ & d_1 = \text{coeff}(q_0^2, 5) = 0 , \\ & c_2 = \tfrac{1}{2}(\text{coeff}(f, 5) - d_1) = 0 , \\ & q_1^0(x) = 1, \quad q_1^1(x) = x^3, \quad q_1^2 = x^6 \\ k = 2:\ & d_2 = \text{coeff}(q_1^2, 4) = 0 , \\ & c_1 = \tfrac{1}{2}(\text{coeff}(f, 4) - d_2) = 3 , \\ & q_2^0(x) = 1, \quad q_2^1(x) = x^3 + 3x, \quad q_2^2 = x^6 + 6x^4 + 9x^2 . \end{aligned}$$

Our candidate for $h$ therefore is $h(x) = q_2(x) = x^3 + 3x$.

The linear system for $g$ in step (4) is trivial

$$\begin{pmatrix} 1 & 0 & 0 \\ 0 & 1 & 0 \\ 0 & 0 & 1 \end{pmatrix} \cdot \begin{pmatrix} b_0 \\ b_1 \\ b_2 \end{pmatrix} = \begin{pmatrix} -5 \\ 1 \\ 1 \end{pmatrix} ,$$

so the candidate for $g$ is $g(x) = b_2x^2 + b_1x + b_0 = x^2 + x - 5$.

Indeed, the consistency check in step (5) works, and we get the decomposition

$$f(x) = g(h(x)) ,$$

where

$$g(x) = x^2 + x - 5 \quad \text{and} \quad h(x) = x^3 + 3x .$$

By inspection of the execution of DECOMPOSE on this example we notice, that the linear and quadratic coefficients of $f$ have never been used. So we get the same candidates for $h$ and $g$ by starting from, e.g., the polynomial

$$\tilde{f}(x) = x^6 + 6x^4 + x^3 + x^2 + 3x - 5 .$$

In this case, however, the consistency check does not work, and we find that $\tilde{f}$ cannot be decomposed into components of degrees 2 and 3, respectively.

### Exercises

1. Prove Lemma 6.1.2.

2. Decompose the rational polynomial $f(x) = x^8 - 4x^7 + 6x^6 - 4x^5 + 3x^4 - 4x^3 + 3x^2 - x + 2$ into components $g, h$ of degrees 4 and 2, respectively.

## 6.2 Bibliographic notes

One of the first investigations of the decomposition problem can be found in Ritt (1922). In Barton and Zippel (1985) the decomposition problem is reduced to factorization of polynomials, but the resulting algorithm is exponential in the degree of the input $f$. More or less at the same time as by Kozen and Landau (1989) another polynomial-time algorithm for decomposition was proposed by Gutierrez et al. (1988). A very thorough analysis of the dependence on the particular ground field $K$ can be found in von zur Gathen (1990a, b). Meanwhile, there are also decomposition algorithms for rational functions, e.g., Zippel (1991), Gutierrez and Recio (1992). Related topics are investigated in Weiß (1992). See also Alagar and Thanh (1985), Brackx et al. (1989), Lidl (1985), von zur Gathen et al. (1987).

# 7 Linear algebra – solving linear systems

## 7.1 Bareiss's algorithm

Systems of linear equations over a field $K$ can be solved by various methods, e.g., by Gaussian elimination or Cramer's rule. But if we start with a system over the integers, we will immediately introduce rational numbers, whose arithmetic operations are clearly more costly than the corresponding operations on integers. So for the same reason as in the computation of gcds of polynomials or factorization of polynomials, we are interested in a method for solving systems of linear equations which avoids computation with rational numbers as much as possible. Such a method for fraction free Gaussian elimination is Bareiss's algorithm, as described in Bareiss (1968).

### Integer preserving Gaussian elimination

Let us first remind ourselves of the process of Gaussian elimination for solving systems of linear equations. Let $I$ be an integral domain, $K = Q(I)$ its quotient field. We assume that we are given a system of linear equations with coefficient matrix $A$ and right-hand side $b$ over $I$, i.e., for some $n \in \mathbb{N}$, $A = (a_{ij})_{1 \leq i,j \leq n} \in I_n^n$, $b = (b_1, \ldots, b_n)^T \in I_n$, and we have to solve

$$A \cdot x = b \; . \tag{7.1.1}$$

The solutions $x = (x_1, \ldots, x_n)^T$ are to be found over $K$. The system (7.1.1) will have a unique solution if and only if $A$ is non-singular, i.e., $\det(A) \neq 0$. Let $\hat{A} = (A, b)$ denote the extended coefficient matrix of the system. We will also write $a_{i,n+1}$ for $b_i$, $1 \leq i \leq n$.

The integer preserving Gaussian elimination algorithm proceeds in $n - 1$ steps, determining matrices $\hat{A} = A^{[0]}, A^{[1]}, \ldots, A^{[n-1]}$, such that all these extended matrices have the same solutions. In step $(k)$ we start from a matrix of the form $A^{[k-1]}$, such that $a_{kk}^{[k-1]} \neq 0$, and by elementary linear row operations we transform $A^{[k-1]}$ into a matrix $A^{[k]}$, where

$$A^{[k-1]} = \begin{pmatrix} a_{11}^{[k-1]} & 0 & \cdots & 0 & a_{1k}^{[k-1]} & \cdots & a_{1n}^{[k-1]} & a_{1,n+1}^{[k-1]} \\ 0 & a_{22}^{[k-1]} & \ddots & \vdots & a_{2k}^{[k-1]} & \cdots & a_{2n}^{[k-1]} & a_{2,n+1}^{[k-1]} \\ 0 & \ddots & \ddots & 0 & \vdots & & \vdots & \vdots \\ 0 & \cdots & 0 & a_{k-1,k-1}^{[k-1]} & a_{k-1,k}^{[k-1]} & \cdots & a_{k-1,n}^{[k-1]} & a_{k-1,n+1}^{[k-1]} \\ 0 & \cdots & 0 & 0 & a_{kk}^{[k-1]} & \cdots & a_{k,n}^{[k-1]} & a_{k,n+1}^{[k-1]} \\ \vdots & & \vdots & \vdots & \vdots & \ddots & \vdots & \vdots \\ 0 & \cdots & 0 & 0 & a_{nk}^{[k-1]} & \cdots & a_{nn}^{[k-1]} & a_{n,n+1}^{[k-1]} \end{pmatrix},$$

$$A^{[k]} = \begin{pmatrix} a_{11}^{[k]} & 0 & \cdots & 0 & a_{1,k+1}^{[k]} & \cdots & a_{1n}^{[k]} & a_{1,n+1}^{[k]} \\ 0 & a_{22}^{[k]} & \ddots & \vdots & a_{2,k+1}^{[k]} & \cdots & a_{2n}^{[k]} & a_{2,n+1}^{[k]} \\ 0 & \ddots & \ddots & 0 & \vdots & & \vdots & \vdots \\ 0 & \cdots & 0 & a_{kk}^{[k]} & a_{k,k+1}^{[k]} & \cdots & a_{k,n}^{[k]} & a_{k,n+1}^{[k]} \\ 0 & \cdots & 0 & 0 & a_{k+1,k+1}^{[k]} & \cdots & a_{k+1,n}^{[k]} & a_{k+1,n+1}^{[k]} \\ \vdots & & \vdots & \vdots & \vdots & \ddots & \vdots & \vdots \\ 0 & \cdots & 0 & 0 & a_{n,k+1}^{[k]} & \cdots & a_{nn}^{[k]} & a_{n,n+1}^{[k]} \end{pmatrix}.$$

We might have to do pivoting if $a_{k+1,k+1}^{[k]} = 0$. In the transformation from $A^{[k-1]}$ to $A^{[k]}$ the row $k$ remains unchanged. For all the other elements, i.e., for $1 \leq j \leq n+1$, $i \in \{1, \ldots, k-1, k+1, \ldots, n\}$, the transformation formula

$$a_{ij}^{[k]} := a_{kk}^{[k-1]} \cdot a_{ij}^{[k-1]} - a_{ik}^{[k-1]} \cdot a_{kj}^{[k-1]} = \begin{vmatrix} a_{kk}^{[k-1]} & a_{kj}^{[k-1]} \\ a_{ik}^{[k-1]} & a_{ij}^{[k-1]} \end{vmatrix} \tag{7.1.2}$$

is applied, i.e., we get the $i$-th row in $A^{[k]}$ by multiplying the $i$-th row in $A^{[k-1]}$ by $a_{kk}^{[k-1]}$ and subtracting $a_{ik}^{[k-1]}$ times the $k$-th row of $A^{[k-1]}$.

From the matrix $A^{[n-1]}$ the solution can be read off immediately, namely $x_i = a_{i,n+1}^{[n-1]} / a_{ii}^{[n-1]}$.

If the coefficient matrix $A$ is singular we can still proceed as above, but the diagonalization process will stop after $k < n-1$ steps. Then (7.1.1) is solvable if and only if

$$a_{k+1,n+1}^{[k]} = \ldots = a_{n,n+1}^{[k]} = 0 .$$

A basis for the solutions of the homogeneous system as well as a particular solution of the inhomogeneous system can now obviously be read off from $A^{[k]}$.

The problem with Gaussian elimination is that the elements of the intermediate matrices grow considerably. For instance, if $I = \mathbb{Z}$ and $B$ is a bound for the absolute values of the elements in $\hat{A}$, then $B^{k+1}$ is a bound for the absolute values of $A^{[k]}$. Of course we could always divide the elements in row $i$ by their gcd. However, this involves a considerable number of gcd computations. The method of Bareiss provides guaranteed factors without having to do additional operations.

### Bareiss's modification of Gaussian elimination

By analyzing the minors and subdeterminants of the matrix $\hat{A}$ it is possible to identify divisors of rows in the elimination process without actually having to compute gcds of the corresponding elements. This observation is the basic new idea in the method of Bareiss. We will only describe one of the various possible approaches discussed in Bareiss (1968).

We start out with a slight change of notation. Let $A, b, \hat{A}, n$ be as above, i.e., we are interested in solving the system (7.1.1). Let

$$a_{ij}^{[0]} = a_{ij} \quad \text{for } 1 \leq i \leq n, \quad 1 \leq j \leq n+1 \ .$$

For $k = 1, \ldots, n-1$, $k < i \leq n$, $k < j \leq n+1$, we define

$$a_{ij}^{[k]} := \begin{vmatrix} a_{11} & a_{12} & \cdots & a_{1k} & a_{1j} \\ a_{21} & a_{22} & \cdots & a_{2k} & a_{2j} \\ \vdots & \vdots & \ddots & \vdots & \vdots \\ a_{k1} & a_{k2} & \cdots & a_{kk} & a_{kj} \\ a_{i1} & a_{i2} & \cdots & a_{ik} & a_{ij} \end{vmatrix} ,$$

i.e., $a_{ij}^{[k]}$ is the $k$-th minor of $A$ bordered by a row and a column. For these subdeterminants we have Sylvester's identity.

**Theorem 7.1.1** (Sylvester's identity). Let $A$ be a matrix in $I_n^n$ such that the $k$-th minor of $A$ is non-zero. Then

$$|A| \cdot (a_{kk}^{[k-1]})^{n-k-1} = \begin{vmatrix} a_{k+1,k+1}^{[k]} & \cdots & a_{k+1,n}^{[k]} \\ \vdots & & \vdots \\ a_{n,k+1}^{[k]} & \cdots & a_{nn}^{[k]} \end{vmatrix} .$$

*Proof.* Let $A_{11}$ be the upper left submatrix of $A$ with $k$ rows and columns. By our assumption $|A_{11}| \neq 0$. For some matrices $A_{12}, A_{21}, A_{22}$ we can write $A$ in block form as

$$A = \begin{pmatrix} A_{11} & A_{12} \\ A_{21} & A_{22} \end{pmatrix} = \begin{pmatrix} A_{11} & 0 \\ A_{21} & I \end{pmatrix} \cdot \begin{pmatrix} I & A_{11}^{-1} A_{12} \\ 0 & A_{22} - A_{21} A_{11}^{-1} A_{12} \end{pmatrix} .$$

So

$$|A| = |A_{11}| \cdot |A_{22} - A_{21}A_{11}^{-1}A_{12}| \tag{7.1.3}$$

and therefore

$$|A| \cdot |A_{11}|^{n-k-1} = |\,|A_{11}| \cdot (A_{22} - A_{21}A_{11}^{-1}A_{12})\,| \,, \tag{7.1.4}$$

because the determinant on the right-hand side is of order $n-k$. Now if we let $a, b$ be the vectors such that

$$a_{ij}^{[k]} = \begin{vmatrix} A_{11} & b \\ a & a_{ij} \end{vmatrix}$$

and apply (7.1.3) to each $a_{ij}^{[k]}$, we obtain

$$a_{ij}^{[k]} = |A_{11}| \cdot \left(a_{ij} - \sum_{r=1}^{k} \sum_{s=1}^{k} a_{ir}(A_{11}^{-1})_{rs}a_{sj}\right) \quad \text{for } k < i, j \le n \,.$$

Since $|A_{11}| = a_{kk}^{[k-1]}$, (7.1.4) takes the form

$$|A| \cdot (a_{kk}^{[k-1]})^{n-k-1} = \begin{vmatrix} a_{k+1,k+1}^{[k]} & \cdots & a_{k+1,n}^{[k]} \\ \vdots & & \vdots \\ a_{n,k+1}^{[k]} & \cdots & a_{nn}^{[k]} \end{vmatrix} \,. \qquad \square$$

**Corollary.** Let $A, k$ be as in the theorem. Let $l \in \mathbb{N}$ such that $0 < l < k$ and the $l$-th minor of $A$ is non-zero. Then for $i, j$ $(k < i, j \le n)$ we have

$$a_{ij}^{[k]} = \frac{1}{\left(a_{ll}^{[l-1]}\right)^{k-l}} \cdot \underbrace{\begin{vmatrix} a_{l+1,l+1}^{[l]} & \cdots & a_{l+1,k}^{[l]} & a_{l+1,j}^{[l]} \\ \vdots & & \vdots & \vdots \\ a_{k,l+1}^{[l]} & \cdots & a_{kk}^{[l]} & a_{kj}^{[l]} \\ a_{i,l+1}^{[l]} & \cdots & a_{ik}^{[l]} & a_{ij}^{[l]} \end{vmatrix}}_{D_{ij}} \,,$$

i.e., $(a_{ll}^{[l-1]})^{k-l}$ is a divisor of the determinant $D_{ij}$.

*Proof.* $a_{ij}^{[k]}$ is a determinant of size $k+1$, so we can apply the theorem with minor of size $l$, leading to

$$a_{ij}^{[k]} \cdot \left(a_{ll}^{[l-1]}\right)^{k-l} = \begin{vmatrix} a_{l+1,l+1}^{[l]} & \cdots & a_{l+1,k}^{[l]} & a_{l+1,j}^{[l]} \\ \vdots & & \vdots & \vdots \\ a_{k,l+1}^{[l]} & \cdots & a_{kk}^{[l]} & a_{kj}^{[l]} \\ a_{i,l+1}^{[l]} & \cdots & a_{ik}^{[l]} & a_{ij}^{[l]} \end{vmatrix} . \qquad \square$$

In particular, for $l = k - 1$ the statement in the corollary reduces to

$$a_{ij}^{[k]} = \frac{1}{a_{k-1,k-1}^{[k-2]}} \cdot \begin{vmatrix} a_{kk}^{[k-1]} & a_{kj}^{[k-1]} \\ a_{ik}^{[k-1]} & a_{ij}^{[k-1]} \end{vmatrix} . \tag{7.1.5}$$

If we let

$$a_{00}^{[-1]} = 1, \quad \text{and} \quad a_{ij}^{[0]} = a_{ij} \text{ for } 1 \le i, j \le n , \tag{7.1.6}$$

then (7.1.5) is basically the same as the transformation formula (7.1.2), only that now we know a factor of the determinant and we divide it out. In fact, (7.1.5) also holds for $1 \le j < k$, i.e., we can diagonalize the coefficient matrix of the linear system in this way. As in Gaussian elimination we might have to do pivoting to keep the elements of the diagonal non-zero. Again we can consider the extended coefficient matrix $\hat{A}$ and proceed as in the Gaussian elimination process.

*Example 7.1.1.* Let us demonstrate these ideas by solving the linear system

$$\begin{pmatrix} 3 & -5 & 7 & 1 \\ 2 & -1 & 4 & 3 \\ 1 & 5 & 5 & 6 \\ -2 & 0 & 3 & 7 \end{pmatrix} \cdot \begin{pmatrix} x_1 \\ x_2 \\ x_3 \\ x_4 \end{pmatrix} = \begin{pmatrix} 1 \\ 1 \\ 1 \\ 1 \end{pmatrix} .$$

The elimination according to the Gauss process (7.1.2) and the Bareiss process (7.1.5), (7.1.6) create the following transformations:

Gauss | Bareiss

$$a_{00}^{[-1]} = 1$$

$$A^{[0]} : \begin{pmatrix} 3 & -5 & 7 & 1 & \vdots & 1 \\ 2 & -1 & 4 & 3 & \vdots & 1 \\ 1 & 5 & 5 & 6 & \vdots & 1 \\ -2 & 0 & 3 & 7 & \vdots & 1 \end{pmatrix} \qquad \begin{pmatrix} 3 & -5 & 7 & 1 & \vdots & 1 \\ 2 & -1 & 4 & 3 & \vdots & 1 \\ 1 & 5 & 5 & 6 & \vdots & 1 \\ -2 & 0 & 3 & 7 & \vdots & 1 \end{pmatrix} \qquad a_{11}^{[0]} = 3$$

$$A^{[1]}: \quad \left(\begin{array}{cccc:c} 3 & -5 & 7 & 1 & 1 \\ 0 & 7 & -2 & 7 & 1 \\ 0 & 20 & 8 & 17 & 2 \\ 0 & -10 & 23 & 23 & 5 \end{array}\right) \quad \left(\begin{array}{cccc:c} 3 & -5 & 7 & 1 & 1 \\ 0 & 7 & -2 & 7 & 1 \\ 0 & 20 & 8 & 17 & 2 \\ 0 & -10 & 23 & 23 & 5 \end{array}\right) \quad a_{22}^{[1]} = 7$$

$$A^{[2]}: \quad \left(\begin{array}{cccc:c} 21 & 0 & 39 & 42 & 12 \\ 0 & 7 & -2 & 7 & 1 \\ 0 & 0 & 96 & -21 & -6 \\ 0 & 0 & 141 & 231 & 45 \end{array}\right) \quad \left(\begin{array}{cccc:c} 7 & 0 & 13 & 14 & 4 \\ 0 & 7 & -2 & 7 & 1 \\ 0 & 0 & 32 & -7 & -2 \\ 0 & 0 & 47 & 77 & 15 \end{array}\right) \quad a_{33}^{[2]} = 32$$

$$A^{[3]}: \quad \left(\begin{array}{cccc:c} 2016 & 0 & 0 & 4851 & 1386 \\ 0 & 672 & 0 & 630 & 84 \\ 0 & 0 & 96 & -21 & -6 \\ 0 & 0 & 0 & 25137 & 5166 \end{array}\right) \left(\begin{array}{cccc:c} 32 & 0 & 0 & 77 & 22 \\ 0 & 32 & 0 & 30 & 4 \\ 0 & 0 & 32 & -7 & -2 \\ 0 & 0 & 0 & 399 & 82 \end{array}\right) \quad a_{44}^{[3]} = 399$$

$$A^{[4]}: \quad \left(\begin{array}{cccc:c} 50676192 & 0 & 0 & 0 & 9779616 \\ 0 & 16892064 & 0 & 0 & -1143072 \\ 0 & 0 & 2413152 & 0 & -42336 \\ 0 & 0 & 0 & 25137 & 5166 \end{array}\right)$$

$$\left(\begin{array}{cccc:c} 399 & 0 & 0 & 0 & 77 \\ 0 & 399 & 0 & 0 & -27 \\ 0 & 0 & 399 & 0 & -7 \\ 0 & 0 & 0 & 399 & 82 \end{array}\right)$$

So the solution is $x = (1/399) \cdot (77, -27, -7, 82)$. □

## Exercises

1. Let $A$ be a square matrix of size $n$ over $\mathbb{Z}$. Prove that the statement of Theorem 7.1.1 holds also if the $k$-th minor of $A$ is zero.
2. Apply Bareiss's algorithm to solve the system

$$\begin{pmatrix} 2 & -1 & 4 & 1 \\ 1 & -2 & 0 & 1 \\ 3 & 3 & -4 & -1 \\ 2 & -1 & 1 & 2 \end{pmatrix} \begin{pmatrix} x_1 \\ x_2 \\ x_3 \\ x_4 \end{pmatrix} = \begin{pmatrix} 6 \\ 0 \\ 1 \\ 4 \end{pmatrix}.$$

3. Why does (7.1.5) also hold for $1 \leq j \leq k$?

## 7.2 Hankel matrices

In this section we consider special systems of linear equations, so-called Hankel (or Toeplitz) systems, which have many interesting applications, e.g., to Padé approximation or to problems in signal processing. However, we will only be concerned with the application of Hankel systems to problems in computer algebra. These Hankel or Toeplitz systems (and the corresponding matrices) are generated by linear recurrence relations, which can be exploited for speeding up the solution of the corresponding linear systems. In our description of Hankel matrices we will mainly follow the approach taken by Sendra and Llovet (1989; 1992a, b), Sendra (1990a, b). The classical results on Hankel matrices can be found in books on linear algebra and theory of matrices such as Gantmacher (1977) or Heinig and Rost (1984).

Throughout this section let $I$ be a ufd and $K = Q(I)$ its quotient field.

*Definition 7.2.1.* Let $\mathcal{D} = (d_i)_{i \in \mathbb{N}}$ be a sequence of elements of $I$. The *infinite Hankel matrix generated by* $\mathcal{D}$ is

$$H_\infty(\mathcal{D}) = (h_{ij})_{i,j \in \mathbb{N}} ,$$

where

$$h_{ij} = d_{i+j-1} \quad \text{for } i, j \in \mathbb{N} .$$

The *finite Hankel matrix of order* $n$ ($\in \mathbb{N}$) *generated by* $\mathcal{D}$, $H_n(\mathcal{D})$, is the $n \times n$ principal submatrix of $H_\infty$, i.e.,

$$H_n(\mathcal{D}) = \begin{pmatrix} d_1 & d_2 & d_3 & \cdots & d_n \\ d_2 & d_3 & & & d_{n+1} \\ d_3 & & & & \vdots \\ \vdots & & & & \vdots \\ d_n & d_{n+1} & \cdots & \cdots & d_{2n-1} \end{pmatrix} .$$

A (finite or infinite) *Hankel matrix* is a (finite or infinite) Hankel matrix $H$ generated by some sequence $\mathcal{D}$ over $I$.

Since only the first $2n-1$ elements of a sequence $\mathcal{D}$ are necessary for specifying a Hankel matrix of order $n$, we sometimes speak of the $n \times n$ Hankel matrix generated by a finite sequence $(d_1, \ldots, d_{2n-1})$.

*Definition 7.2.2.* Let $H$ be an infinite Hankel matrix, and $H_m$ its $m \times m$ principal submatrix. If $H_m$ is non-singular, then the *$m$-th fundamental vector* $\omega^{(m,H)} \in K^m$ of $H$ is the solution of the linear system

$$(x_1, \ldots, x_m) \cdot H_m = (d_{m+1}, \ldots, d_{2m}) .$$

When $H$ is clear from the context, we drop it from the index of $\omega^{(m,H)}$ and write simply $\omega^{(m)}$.

$H$ is called *proper* iff it has finite rank.

In abuse of notation, we allow ourselves to write $(x, a)$ for the vector $(x_1, \ldots, x_m, a)$ if $x = (x_1, \ldots, x_m)$. In particular, we will use this notation in connection with fundamental vectors.

**Lemma 7.2.1.** Let $H_n$ be generated by $(d_1, \ldots, d_{2n-1})$, $H_p$ $(p < n)$ a regular principal submatrix of $H_n$, $\omega^{(p)}$ the $p$-th fundamental vector, and $q$ such that $p < q \leq n$. Then the following statements are equivalent:

a. $(\omega^{(p)}, -1) \cdot (d_i, \ldots, d_{i+p})^T = 0$ for $p < i < q$, and $(\omega^{(p)}, -1) \cdot (d_q, \ldots, d_{q+p})^T \neq 0$,
b. $\det(H_i) = 0$ for $p < i < q$ and $\det(H_q) \neq 0$.

*Proof.* (a) $\Longrightarrow$ (b): Let $\sigma_j = (\omega^{(p)}, -1) \cdot (d_j, \ldots, d_{j+p})^T$ for $0 < j \leq 2q - p - 1$. Since $\omega^{(p)}$ is the fundamental vector of $H_p$, we have $\sigma_j = 0$ for $1 \leq j \leq p$. So if $p + 1 < q$ then the $(p + 1)$-st row of $H_{q-1}$ is a linear combination of the first $p$ rows of $H_{q-1}$, and therefore $\det(H_i) = 0$ for $p < i < q$.

Suppose $\lambda_1, \ldots, \lambda_q \in K$ such that

$$(\lambda_1, \ldots, \lambda_q) \cdot H_q = (0, \ldots, 0) .$$

Then for every $i$, $1 \leq i \leq q - p$, we have

$$\underbrace{\Big(0, \ldots, 0, \overset{\substack{i\text{-th pos.}\\ \downarrow}}{\omega_1^{(p)}}, \ldots, \omega_p^{(p)}, -1, 0, \ldots, 0\Big) \cdot H_q}_{(\sigma_i, \ldots, \sigma_{i+q-1})} \cdot \begin{pmatrix} \lambda_1 \\ \vdots \\ \lambda_q \end{pmatrix} = 0 .$$

So from $\sigma_1 = \ldots = \sigma_{q-1} = 0$ and $\sigma_q \neq 0$ we get $\lambda_{p+1} = \ldots = \lambda_q = 0$. Therefore,

$$(\lambda_1, \ldots, \lambda_p) \cdot H_p = (0, \ldots, 0)$$

and this implies $\lambda_1 = \ldots = \lambda_p = 0$ because $H_p$ is nonsingular. Thus, the rows of $H_q$ are linearly independent, i.e., $\det(H_q) \neq 0$.

(b) $\Longrightarrow$ (a): This is left to the reader. □

**Theorem 7.2.2.** Let $H = H_\infty(\mathcal{D})$, where $\mathcal{D} = (d_i)_{i \in \mathbb{N}}$. Then the following are equivalent:

a. $H$ is proper of rank $r$ $(r > 0)$,
b. there exist $a_1, \ldots, a_r \in K$ such that $d_j = a_1 d_{j-r} + \ldots + a_r d_{j-1}$ for $j > r$, and $r$ is the smallest integer with this property,
c. $\det(H_r) \neq 0$ and $\det(H_i) = 0$ for $i > r$.

*Proof.* (a) $\Longrightarrow$ (b): Let $\operatorname{rank}(H) = r < \infty$. Then for some integer $p, 0 < p \leq r$, the first $p$ columns of $H$ are linearly independent over $K$ but the $(p+1)$-st column depends linearly on the previous ones. That is, for some $a_1, \ldots, a_p \in K$ we have

$$\begin{pmatrix} d_{p+1} \\ d_{p+2} \\ \vdots \end{pmatrix} = \sum_{i=1}^{p} a_i \cdot \begin{pmatrix} d_i \\ d_{i+1} \\ \vdots \end{pmatrix},$$

or, in other words, the linear recurrence relation

$$d_j = \sum_{i=1}^{p} a_i d_{j-p-1+i} \quad \text{for } j > p$$

holds. From this and the Hankel structure of $H$ one deduces that every column of $H$ is a linear combination of the first $p$ columns. So every submatrix of order greater than $p$ is singular and hence $p$ must be equal to the rank of $H$.

(b) $\Longrightarrow$ (c): Every minor of order greater than $r$ must vanish. In particular, $\det(H_i) = 0$ for $i > r$. On the other hand, let $p \leq r$ be the greatest integer such that $\det(H_p) \neq 0$, and let $\omega^{(p)}$ be the $p$-th fundamental vector of $H$. Then, by Lemma 7.2.1, we must have $(\omega^{(p)}, -1) \cdot (d_i, \ldots, d_{i+p})^T = 0$ for $i > 0$. But $r$ is the smallest integer with this property, so $p = r$ and therefore $\det(H_r) \neq 0$.

(c) $\Longrightarrow$ (a): Let $\omega^{(r)}$ be the $r$-th fundamental vector. Then, because of $\det(H_i) = 0$ for $i > r$ and Lemma 7.2.1 we have $(\omega^{(r)}, -1) \cdot (d_i, \ldots, d_{i+r})^T = 0$ for $i > 0$. Hence, every column of $H$ is a linear combination of the first $r$ columns. Furthermore, since $\det(H_r) \neq 0$, $H$ is proper of rank $r$. □

*Definition 7.2.3.* Let $H$ be an infinite Hankel matrix. If $\operatorname{rank}(H) = r < \infty$, then $\omega^{(r,H)} = \omega^{(H)}$ is called the *fundamental vector* of $H$.

If the infinite Hankel matrix $H$ is proper of rank $r$, the fundamental vector $\omega^{(H)}$ is the vector $(a_1, \ldots, a_r)$ of Theorem 7.2.2 (b). On the other hand, by Theorem 7.2.2 (b) we know that all the entries $d_j$ of the generating sequence $\mathcal{D} = (d_j)_{j \in \mathbb{N}}$ are determined by the fundamental vector $\omega^{(H)}$ and the first $r$ entries. So all the information on $H$ is already contained in its $r \times (r+1)$ principal submatrix and also in $H_{r+1}$.

**Theorem 7.2.3.** Let $\mathcal{D} = (d_1, \ldots, d_{2n-1})$ be a finite sequence over $I$. Let $H_n$ be the finite Hankel matrix of order $n$ generated by $\mathcal{D}$.

a. The sequence of fundamental vectors of $H_n$, its determinant and its rank can be computed in $\mathcal{O}(n^2)$ arithmetic operations in $I$.
b. Let $I = \mathbb{Z}[x_1, \ldots, x_r]$, $D$ the maximal degree in any variable of the $d_i$'s and $L$ the maximum norm of the $d_i$'s. The complexity of computing the determinant and the rank of $H_n$ is $\mathcal{O}((n^{r+3}D^r + n^{r+1}D^{r+1})(\log n)(\log^2 L))$.

*Proof.* See Sendra (1990a) and Sendra and Llovet (1992b). Modular methods are used in the proof of (b). □

Solutions of Hankel linear systems and determination of signatures can be achieved with algorithms of similar complexity.

## 7.3 Application of Hankel matrices to polynomial problems

Proper Hankel matrices may be put in a 1–1 correspondence with proper rational functions. This relation provides the basis for the application to polynomial problems. Again we assume that $I$ is a ufd and $K$ its quotient field.

**Theorem 7.3.1.** Let $H$ be the infinite Hankel matrix generated by $\mathcal{D} = (d_i)_{i\in\mathbb{N}}$, $d_i \in K$. Then $H$ is proper if and only if there exists a rational function $R(x) = g(x)/f(x)$ in $K(x)$, $f$, $g$ relatively prime, $\deg(g) < \deg(f)$ (i.e., $R(x)$ is proper), which has the power series expansion

$$R(x) = \sum_{i\in\mathbb{N}} d_i x^{-i} .$$

Furthermore, if $H$ is proper then $\operatorname{rank}(H) = \deg(f)$.

*Proof.* If $\operatorname{rank}(H) = r < \infty$, by Theorem 7.2.2 there exist $a_1, \ldots, a_r \in K$ such that $d_j = \sum_{i=1}^{r} a_i d_{j-r-1+i}$ for $j > r$. Then the polynomials defining $R = g/f$ are

$$\begin{aligned} f(x) &= x^r - a_r x^{r-1} - \ldots - a_1 , \\ g(x) &= d_1 x^{r-1} + (d_2 - d_1 a_r) x^{r-2} + \ldots + (d_r - d_{r-1} a_r - \ldots - d_1 a_2) . \end{aligned}$$

Conversely, let $R = g/f$, $\deg(f) = n$, $f = f_n x^n + \cdots + f_0$. So

$$g = R \cdot f = \Big(\sum_{i\in\mathbb{N}} d_i x^{-i}\Big) \cdot f .$$

By equating coefficients of like powers of $x$ on both sides of this equation we obtain

$$d_j = -\Big(\frac{f_0}{f_n} d_{j-n} + \ldots + \frac{f_{n-1}}{f_n} d_{j-1}\Big) \quad \text{for } j > n .$$

Furthermore, if one assumes that there exist $a_1, \ldots, a_s \in K$, $s < n$, that satisfy $d_j = \sum_{i=1}^{s} a_i d_{j-s-1+i}$ for $j > s$, then one has

$$\begin{aligned} &(x^s - a_s x^{s-1} - \ldots - a_1) \frac{g(x)}{f(x)} \\ &= d_1 x^{s-1} + (d_2 - a_s d_1) x^{s-2} + \ldots + (d_s - a_s d_{s-1} - \ldots - a_2 d_1) . \end{aligned}$$

So, since $g$ and $f$ are relatively prime, $f(x)$ must divide $x^s - a_s x^{s-1} - \ldots - a_1$, which is impossible because $s < n$. Thus, by Theorem 7.2.2, $\text{rank}(H) = \deg(f)$. □

**Corollary.** If $R(x) = g(x)/f(x)$, $\gcd(f, g) = 1$, $\deg(g) < \deg(f)$, is a proper rational function and $f(x) = f_r x^r + \ldots + f_0$, then the Hankel matrix $H$ associated with $R$ is defined by $(d_i)_{i\in\mathbb{N}}$, where $R(x) = \sum_{i\in\mathbb{N}} d_i x^{-i}$, and the fundamental vector $\omega^{(H)}$ of $H$ is $(-f_0/f_r, \ldots, -f_{r-1}/f_r)$. Conversely, if the proper Hankel matrix $H$ is of rank $r$ and generated by $(d_i)_{i\in\mathbb{N}}$, then $R(x)$ can be expressed as

$$R(x) = \frac{d_1 x^{r-1} + (d_2 - d_1 a_r)x^{r-2} + \ldots + (d_r - d_{r-1}a_r - \ldots - d_1 a_2)}{x^r - a_r x^{r-1} - \ldots - a_1},$$

where $(a_1, \ldots, a_r)$ is the fundamental vector of $H$.

So there exists a bijection

$$\varphi\colon\ \mathcal{R}_p \longrightarrow \mathcal{H}_p$$

between the set $\mathcal{R}_p$ of proper rational functions over $K$ and the set $\mathcal{H}_p$ of proper Hankel matrices over $K$.

*Example 7.3.1.* Let $\mathcal{D} = (d_i) = (1, 1, 1, 6, 21, \ldots)$ generate a proper Hankel matrix $H$ of rank 3 over $\mathbb{Z}$ with fundamental vector $\omega = (a_1, a_2, a_3) = (1, 2, 3)$, i.e.,

$$H_3 = \begin{pmatrix} 1 & 1 & 1 \\ 1 & 1 & 6 \\ 1 & 6 & 21 \end{pmatrix}.$$

The corresponding rational function is

$$\begin{aligned} R(x) &= \frac{g(x)}{f(x)} \\ &= \frac{d_1 x^2 + (d_2 - d_1 a_3)x + (d_3 - d_2 a_3 - d_1 a_2)}{x^3 - a_3 x^2 - a_2 x - a_1} = \frac{x^2 - 2x - 4}{x^3 - 3x^2 - 2x - 1}. \end{aligned}$$

On the other hand, starting from $R(x)$ and expanding it we get

$$g(x) = \Big(\sum_{i\in\mathbb{N}} d_i x^{-i}\Big) \cdot f(x),$$

i.e.,

$$\begin{aligned} (x^2 - 2x - 4) = (x^{-1} + x^{-2} + x^{-3} + 6x^{-4} + 21x^{-5} + \ldots) \cdot \\ \cdot (x^3 - 3x^2 - 2x - 1), \end{aligned}$$

and the fundamental vector

$$(a_1, a_2, a_3) = (-f_0/f_3, -f_1/f_3, -f_2/f_3) = (1, 2, 3) .$$

In fact,

$$x^5 \cdot g(x) = (x^4 + x^3 + x^2 + 6x + 21) \cdot f(x) + (76x^2 + 48x + 21) ,$$

so we get the generating sequence of $H_3$ as the coefficients of this quotient.

*Definition 7.2.3.* If $H$ and $R$ are as in Theorem 7.3.1, then $R$ is called the *rational function associated with* $H$, and conversely $H$ is called the *proper Hankel matrix associated with* $R$.

The bijection $\varphi$: $\mathcal{R}_p \longrightarrow \mathcal{H}_p$ is the basis for various applications of Hankel matrix computations to solving problems for polynomials.

If $f, g \in K[x]$ and no restriction on degrees is imposed on $f$ and $g$, we can still associate a proper Hankel matrix $H$ with $g/f$ by letting $H$ be generated by $(d_i)_{i\in\mathbb{N}}$, where

$$\frac{g(x)}{f(x)} = \sum_{i=0}^{s} b_i x^i + \sum_{i\in\mathbb{N}} d_i x^{-i} ,$$

i.e., by the asymptotic part of the expansion, or by the expansion of $\mathrm{rem}(g, f)/f$.

As in Example 7.3.1 we can compute the generating sequence $(d_1, \ldots, d_{2n-1})$ for the Hankel matrix associated with the proper rational function $g/f$ as the coefficients of $\mathrm{quot}(x^{2n-1} \cdot g, f)$, where $n = \deg(f)$.

Computation of resultants

**Theorem 7.3.2.** a. Let $f, g \in K[x]$, $\deg(g) = m \leq \deg(f) = n$, $0 < n$, and let $f_n = \mathrm{lc}(f)$. Then

$$\mathrm{res}_x(f, g) = (-1)^{\frac{n(n+3)}{2} + m(n+1)} f_n^{n+m} \det(H_n) ,$$

where $H_n$ is the $n \times n$ principal submatrix of $\varphi(g(x)/f(x))$.

b. Let $f, g \in I[x]$ and $m, n, f_n$ as above. Then

$$\deg(H_n^*) = (-1)^{\frac{n(n+3)}{2} + m(n+1)} f_n^{(n+m)(n-1)} \mathrm{res}_x(f, g) ,$$

where $H_n^*$ is the $n \times n$ principal submatrix of $\varphi(f_n^{n+m} g(x)/f(x))$.

*Proof.* a. Let $f(x) = f_n \prod_{i=1}^{n}(x - \alpha_i)$, $g(x) = g_m \prod_{j=1}^{m}(x - \beta_j)$, and $C$ the companion matrix of $\tilde{f} = (1/\mathrm{lc}(f))f(x)$, i.e. $C = (c_{ij}) \in K^{n\times n}$ with $c_{i+1\,i} = 1$ for $1 \leq i < n$, $c_{in} = -a_{i-1}$ for $1 \leq i \leq n$, and all other entries are 0, where

$\tilde{f} = x^n + a_{n-1}x^{n-1} + \ldots + a_0$. Then, by Exercise 1, we have

$$\begin{aligned}\deg(H_n) &= (-1)^{\frac{n(n+3)}{2}} f_n^{-n} g_m^n \prod_{j=1}^{m} \deg(C - \beta_j I) \\ &= (-1)^{\frac{n(n+3)}{2}+nm+m} f_n^{-n} g_m^n \prod_{j=1}^{m} \prod_{i=1}^{n} (\alpha_i - \beta_j) \\ &= (-1)^{\frac{n(n+3)}{2}+nm+m} f_n^{-n-m} \mathrm{res}_x(f, g) \ .\end{aligned}$$

b. This extension is not difficult and is left to the reader. □

So we get the following algorithm for computing resultants.

**Algorithm RES_H**(in: $f, g$; out: $r$);
[$f, g \in I[x]$, $\deg(g) \le \deg(f)$, $0 < \deg(f)$; $r = \mathrm{res}_x(f, g)$.]
1. $n := \deg(f)$; $m := \deg(g)$;
2. $H_n^* := n \times n$ principal submatrix of $\varphi(\mathrm{lc}(f)^{n+m} g/f)$;
3. $D := \det(H_n^*)$;
4. if $D = 0$
 then $r := 0$
 else $r := (-1)^{\frac{n(n+3)}{2}+m(n+1)} \mathrm{lc}(f)^{(1-n)(n+m)} D$;
 return.

From the complexity bounds in the previous section we immediately get a complexity bound for RES_H.

**Theorem 7.3.3.** The number of arithmetic operations in RES_H is dominated by $n^2$, where $n = \deg(f)$.

Computation of greatest common divisors

*Definition 7.3.1.* Let $f, g \in I[x]$, $\deg(g) \le \deg(f) = n$, $f \ne 0$. Let $H = \varphi(g/f)$, and let $0 = n_0 < n_1 < \cdots < n_t (\le n)$ be the indices of principal submatrices $H_{n_i}$ of $H$ such that $\det(H_{n_i}) \ne 0$ for $1 \le i \le t$. Let the $n_i$-th fundamental vector be $\omega^{(n_i)} = (a_{i\,1}, \ldots, a_{i\,n_i})$ for $1 \le i \le t$ and let

$$p_{n_0}(x) = 1 \quad \text{and} \quad p_{n_i}(x) = x^{n_i} - a_{i\,n_i} x^{n_i - 1} - \ldots - a_{i\,1} \quad \text{for } 1 \le i \le t \ .$$

The *polynomial Hankel sequence* of $f$ and $g$ is the sequence of polynomials $p_{n_0}(x), \ldots, p_{n_t}(x)$.

**Theorem 7.3.4.** Let $f, g \in K[x]$, $m = \deg(g) \le \deg(f) = n$, $0 < n$. Let $H = \varphi(g/f)$ and let $H$ be generated by the sequence $(d_i)_{i \in \mathbb{N}}$. Let $p_{n_0}(x), \ldots, p_{n_t}(x)$ be the polynomial Hankel sequence of $f$ and $g$, $p_{n_{t-1}}(x) = b_{n_{t-1}} x^{n_{t-1}} + \ldots + b_0$, $a = \mathrm{lc}(f) \cdot (b_0, \ldots, b_{n_{t-1}}) \cdot (d_{n_t}, \ldots, d_{n_t + n_{t-1}})^T$, $d_0 = \mathrm{lc}(g)/\mathrm{lc}(f)$ if $m = n$ and

$d_0 = 0$ otherwise. Then

$$f(x) = \mathrm{lc}(f) \cdot p_{n_t}(x) \cdot \gcd(f, g)$$

(where $\gcd(f, g)$ is assumed to be monic) and the polynomials

$$v(x) = \frac{1}{a} p_{n_{t-1}}(x), \qquad u(x) = -\frac{1}{a} \sum_{j=0}^{n_{t-1}} b_j \sum_{i=0}^{j} d_i x^{j-i}$$

satisfy the Bezout equality

$$u(x) f(x) + v(x) g(x) = \gcd(f, g) \ .$$

The proof of this theorem can be found in Sendra and Llovet (1992a). So we get the following algorithm for solving the extended gcd problem.

**Algorithm EGCD_H**(in: $f, g$; out: $h, u, v$);
[$f, g \in K[x]$, $\deg(g) \leq \deg(f)$, $0 < \deg(f)$;
$h = \gcd(f, g)$, $u, v \in K[x]$, such that $h = u \cdot f + v \cdot g$.]
1. $n := \deg(f)$; $m := \deg(g)$;
2. $H_n :=$ $n \times n$ principal submatrix of $\varphi(g/f)$ generated by $(d_i)_{i \in \mathbb{N}}$;
3. determine the polynomial Hankel sequence $p_{n_0}, \ldots, p_{n_t}$ of $f$ and $g$ from $H_n$;
4. $a := \mathrm{lc}(f) \cdot (b_0, \ldots, b_{n_{t-1}}) \cdot (d_{n_t}, \ldots, d_{n_t + n_{t-1}})^T$,
   where $p_{n_{t-1}}(x) = b_{n_{t-1}} x^{n_{t-1}} + \ldots + b_0$;
5. $v := \frac{1}{a} p_{n_{t-1}}(x)$;
   $u := -\frac{1}{a} \sum_{j=0}^{n_{t-1}} b_j \sum_{i=0}^{j} d_i x^{j-i}$;
   $h := \mathrm{quot}(f, \mathrm{lc}(f) \cdot p_{n_t})$;
   return.

**Theorem 7.3.5.** The number of arithmetic operations in EGCD_H is dominated by $n^2$, where $n = \deg(f)$.

### Exercises

1. Let $f, g \in I[x]$, $\deg(g) \leq \deg(f) = n$, $0 < n$. Let $H_n$ and $M_n$ be the $n \times n$ principal submatrices of $\varphi(g/f)$ and $\varphi(1/f)$, respectively, and let $C$ be the companion matrix of $\tilde{f}(x) = \frac{1}{\mathrm{lc}(f)} f(x)$. Then

$$H_n = M_n \cdot g(C) \ .$$

2. Show that the sequence $(d_1, \ldots, d_{2n-1})$ generating the $n \times n$ principal submatrix of $\varphi(g/f)$ in the corollary to Theorem 7.3.1 can be computed by dividing a properly scaled $g$ by $f$, i.e.,

$$\mathrm{quot}(x^{2n-1} \cdot g, f) = d_1 x^{2n-2} + \ldots + d_{2n-1} \ .$$

3. Compute the finite Hankel matrix and the fundamental vector associated with

$$R(x) = \frac{x^2 + 3x + 2}{x^3 - x^2 - x - 2} \, .$$

## 7.4 Bibliographic notes

An efficient algorithm for solving sparse linear systems over a finite field was designed by D. Wiedemann (1986).

In the computation of polynomial gcd by Hankel matrices one can use ideas similar to subresultants for bounding the coefficients (Gemignani 1994). Several variants of resultants can also be computed by Hankel methods (Hong and Sendra 1996). Moreover, Hankel matrices can be employed for computing Padé approximations (Brent et al. 1980), determining the number of real roots of polynomials (Gantmacher 1977, Heinig and Rost 1984, Llovet et al. 1992) and for factorization of polynomials. Issues of parallel computation are considered in Bini and Pan (1993).

# 8 The method of Gröbner bases

## 8.1 Reduction relations

Many of the properties that are important for Gröbner bases can be developed in the frame of binary relations on arbitrary sets, so-called reduction relations (Huet 1980). The theory of reduction relations forms a common basis for the theory of Gröbner bases, word problems in finitely presented groups, term rewriting systems, and lambda calculus.

*Definition 8.1.1.* Let $M$ be a set and $\longrightarrow$ a binary relation on $M$, i.e., $\longrightarrow \subseteq M \times M$. We call $\longrightarrow$ a *reduction relation* on $M$. Instead of $(a, b) \in \longrightarrow$ we usually write $a \longrightarrow b$ and say that *$a$ reduces to $b$*.

Given reduction relations $\longrightarrow$ and $\longrightarrow'$ on $M$, we define operations on $M \times M$ for constructing new reduction relations.

- $\longrightarrow \circ \longrightarrow'$ (or just $\longrightarrow\longrightarrow'$), the *composition* of $\longrightarrow$ and $\longrightarrow'$, is the reduction relation defined as $a \longrightarrow\longrightarrow' b$ iff there exists a $c \in M$ such that $a \longrightarrow c \longrightarrow' b$;
- $\longrightarrow^{-1}$ (or just $\longleftarrow$), the *inverse relation* of $\longrightarrow$, is the reduction relation defined as $a \longleftarrow b$ iff $b \longrightarrow a$;
- $\longrightarrow_{\text{sym}}$ (or just $\longleftrightarrow$), the *symmetric closure* of $\longrightarrow$, is the reduction relation defined as $\longrightarrow \cup \longleftarrow$, i.e., $a \longleftrightarrow b$ iff $a \longrightarrow b$ or $a \longleftarrow b$;
- $\longrightarrow^i$, the *$i$-th power* of $\longrightarrow$, is the reduction relation defined inductively for $i \in \mathbb{N}_0$ as
  $\longrightarrow^0 :=$ id (identity relation on $M$), i.e., $a \longrightarrow^0 b$ iff $a = b$, and
  $\longrightarrow^i := \longrightarrow \longrightarrow^{i-1}$ for $i \geq 1$.
  So $a \longrightarrow^i b$ if and only if there exist $c_0, \ldots, c_i$ such that $a = c_0 \longrightarrow c_1 \longrightarrow \ldots \longrightarrow c_i = b$. In this case we say that *$a$ reduces to $b$ in $i$ steps*;
- $\longrightarrow^+ := \bigcup_{i=1}^{\infty} \longrightarrow^i$, the *transitive closure* of $\longrightarrow$;
- $\longrightarrow^* := \bigcup_{i=0}^{\infty} \longrightarrow^i$, the *reflexive-transitive closure* of $\longrightarrow$;
- $\longleftrightarrow^*$ is the *reflexive-transitive-symmetric closure* of $\longrightarrow$.

In the sequel we will always assume that the set $M$ is recursively enumerable and the reduction relation $\longrightarrow$ is recursive, i.e., for given $x, y \in M$ we can decide whether $x \longrightarrow y$.

$\longleftrightarrow^*$ is an equivalence relation on $M$ and $M_{/\longleftrightarrow^*}$ is the set of equivalence classes modulo $\longleftrightarrow^*$. One of the main problems in connection with reduction relations is to decide $\longleftrightarrow^*$, i.e., to determine for $a, b \in M$ whether $a \longleftrightarrow^* b$; or, in other words, whether $a$ and $b$ belong to the same equivalence class. We call this problem the *equivalence problem* for the reduction relation $\longrightarrow$.

*Example 8.1.1.* a. One well known version of the equivalence problem is the word problem for groups. A *free presentation* of a group is a set $X$ of *generators* together with a set $R$ of words (strings) in the generators, called *relators*. Words are formed by concatenating symbols $x$ or $x^{-1}$ for $x \in X$. Such a presentation is usually written as $\langle X|R \rangle$ and it denotes the group $F(X)$ modulo $\langle R \rangle$, $F(X)_{/\langle R \rangle}$, where $F(X)$ is the free group generated by $X$ and $\langle R \rangle$ is the smallest normal subgroup of $F(X)$ which contains $R$. In more concrete terms, we think of $\langle X|R \rangle$ as the group obtained from $F(X)$ by forcing all words in $R$ to be equal to the identity together with all consequences of these equations.

For example, consider the group

$$G = \langle \ \{a, b\} \mid \{a^2, b^2, aba^{-1}b^{-1}\} \ \rangle \ .$$

The first relator tells us that we can replace $a^m$ by 1 if $m$ is even and by $a$ if $m$ is odd. Similarly for powers of $b$. The third relator tells us that $a$ and $b$ commute so that we can collect all powers of $a$ and then all powers of $b$ in a word. Thus, every element of $G$ is equal to one of

$$1, a, b, ab$$

and it can be shown that these are distinct.

The *word problem for freely presented groups* is:
given: a presentation $\langle X|R \rangle$ and words $u, v \in F(X)$;
decide: $u \stackrel{?}{=} v$ in $\langle X|R \rangle$.

Actually this definition looks as though the problem were about the presentation of the group rather than the group itself. But, in fact, if we insist that the presentations considered must be effectively given, i.e., both $X$ and $R$ are recursively enumerable, then the decidability is independent of the presentation. It is not very hard to show that the problem is undecidable in general. It is much harder to show that the same is true even if we consider only finite presentations, i.e., both $X$ and $R$ are finite sets.

b. Another example is from polynomial ideal theory and it will lead us to the introduction of Gröbner bases. Consider the polynomial ring $K[x_1, \ldots, x_n]$, $K$ a field, and let $I = \langle p_1, \ldots, p_m \rangle$ be the ideal generated by $p_1, \ldots, p_m$ in $K[x_1, \ldots, x_n]$. The *main problem in polynomial ideal theory* according to van der Waerden is:
given: generators $p_1, \ldots, p_m$ for an ideal $I$ in $K[x_1, \ldots, x_n]$, and polynomials $f, g \in K[x_1, \ldots, x_n]$,
decide: whether $f \equiv g \pmod{I}$, or equivalently, whether $f$ and $g$ represent the same element of the factor ring $K[x_1, \ldots, x_n]_{/I}$.
Later we will introduce a reduction relation $\longrightarrow$ such that $\longleftrightarrow^* = \equiv_I$, so again the problem is to decide the equivalence problem of a reduction relation.

Let us introduce some more useful notations for abbreviating our arguments about reduction relations.

*Definition 8.1.2.*

- $x \longrightarrow$ means $x$ is *reducible*, i.e., $x \longrightarrow y$ for some $y$;
- $\underline{x}_{\longrightarrow}$ means $x$ is *irreducible* or *in normal form* w.r.t. $\longrightarrow$. We omit mentioning the reduction relation if it is clear from the context;
- $x \downarrow y$ means that $x$ and $y$ have a *common successor*, i.e., $x \longrightarrow z \longleftarrow y$ for some $z$;
- $x \uparrow y$ means that $x$ and $y$ have a *common predecessor*, i.e., $x \longleftarrow z \longrightarrow y$ for some $z$;
- *$x$ is a $\longrightarrow$-normal form* of $y$ iff $y \longrightarrow^* \underline{x}$.

In the sequel we will always assume that we can decide whether $x \in M$ is reducible and if so compute a $y$ such that $x \longrightarrow y$. Based on these assumptions about the decidability of the reduction relation we will establish that the equivalence problem for $\longrightarrow$ can be decided if $\longrightarrow$ has two basic properties, namely the Church–Rosser property and the termination property.

*Definition 8.1.3.* a. $\longrightarrow$ is *Noetherian* or has the *termination property* iff every reduction sequence terminates, i.e., there is no infinite sequence $x_1, x_2, \ldots$ in $M$ such that $x_1 \longrightarrow x_2 \longrightarrow \ldots$.

b. $\longrightarrow$ is *Church–Rosser* or has the *Church–Rosser property* iff $a \longleftrightarrow^* b$ implies $a \downarrow_* b$.

Whenever a set $M$ is equipped with a Noetherian relation $\longrightarrow$ we can apply the *principle of Noetherian induction* for proving that a predicate $P$ holds for all $x \in M$:

if for all $x \in M$

$$[\text{for all } y \in M\colon (x \longrightarrow y) \Longrightarrow P(y)] \Longrightarrow P(x)$$

then

$$\text{for all } x \in M\colon P(x) .$$

A correctness proof of this principle can be found in (Cohn 1974).

**Theorem 8.1.1.** Let $\longrightarrow$ be Noetherian and Church–Rosser. Then the equivalence problem for $\longrightarrow$ is decidable.

*Proof.* Let $x, y \in M$. Let $\tilde{x}, \tilde{y}$ be normal forms of $x, y$, respectively (by Noetherianity every sequence of reductions leads to a normal form after finitely many steps). Obviously $x \longleftrightarrow^* y$ if and only if $\tilde{x} \longleftrightarrow^* \tilde{y}$. By the Church–Rosser property $\tilde{x} \longleftrightarrow^* \tilde{y}$ if and only if $\tilde{x} \downarrow_* \tilde{y}$. Since $\tilde{x}$ and $\tilde{y}$ are irreducible, $\tilde{x} \downarrow_* \tilde{y}$ if and only if $\tilde{x} = \tilde{y}$.

Summarizing we have $x \longleftrightarrow^* y$ if and only if $\tilde{x} = \tilde{y}$. □

Theorem 8.1.1 cannot be reversed, i.e., the equivalence problem for $\longrightarrow$ could be decidable although $\longrightarrow$ is not Noetherian or $\longrightarrow$ is not Church–Rosser.

*Example 8.1.2.* a. Let $M = \mathbb{N}$ and $\longrightarrow\ = \{(n, n+1) \mid n \in \mathbb{N}\}$. Obviously the equivalence problem for $\longrightarrow$ is decidable, but $\longrightarrow$ is not Noetherian.

b. Let $M = \{a, b, c\}$ and $\longrightarrow = \{(a, b), (a, c)\}$. So

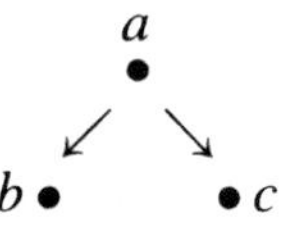


Obviously the equivalence problem for $\longrightarrow$ is decidable, but $\longrightarrow$ is not Church–Rosser.

So if $\longrightarrow$ is Noetherian and Church–Rosser then we have a *canonical simplifier* for $M_{/\longleftrightarrow^*}$, i.e., a function which for every equivalence class computes a unique representative in that equivalence class. For $x \in M$ any normal form of $x$ can be taken as the simplified form of $x$, since all these normal forms are equal.

*Example 8.1.3.* a. Let $H$ be the commutative semigroup generated by $a, b, c, f, s$ modulo the relations

$$as = c^2 s, \quad bs = cs, \quad s = f \ . \tag{E}$$

Consider the reduction relation $\longrightarrow$ given by

$$s \longrightarrow f, \quad cf \longrightarrow bf, \quad b^2 f \longrightarrow af$$

and if $u \longrightarrow v$ then $ut \longrightarrow vt$ for all words $u, v, t$.

$\longrightarrow$ is Church–Rosser and Noetherian and $\longleftrightarrow^* = =_{(E)}$. So, for example, we can discover that $a^3bcf^3 =_{(E)} a^2b^4fs^2$ by computing the normal forms of both words, which turn out to be equal.

b. Let $I$ be the ideal in $\mathbb{Q}[x, y]$ generated by

$$x^3 - x^2, \quad x^2 y - x^2 \ .$$

Let $\longrightarrow$ be defined on $\mathbb{Q}[x, y]$ in such a way that every occurrence of $x^3$ or $x^2 y$ can be replaced by $x^2$. Then $\longrightarrow$ is Church–Rosser and Noetherian. Thus, we can decide whether $f \equiv g \pmod{I}$ for arbitrary $f, g \in \mathbb{Q}[x, y]$, i.e., we can compute in $\mathbb{Q}[x, y]_{/I}$.

Checking whether the Church–Rosser property and the Noetherian property are satisfied for a given reduction relation is not an easy task. Fortunately, in the situation of polynomial ideals Noetherianity is always satisfied as we will see later. Our goal now is to reduce the problem of checking the Church–Rosser property to checking simpler properties.

*Definition 8.1.4.* a. $\longrightarrow$ is *confluent* iff $x \uparrow^* y$ implies $x \downarrow_* y$, or graphically every diamond of the following form can be completed:

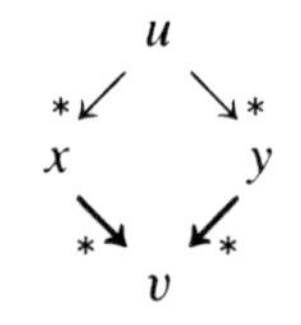

b. $\longrightarrow$ is *locally confluent* iff $x \uparrow y$ implies $x \downarrow_* y$, or graphically every diamond of the following form can be completed:

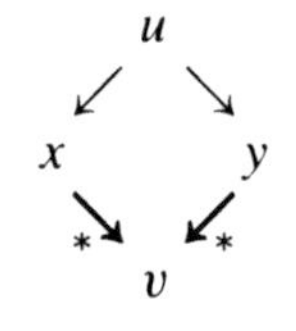

**Theorem 8.1.2.** a. $\longrightarrow$ is Church–Rosser if and only if $\longrightarrow$ is confluent.

b. (Newman lemma) Let $\longrightarrow$ be Noetherian. Then $\longrightarrow$ is confluent if and only if $\longrightarrow$ is locally confluent.

*Proof.* a. If $\longrightarrow$ is Church–Rosser then it is obviously confluent. So let us assume that $\longrightarrow$ is confluent. Suppose that $x \longleftrightarrow^* y$ in $n$ steps, i.e., $x \longleftrightarrow^n y$. We use induction on $n$. The case $n = 0$ is immediate. For $n > 0$ there are two possible situations:

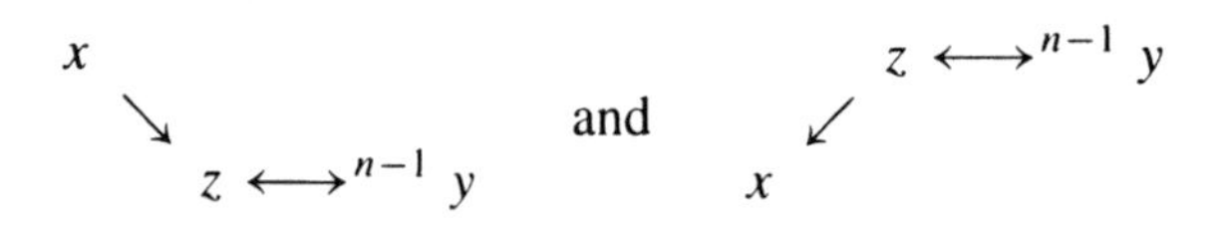

for some $z$. In the first case by the induction hypothesis there is a $u$ such that

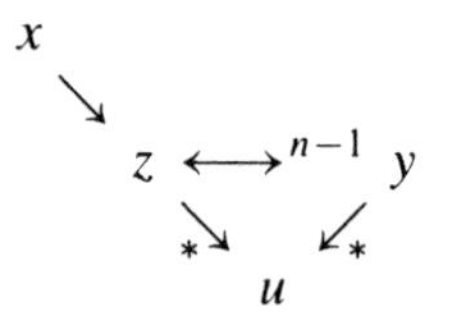

and in the second case by the induction hypothesis and by confluence there are $u, v$ such that

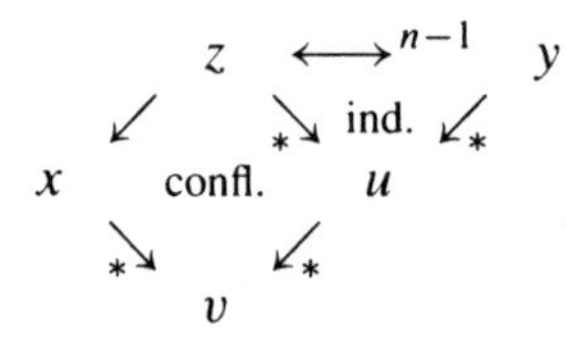


In either case $x \downarrow_* y$.

b. Confluence obviously implies local confluence. So assume that $\longrightarrow$ is locally confluent. We use Noetherian induction on the Noetherian ordering $\longrightarrow$. The induction hypothesis is

"for all $z$ with $z_0 \longrightarrow z$ and for all $x', y'$ with $x' \longleftarrow^* z \longrightarrow^* y'$ we have $x' \downarrow_* y'$."

Now assume that $x \longleftarrow^* z_0 \longrightarrow^* y$. The cases $x = z_0$, $y = z_0$ are obvious. So consider

$$x \longleftarrow^* x_1 \longleftarrow z_0 \longrightarrow y_1 \longrightarrow^* y\ .$$

By local confluence and the induction hypothesis there are $u$, $v$, $w$ such that

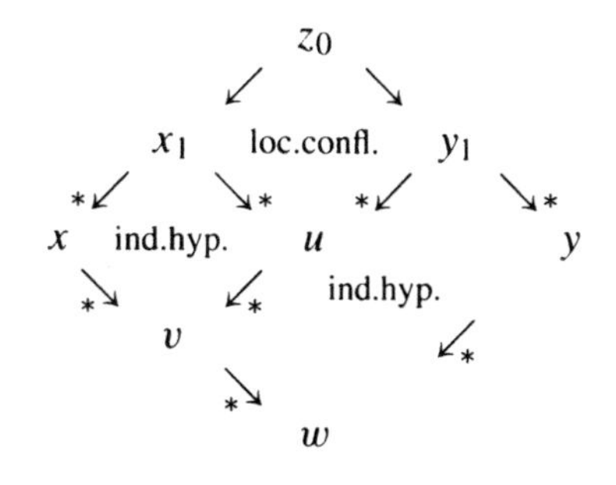


So $x \downarrow_* y$. □

If we drop the requirement of Noetherianity in Theorem 8.1.2 (b) then the statement does not hold any more, as can be seen from the counterexample

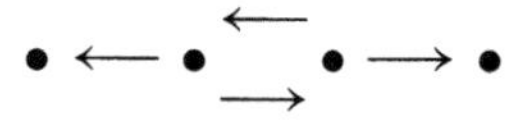

*Definition 8.1.5.* Let $\longrightarrow$ be a reduction relation on the set $M$ and $>$ a partial ordering on $M$. Let $x, y, z \in M$. $x$ and $y$ are *connected (w.r.t.* $\longrightarrow$*) below (w.r.t.* $>$*)* $z$ iff there are $w_1, \ldots, w_n \in M$ such that $x = w_1 \longleftrightarrow \ldots \longleftrightarrow w_n = y$ and $w_i < z$ for all $1 \leq i \leq n$. We use the notation $x \longleftrightarrow^*_{(<z)} y$.

**Theorem 8.1.3** (Refined Newman lemma). Let $\longrightarrow$ be a reduction relation on $M$ and $>$ a partial Noetherian ordering on $M$ such that $\longrightarrow \subseteq >$. Then $\longrightarrow$ is confluent if and only if for all $x, y, z$ in $M$:

$$x \longleftarrow z \longrightarrow y \quad \text{implies} \quad x \longleftrightarrow^*_{(<z)} y\ .$$

*Proof.* Confluence obviously implies connectedness. So now let us assume that the connectedness property holds. We use Noetherian induction on $>$ with the induction hypothesis

$$\text{for all } \tilde{x}, \tilde{y}, \tilde{z}\text{: if } \tilde{z} < z \text{ and } \tilde{x} \longleftarrow^* \tilde{z} \longrightarrow^* \tilde{y} \text{ then } \tilde{x} \downarrow_* \tilde{y}\ . \qquad \text{(IH 1)}$$

Now consider the situation $x \longleftarrow^* z \longrightarrow^* y$. If $x = z$ or $y = z$ then we are done. Otherwise we have

$$x \longleftarrow^* x_1 \longleftarrow z \longrightarrow y_1 \longrightarrow^* y\ .$$

By the assumption of connectedness there are $u_1, \ldots, u_n < z$ such that

$$x_1 = u_1 \longleftrightarrow \ldots \longleftrightarrow u_n = y_1\ .$$

We use induction on $n$ to show that for all $n$ and all $u_1, \ldots, u_n \in M$:

$$\begin{array}{l} \text{if } u_1 \longleftrightarrow \ldots \longleftrightarrow u_n \text{ and } u_i < z \text{ for all } 1 \le i \le n\ , \\ \text{then } u_1 \downarrow_* u_n\ . \end{array} \qquad (8.1.1)$$

The case $n = 1$ is clear. So we formulate induction hypothesis 2:

$$(8.1.1) \text{ holds for some fixed } n\ . \qquad \text{(IH 2)}$$

For the induction step let $u_1, \ldots, u_{n+1} \in M$ such that $u_i < z$ for $1 \le i \le n+1$ and $u_1 \longleftrightarrow \ldots \longleftrightarrow u_{n+1}$. We distinguish two cases in which the existence of a common successor $v$ to $u_1$ and $u_{n+1}$ can be shown by the following diagrams:

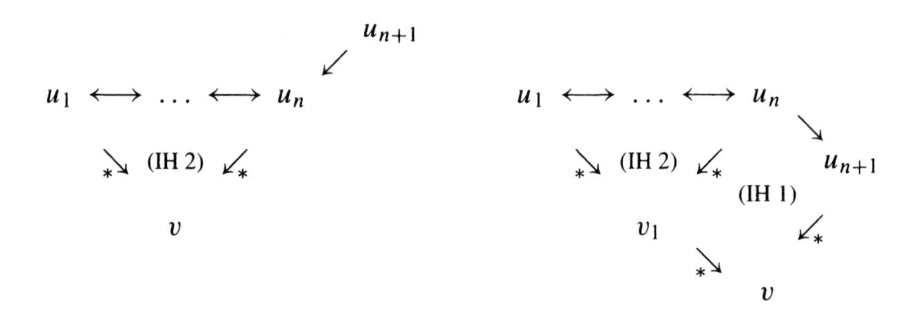


This proves (8.1.1). The proof of the theorem can now be completed by the diagram

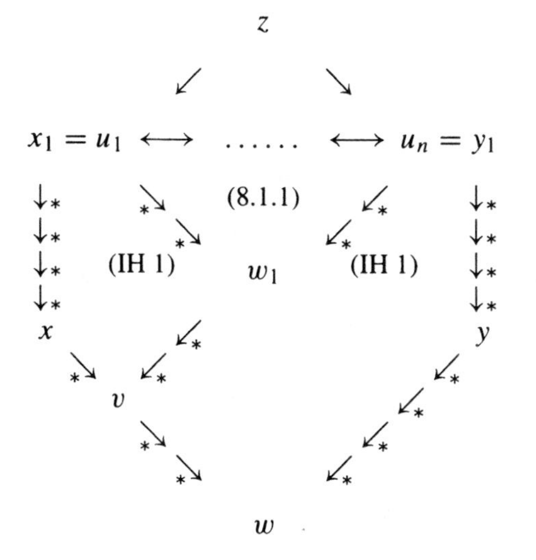


□

**Exercises**

1. If the reduction relation $\longrightarrow$ on the set $M$ is Noetherian, does that mean that $R(x) = \{y \mid x \longrightarrow y\}$ is finite for every $x$?
2. Give another example of a locally confluent reduction relation which is not confluent.

## 8.2 Polynomial reduction and Gröbner bases

The theory of Gröbner bases was developed by B. Buchberger (1965, 1970, 1985b). Gröbner bases are very special and useful bases for polynomial ideals. The Buchberger algorithm for constructing Gröbner bases is at the same time a generalization of Euclid's gcd algorithm and of Gauss's triangularization algorithm for linear systems.

Let $R$ be a commutative ring with 1 and $R[X] = R[x_1, \ldots, x_n]$ the polynomial ring in $n$ indeterminates over $R$. If $F$ is any subset of $R[X]$ we write $\langle F \rangle$ or ideal$(F)$ for the ideal generated by $F$ in $R[X]$. By $[X]$ we denote the monoid (under multiplication) of *power products* $x_1^{i_1} \ldots x_n^{i_n}$ in $x_1, \ldots, x_n$. $1 = x_1^0 \ldots x_n^0$ is the unit element in the monoid $[X]$. lcm$(s, t)$ denotes the least common multiple of the power products $s, t$.

Commutative rings with 1 in which the *basis condition* holds, i.e., in which every ideal has a finite basis, are usually called *Noetherian rings*. This notation is motivated by the following lemma.

**Lemma 8.2.1.** In a Noetherian ring there are no infinitely ascending chains of ideals.

**Theorem 8.2.2** (Hilbert's basis theorem). If $R$ is a Noetherian ring then also the univariate polynomial ring $R[x]$ is Noetherian.

*Proof.* See van der Waerden (1970: chap. 15). □

Hilbert's basis theorem implies that the multivariate polynomial ring $K[X]$ is Noetherian, if $K$ is a field. So every ideal $I$ in $K[X]$ has a finite basis, and if we are able to effectively compute with finite bases then we are dealing with all the ideals in $K[X]$.

Before we can define the reduction relation on the polynomial ring, we have to introduce an ordering of the power products with respect to which the reduction relation should be decreasing.

*Definition 8.2.1.* Let $<$ be an ordering on $[X]$ that is compatible with the monoid structure, i.e.,
a. $1 = x_1^0 \dots x_n^0 < t$ for all $t \in [X] \setminus \{1\}$, and
b. $s < t \Longrightarrow su < tu$ for all $s, t, u \in [X]$.
We call such an ordering $<$ on $[X]$ an *admissible ordering*.

*Example 8.2.1.* We give some examples of frequently used admissible orderings on $[X]$.

a. The *lexicographic ordering* with $x_{\pi(1)} > x_{\pi(2)} > \dots > x_{\pi(n)}$, $\pi$ a permutation of $\{1, \dots, n\}$: $x_1^{i_1} \dots x_n^{i_n} <_{\text{lex},\pi} x_1^{j_1} \dots x_n^{j_n}$ iff there exists a $k \in \{1, \dots, n\}$ such that for all $l < k$ $i_{\pi(l)} = j_{\pi(l)}$ and $i_{\pi(k)} < j_{\pi(k)}$.

If $\pi = \text{id}$, we get the usual lexicographic ordering $<_{\text{lex}}$.

b. The *graduated lexicographic ordering* w.r.t. the permutation $\pi$ and the weight function $w$: $\{1, \dots, n\} \to \mathbb{R}^+$: for $s = x_1^{i_1} \dots x_n^{i_n}$, $t = x_1^{j_1} \dots x_n^{j_n}$ we define $s <_{\text{glex},\pi,w} t$ iff

$$\left(\sum_{k=1}^{n} w(k) i_k < \sum_{k=1}^{n} w(k) j_k\right) \quad \text{or}$$
$$\left(\sum_{k=1}^{n} w(k) i_k = \sum_{k=1}^{n} w(k) j_k \ \text{and} \ s <_{\text{lex},\pi} t\right) .$$

We get the usual graduated lexicographic ordering $<_{\text{glex}}$ by setting $\pi = \text{id}$ and $w = 1_{\text{const}}$.

c. The *graduated reverse lexicographic ordering*: we define $s <_{\text{grlex}} t$ iff

$$\deg(s) < \deg(t) \quad \text{or}$$
$$(\deg(s) = \deg(t) \ \text{and} \ t <_{\text{lex},\pi} s, \ \text{where} \ \pi(j) = n - j + 1) .$$

d. The *product ordering* w.r.t. $i \in \{1, \ldots, n-1\}$ and the admissible orderings $<_1$ on $X_1 = [x_1, \ldots, x_i]$ and $<_2$ on $X_2 = [x_{i+1}, \ldots, x_n]$: for $s = s_1 s_2, t = t_1 t_2$, where $s_1, t_1 \in X_1, s_2, t_2 \in X_2$, we define $s <_{\mathrm{prod}, i, <_1, <_2} t$ iff

$$s_1 <_1 t_1 \quad \text{or} \quad (s_1 = t_1 \text{ and } s_2 <_2 t_2) \ .$$

A complete classification of admissible orderings is given in Robbiano (1985, 1986).

**Lemma 8.2.3.** Let $<$ be an admissible ordering on $[X]$.
a. If $s, t \in [X]$ and $s$ divides $t$ then $s \leq t$.
b. $<$ (or actually $>$) is Noetherian, and consequently every subset of $[X]$ has a smallest element.

*Proof.* a. For some $u$ we have $su = t$. By the admissibility of $<$, $s = 1s \leq us = t$.

b. Let $s_1 > s_2 > \ldots$ be a sequence of decreasing elements in $[X]$. Let $K$ be any field. So the sequence of ideals $\langle s_1 \rangle \subset \langle s_1, s_2 \rangle \subset \ldots$ in $K[X]$ is increasing. But $K[X]$ is Noetherian, thus the sequence has to be finite. □

Throughout this chapter let $R$ be a commutative ring with 1, $K$ a field, $X$ a set of variables, and $<$ an admissible ordering on $[X]$.

*Definition 8.2.2.* Let $s$ be a power product in $[X]$, $f$ a non-zero polynomial in $R[X]$, $F$ a subset of $R[X]$.

By $\mathrm{coeff}(f, s)$ we denote the coefficient of $s$ in $f$.
$\mathrm{lpp}(f) := \max_<\{t \in [X] \mid \mathrm{coeff}(f, t) \neq 0\}$ (*leading power product* of $f$),
$\mathrm{lc}(f) := \mathrm{coeff}(f, \mathrm{lpp}(f))$ (*leading coefficient* of $f$),
$\mathrm{in}(f) := \mathrm{lc}(f)\mathrm{lpp}(f)$ (*initial* of $f$),
$\mathrm{red}(f) := f - \mathrm{in}(f)$ (*reductum* of $f$),
$\mathrm{lpp}(F) := \{\mathrm{lpp}(f) \mid f \in F \setminus \{0\}\}$,
$\mathrm{lc}(F) := \{\mathrm{lc}(f) \mid f \in F \setminus \{0\}\}$,
$\mathrm{in}(F) := \{\mathrm{in}(f) \mid f \in F \setminus \{0\}\}$,
$\mathrm{red}(F) := \{\mathrm{red}(f) \mid f \in F \setminus \{0\}\}$.

If $I$ is an ideal in $R[X]$, then $\mathrm{lc}(I) \cup \{0\}$ is an ideal in $R$. However, $\mathrm{in}(F) \cup \{0\}$ in general is not an ideal in $R[X]$.

*Definition 8.2.3.* Any admissible ordering $<$ on $[X]$ induces a partial ordering $\ll$ on $R[X]$, the *induced ordering*, in the following way:
$f \ll g$ iff $f = 0$ and $g \neq 0$ or
$f \neq 0$, $g \neq 0$ and $\mathrm{lpp}(f) < \mathrm{lpp}(g)$ or
$f \neq 0$, $g \neq 0$, $\mathrm{lpp}(f) = \mathrm{lpp}(g)$ and $\mathrm{red}(f) \ll \mathrm{red}(g)$.

**Lemma 8.2.4.** $\ll$ (or actually $\gg$) is a Noetherian partial ordering on $R[X]$.

One of the central notions of the theory of Gröbner bases is the concept of polynomial reduction.

*Definition 8.2.4.* Let $f, g, h \in K[X]$, $F \subseteq K[X]$. We say that $g$ *reduces to* $h$ *w.r.t.* $f$ ($g \longrightarrow_f h$) iff there are power products $s, t \in [X]$ such that $s$ has a non-vanishing coefficient $c$ in $g$ ($\text{coeff}(g, s) = c \neq 0$), $s = \text{lpp}(f) \cdot t$, and

$$h = g - \frac{c}{\text{lc}(f)} \cdot t \cdot f \ .$$

If we want to indicate which power product and coefficient are used in the reduction, we write

$$g \longrightarrow_{f,b,t} h, \quad \text{where } b = \frac{c}{\text{lc}(f)} \ .$$

We say that $g$ *reduces to* $h$ *w.r.t.* $F$ ($g \longrightarrow_F h$) iff there is $f \in F$ such that $g \longrightarrow_f h$.

*Example 8.2.2.* Let $F = \{\ldots, f = x_1x_3 + x_1x_2 - 2x_3, \ldots\}$ in $\mathbb{Q}[x_1, x_2, x_3]$, and $g = x_3^3 + 2x_1x_2x_3 + 2x_2 - 1$. Let $<$ be the graduated lexicographic ordering with $x_1 < x_2 < x_3$.

Then $g \longrightarrow_F x_3^3 - 2x_1x_2^2 + 4x_2x_3 + 2x_2 - 1 =: h$, and in fact $g \longrightarrow_{f,2,x_2} h$.

As an immediate consequence of this definition we get that the reduction relation $\longrightarrow$ is (nearly) compatible with the operations in the polynomial ring. Moreover, the reflexive-transitive-symmetric closure of the reduction relation $\longrightarrow_F$ is equal to the congruence modulo the ideal generated by $F$. The proofs are not difficult and are left to the reader.

**Lemma 8.2.5.** Let $a \in K^*$, $s \in [X]$, $F \subseteq K[X]$, $g_1, g_2, h \in K[X]$.

a. $\longrightarrow_F \subseteq \gg$,
b. $\longrightarrow_F$ is Noetherian,
c. if $g_1 \longrightarrow_F g_2$ then $a \cdot s \cdot g_1 \longrightarrow_F a \cdot s \cdot g_2$,
d. if $g_1 \longrightarrow_F g_2$ then $g_1 + h \downarrow_F^* g_2 + h$.

**Theorem 8.2.6.** Let $F \subseteq K[X]$. The ideal congruence modulo $\langle F \rangle$ equals the reflexive-transitive-symmetric closure of $\longrightarrow_F$, i.e., $\equiv_{\langle F \rangle} = \longleftrightarrow_F^*$.

So the congruence $\equiv_{\langle F \rangle}$ can be decided if $\longrightarrow_F$ has the Church–Rosser property. Of course, this is not the case for an arbitrary set $F$. Such distinguished sets (bases for polynomial ideals) are called Gröbner bases. In the literature sometimes the terms "standard basis" or "canonical basis" are used.

*Definition 8.2.5.* A subset $F$ of $K[X]$ is a *Gröbner basis* (for $\langle F \rangle$) iff $\longrightarrow_F$ is Church–Rosser.

A Gröbner basis of an ideal $I$ in $K[X]$ is by no means uniquely defined. In fact, whenever $F$ is a Gröbner basis for $I$ and $f \in I$, then also $F \cup \{f\}$ is a Gröbner basis for $I$.

### Exercises

1. Prove Lemma 8.2.4.
2. Prove Lemma 8.2.5.
3. Prove Theorem 8.2.6.

## 8.3 Computation of Gröbner bases

For testing whether a given basis $F$ of an ideal $I$ is a Gröbner basis it suffices to test for local confluence of the reduction relation $\longrightarrow_F$. This, however, does not yield a decision procedure, since there are infinitely many situations $f \uparrow_F g$. However, Buchberger (1965) has been able to reduce this test for local confluence to just testing a finite number of situations $f \uparrow_F g$. For that purpose he has introduced the notion of subtraction polynomials, or S-polynomials for short.

*Definition 8.3.1.* Let $f, g \in K[X]^*$, $t = \text{lcm}(\text{lpp}(f), \text{lpp}(g))$. Then

$$\text{cp}(f, g) = \left(t - \frac{1}{\text{lc}(f)} \cdot \frac{t}{\text{lpp}(f)} \cdot f,\ t - \frac{1}{\text{lc}(g)} \cdot \frac{t}{\text{lpp}(g)} \cdot g\right)$$

is the *critical pair* of $f$ and $g$. The difference of the elements of $\text{cp}(f, g)$ is the *S-polynomial* $\text{spol}(f, g)$ of $f$ and $g$.

If $\text{cp}(f, g) = (h_1, h_2)$ then we can depict the situation graphically in the following way:

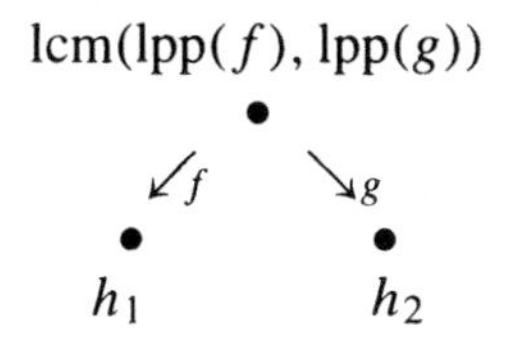


The critical pairs of elements of $F$ describe exactly the essential branchings of the reduction relation $\longrightarrow_F$.

**Theorem 8.3.1** (Buchberger's theorem). Let $F$ be a subset of $K[X]$.

a. $F$ is a Gröbner basis if and only if $g_1 \downarrow_F^* g_2$ for all critical pairs $(g_1, g_2)$ of elements of $F$.
b. $F$ is a Gröbner basis if and only if $\text{spol}(f, g) \longrightarrow_F^* 0$ for all $f, g \in F$.

*Proof.* a. Obviously, if $F$ is a Gröbner basis then $g_1 \downarrow_F^* g_2$ for all critical pairs $(g_1, g_2)$ of $F$.

On the other hand, assume that $g_1 \downarrow_F^* g_2$ for all critical pairs $(g_1, g_2)$. By the refined Newman lemma (Theorem 8.1.3) it suffices to show $h_1 \longleftrightarrow^*_{F(\ll h)} h_2$ for all $h, h_1, h_2$ such that $h_1 \longleftarrow_F h \longrightarrow_F h_2$.

Let $s_1, s_2$ be the power products that are eliminated in the reductions of $h$ to $h_1$ and $h_2$, respectively. That is, there are polynomials $f_1, f_2 \in F$, coefficients $c_1 = \mathrm{coeff}(h, s_1) \neq 0$, $c_2 = \mathrm{coeff}(h, s_2) \neq 0$, and power products $t_1, t_2$ such that

$$s_1 = t_1 \mathrm{lpp}(f_1), \quad h_1 = h - \frac{c_1}{\mathrm{lc}(f_1)} t_1 f_1 \quad \text{and}$$

$$s_2 = t_2 \mathrm{lpp}(f_2), \quad h_2 = h - \frac{c_2}{\mathrm{lc}(f_2)} t_2 f_2 \ .$$

We distinguish two cases, depending on whether or not $s_1 = s_2$.

*Case* $s_1 \neq s_2$: w.l.o.g. assume $s_1 > s_2$. Let $a = \mathrm{coeff}(-(c_1/\mathrm{lc}(f_1))t_1 f_1, s_2)$. Then $\mathrm{coeff}(h_1, s_2) = c_2 + a$ and therefore

$$h_1 \longrightarrow_F h_1 - \frac{c_2 + a}{\mathrm{lc}(f_2)} t_2 f_2 = h - \frac{c_1}{\mathrm{lc}(f_1)} t_1 f_1 - \frac{c_2 + a}{\mathrm{lc}(f2)} t_2 f_2 \ .$$

On the other hand,

$$h_2 \longrightarrow_F h_2 - \frac{c_1}{\mathrm{lc}(f_1)} t_1 f_1 \longrightarrow_F h_2 - \frac{c_1}{\mathrm{lc}(f_1)} t_1 f_1 - \frac{a}{\mathrm{lc}(f_2)} t_2 f_2 =$$
$$h - \frac{c_1}{\mathrm{lc}(f_1)} t_1 f_1 - \frac{c_2 + a}{\mathrm{lc}(f2)} t_2 f_2 \ .$$

Thus, $h_1 \longleftrightarrow^*_{F(\ll h)} h_2$, in fact $h_1 \downarrow_F^* h_2$.

*Case* $s_1 = s_2$: let $s = s_1 = s_2$, $c = \mathrm{coeff}(h, s)$ and $h' = h - cs$. So for some power product $t$ we have $s = t \cdot \mathrm{lcm}(\mathrm{lpp}(f_1), \mathrm{lpp}(f_2))$, and $h_1 = h' + c \cdot t \cdot g_1$, $h_2 = h' + c \cdot t \cdot g_2$, where $(g_1, g_2) = \mathrm{cp}(f_1, f_2)$. By assumption $g_1 \downarrow_F^* g_2$, i.e., there are $p_1, \ldots, p_k$ and $q_1, \ldots, q_l$ such that

$$g_1 = p_1 \longrightarrow_F \ldots \longrightarrow_F p_k = q_l \longleftarrow_F \ldots \longleftarrow_F q_1 = g_2 \ .$$

So, by Lemma 8.2.5 (c),

$$ctg_1 = ctp_1 \longrightarrow_F \ldots \longrightarrow_F ctp_k = ctq_l \longleftarrow_F \ldots \longleftarrow_F ctq_1 = ctg_2 \ .$$

Applying Lemma 8.2.5 (d) we get

$$\begin{aligned} h_1 &= h' + ctp_1 \downarrow_F^* \ldots \downarrow_F^* h' + ctp_k \\ &= h' + ctq_l \downarrow_F^* \ldots \downarrow_F^* h' + ctq_1 = h_2 \ . \end{aligned}$$

All the intermediate polynomials in these reductions are less than $h$ w.r.t. $\ll$. Thus, $h_1 \longleftrightarrow^*_{F(\ll)} h_2$.

b. Every S-polynomial is congruent to 0 modulo $\langle F\rangle$. So by Theorem 8.2.6 $\mathrm{spol}(f, g) \longleftrightarrow^*_F 0$. If $F$ is a Gröbner basis, this implies $\mathrm{spol}(f, g) \longrightarrow^*_F 0$.

On the other hand, assume that $\mathrm{spol}(f, g) \longrightarrow^*_F 0$ for all $f, g \in F$. We use the same notation as in (a). In fact, the whole proof is analogous to the one for (a), except for the case $s_1 = s_2 = s$. So for $h_1 = h' + ctg_1 \longleftarrow_F h \longrightarrow_F h' + ctg_2 = h_2$ we have to show $h_1 \longleftrightarrow^*_{F(\ll h)} h_2$.

$g_1 - g_2$ is the S-polynomial of $f_1, f_2 \in F$, so by the assumption $g_1 - g_2 \longrightarrow^*_F 0$. By Lemma 8.2.5 also $h_1 - h_2 = ct(g_1 - g_2) \longrightarrow^*_F 0$, i.e., for some $p_1, \ldots, p_k$ we have

$$h_1 - h_2 = p_1 \longrightarrow_F \ldots \longrightarrow_F p_k = 0 .$$

Again by Lemma 8.2.5 we get

$$h_1 = p_1 + h_2 \downarrow^*_F \ldots \downarrow^*_F p_k + h_2 = h_2 ,$$

and therefore $h_1 \longleftrightarrow^*_{F(\ll h)} h_2$. □

Buchberger's theorem suggests an algorithm for checking whether a given finite basis is a Gröbner basis: reduce all the S-polynomials to normal forms and check whether they are all 0. In fact, by a simple extension we get an algorithm for constructing Gröbner bases.

**Algorithm GRÖBNER_B**(in: $F$; out: $G$);
[Buchberger algorithm for computing a Gröbner basis. $F$ is a finite subset of $K[X]^*$; $G$ is a finite subset of $K[X]^*$, such that $\langle G\rangle = \langle F\rangle$ and $G$ is a Gröbner basis.]

1. $G := F$;
   $C := \{\{g_1, g_2\} \mid g_1, g_2 \in G, g_1 \neq g_2\}$;
2. while not all pairs $\{g_1, g_2\} \in C$ are marked do
   {choose an unmarked pair $\{g_1, g_2\}$;
   mark $\{g_1, g_2\}$;
   $h :=$ normal form of $\mathrm{spol}(g_1, g_2)$ w.r.t. $\longrightarrow_G$;
   if $h \neq 0$
   then $\{C := C \cup \{\{g, h\} \mid g \in G\}$;
   $G := G \cup \{h\}\}$;
   };
   return.

Every polynomial $h$ constructed in GRÖBNER_B is in $\langle F\rangle$, so $\langle G\rangle = \langle F\rangle$ throughout GRÖBNER_B. Thus, by Theorem 8.3.1 GRÖBNER_B yields a correct result if it stops. The termination of GRÖBNER_B is a consequence of Dickson's lemma (Dickson 1913) which implies that in $[X]$ there is no infinite chain of elements $s_1, s_2, \ldots$ such that $s_i \nmid s_j$ for all $1 \leq i < j$. The leading power products of the polynomials added to the basis form such a sequence in $[X]$, so this sequence must be finite.

**Theorem 8.3.2** (Dickson's lemma). Every $A \subseteq [X]$ contains a finite subset $B$, such that every $t \in A$ is a multiple of some $s \in B$.

*Proof.* We proceed by induction on the number of variables $n$, where $X = \{x_1, \ldots, x_n\}$. For $n = 1$ the statement obviously holds. So let us assume that $n > 1$. We choose any element $t_0 \in A$, say

$$t_0 = x_1^{e_1} \ldots x_n^{e_n} .$$

Now for any $i \in \{1, \ldots, n\}$, $j \in \{0, \ldots, e_i\}$ we consider the set of power products

$$A_{i,j} = \{t \mid t \in A \text{ and } \deg_{x_i}(t) = j\} .$$

Let

$$A'_{i,j} = \{t/x_i^j \mid t \in A_{i,j}\} .$$

The variable $x_i$ does not occur in the elements of $A'_{i,j}$ any more. By the induction hypothesis, there exist finite subsets $B'_{i,j} \subseteq A'_{i,j}$ such that every power product in $A'_{i,j}$ is a multiple of some power product in $B'_{i,j}$. We define

$$B_{i,j} = \{t \cdot x_i^j \mid t \in B'_{i,j}\} .$$

Now every element of $A$ is a multiple of some element of the finite set

$$B = \{t_0\} \cup \bigcup_{i,j} B_{i,j} \subseteq A . \qquad \square$$

The termination of GRÖBNER_B also follows from Hilbert's basis theorem applied to the initial ideals of the sets $G$ constructed in the course of the algorithm, i.e., $\langle \mathrm{in}(G) \rangle$. See Exercise 4.

The algorithm GRÖBNER_B provides a constructive proof of the following theorem.

**Theorem 8.3.3.** Every ideal $I$ in $K[X]$ has a Gröbner basis.

*Example 8.3.1.* Let $F = \{f_1, f_2\}$, with $f_1 = x^2y^2 + y - 1$, $f_2 = x^2y + x$. We compute a Gröbner basis of $\langle F \rangle$ in $\mathbb{Q}[x, y]$ w.r.t. the graduated lexicographic ordering with $x < y$. The following describes one way in which the algorithm GRÖBNER_B could execute (recall that there is a free choice of pairs in the loop):

1. $\mathrm{spol}(f_1, f_2) = f_1 - yf_2 = -xy + y - 1 =: f_3$ is irreducible, so $G := \{f_1, f_2, f_3\}$.
2. $\mathrm{spol}(f_2, f_3) = f_2 + xf_3 = xy \longrightarrow_{f_3} y - 1 =: f_4$, so $G := \{f_1, f_2, f_3, f_4\}$.
3. $\mathrm{spol}(f_3, f_4) = f_3 + xf_4 = y - x - 1 \longrightarrow_{f_4} -x =: f_5$, so $G := \{f_1, \ldots, f_5\}$.

All the other S-polynomials now reduce to 0, so GRÖBNER_B terminates with

$$G = \{x^2y^2 + y - 1, x^2y + x, -xy + y - 1, y - 1, -x\} .$$

In addition to the original definition and the ones given in Theorem 8.3.1, there are many other characterizations of Gröbner bases. We list only a few of them.

**Theorem 8.3.4.** Let $I$ be an ideal in $K[X]$, $F \subseteq K[x]$, and $\langle F \rangle = I$. Then the following are equivalent.
a. $F$ is a Gröbner basis for $I$.
b. $f \longrightarrow_F^* 0$ for every $f \in I$.
c. $f \longrightarrow_F$ for every $f \in I \setminus \{0\}$.
d. For all $g \in I, h \in K[X]$: if $g \longrightarrow_F^* \underline{h}$ then $h = 0$.
e. For all $g, h_1, h_2 \in K[X]$: if $g \longrightarrow_F^* \underline{h_1}$ and $g \longrightarrow_F^* \underline{h_2}$ then $h_1 = h_2$.
f. $\langle \mathrm{in}(F) \rangle = \langle \mathrm{in}(I) \rangle$.

*Proof.* The equivalence of (a), (b), (c), (d), (e) is left to the reader.

Suppose $F$ satisfies (c). Then every non-zero $g$ in $I$ can be reduced, so $\mathrm{in}(g) \in \langle \mathrm{in}(F) \rangle$. Thus (c) implies (f). On the other hand, suppose that $F$ satisfies (f). Let $g \in I \setminus \{0\}$. Then $\mathrm{in}(g) \in \langle \mathrm{in}(F) \rangle$, so $g \longrightarrow_F$. Thus, (f) implies (c). □

The Gröbner basis $G$ computed in Example 8.3.1 is much too complicated. In fact, $\{y - 1, x\}$ is a Gröbner basis for the ideal. There is a general procedure for simplifying Gröbner bases.

**Theorem 8.3.5.** Let $G$ be a Gröbner basis for an ideal $I$ in $K[X]$. Let $g, h \in G$ and $g \neq h$.
a. If $\mathrm{lpp}(g) | \mathrm{lpp}(h)$ then $G' = G \setminus \{h\}$ is also a Gröbner basis for $I$.
b. If $h \longrightarrow_g h'$ then $G' = (G \setminus \{h\}) \cup \{h'\}$ is also a Gröbner basis for $I$.

*Proof.* a. Certainly $\langle G' \rangle \subseteq I$. For $f \in I$ we have $f \longrightarrow_G^* 0$, but in fact we have $f \longrightarrow_{G'}^* 0$, because whenever we could reduce by $h$ we can instead reduce by $g$.

b. $\langle G' \rangle = \langle G \rangle$. If $\mathrm{lpp}(h)$ is reduced then the result follows from (a). Otherwise $\langle \mathrm{in}(G') \rangle = \langle \mathrm{in}(G) \rangle = \langle \mathrm{in}(I) \rangle$. □

Observe that the elimination of basis polynomials described in Theorem 8.3.5 (a) is only possible if $G$ is a Gröbner basis. In particular, we are not allowed to do this during a Gröbner basis computation. Based on Theorem 8.3.5 we can show that every ideal has a unique Gröbner basis after suitable pruning and normalization.

*Definition 8.3.2.* Let $G$ be a Gröbner basis in $K[X]$.
– $G$ is *minimal* iff $\mathrm{lpp}(g) \not| \mathrm{lpp}(h)$ for all $g, h \in G$ with $g \neq h$.

- $G$ is *reduced* iff for all $g, h \in G$ with $g \neq h$ we cannot reduce $h$ by $g$.
- $G$ is *normed* iff $\mathrm{lc}(g) = 1$ for all $g \in G$.

From Theorem 8.3.5 we obviously get an algorithm for transforming any Gröbner basis for an ideal $I$ into a normed reduced Gröbner basis for $I$. No matter from which Gröbner basis of $I$ we start and which path we take in this transformation process, we always reach the same uniquely defined normed reduced Gröbner basis of $I$.

**Theorem 8.3.6.** Every ideal in $K[X]$ has a unique finite normed reduced Gröbner basis.

*Proof.* The existence of such a basis follows from Theorem 8.3.5. Now suppose that $G$ and $G'$ are two such normed reduced Gröbner bases for the ideal $I$. Let

$$G = \{g_1, \ldots, g_m\}, \qquad G' = \{g'_1, \ldots, g'_{m'}\} .$$

So $g_1 \longrightarrow^*_{G'} 0$, in particular $\mathrm{lpp}(g_1)$ can be reduced by some polynomial in $G'$, w.l.o.g. let $g'_1$ be this polynomial, i.e., $\mathrm{lpp}(g'_1) | \mathrm{lpp}(g_1)$. Also $g'_1 \longrightarrow^*_G 0$ and therefore $\mathrm{lpp}(g_k) | \mathrm{lpp}(g'_1)$ for some $k$. Since $G$ is reduced, this is possible only for $k = 1$, i.e., $\mathrm{lpp}(g_1) = \mathrm{lpp}(g'_1)$. Proceeding in this way we obtain $m = m'$ and $\mathrm{lpp}(g_i) = \mathrm{lpp}(g'_i)$ for all $1 \leq i \leq m$ (possibly after reordering the elements of $G'$).

Now consider any $g_i$. We have $g_i \longrightarrow^*_{G'} 0$. Suppose $g_i \neq g'_i$. The only way to eliminate $\mathrm{lpp}(g_i)$ is to use $g'_i$. But $g_i - g'_i \neq 0$ and none of the power products in $g_i - g'_i$ can be reduced modulo $G'$. So $g_i$ cannot be reduced to 0 by $G'$, which is a contradiction. Hence, $g_i = g'_i$ for all $1 \leq i \leq m$. □

Observe that the normed reduced Gröbner basis of an ideal $I$ depends, of course, on the admissible ordering $<$. Different orderings can give rise to different Gröbner bases. However, if we decompose the set of all admissible orderings into sets which induce the same normed reduced Gröbner basis of a fixed ideal $I$, then this decomposition is finite. This leads to the consideration of universal Gröbner bases. A universal Gröbner basis for $I$ is a basis for $I$ which is a Gröbner basis w.r.t. any admissible ordering of the power products. See Mora and Robbiano (1988) and Weispfenning (1989).

If we have a Gröbner basis $G$ for an ideal $I$, then we can compute in the vector space $K[X]_{/I}$ over $K$. The irreducible power products (with coefficient 1) modulo $G$ form a basis of $K[X]_{/I}$. We get that $\dim(K[X]_{/I})$ is the number of irreducible power products modulo $G$. Thus, this number is independent of the particular admissible ordering.

*Example 8.3.2.* Let $I = \langle x^3y - 2y^2 - 1, x^2y^2 + x + y\rangle$ in $\mathbb{Q}[x, y]$. Let $<$ be the graduated lexicographic ordering with $x > y$. Then the normed reduced Gröbner basis of $I$ has leading power products $x^4, x^3y, x^2y^2, y^3$. So there are 9 irreducible power products.

If $<$ is the lexicographic ordering with $x > y$, then the normed reduced

Gröbner basis of $I$ has leading power products $x$ and $y^9$. So again there are 9 irreducible power products.

In fact, $\dim(\mathbb{Q}[x, y]_{/I}) = 9$.

### Exercises

1. Complete the proof of Theorem 8.3.4.
2. Compute the normed reduced Gröbner basis w.r.t. the lexicographic ordering with $x < y$ for the ideal generated by $f_1 = xy^2 + x^2 + x$, $f_2 = x^2y + x$ in $\mathbb{Z}_3[x, y]$.
3. Compute the normed reduced Gröbner basis $G$ for the ideal

$$I = \langle xz - 3x^2 + x + 6x^3 + 1, y^2 + x^2 - 2, x^5 - 6x^3 + x^2 - 1\rangle$$

   in $\mathbb{Q}[x, y, z]$ w.r.t. the lexicographic ordering with $x < y < z$. What is $\dim\mathbb{Q}[x, y, z]_{/I}$?
4. Prove the termination of GRÖBNER_B by Hilbert's basis theorem applied to the *initial ideals* of the bases $G$, i.e., $\langle \mathrm{in}(G)\rangle$, generated in the course of the algorithm. Observe that these initial ideals are homogeneous, i.e., they are generated by a homogeneous basis and with every polynomial $f = f_d + f_{d-1} + \ldots + f_0$ they also contain every form $f_i$ of $f$, where every power product occurring in $f_i$ has degree $i$.

## 8.4 Applications of Gröbner bases

Computation in the vector space of polynomials modulo an ideal

The ring $K[X]_{/I}$ of polynomials modulo the ideal $I$ is a vector space over $K$. A Gröbner basis $G$ provides a basis for this vector space.

**Theorem 8.4.1.** The irreducible power products modulo $G$, viewed as polynomials with coefficient 1, form a basis for the vector space $K[X]_{/I}$ over $K$.

*Proof.* Let $B$ be the set of irreducible power products modulo $G$, viewed as polynomials with coefficient 1. Clearly $B$ generates $K[X]_{/I}$, since every polynomial can be reduced to a linear combination of elements of $B$ with coefficients in $K$. Moreover, $B$ is linearly independent, because any nontrivial linear combination of elements of $B$ is irreducible modulo $G$ and therefore different from 0 in $K[X]_{/I}$. □

Ideal membership

By definition Gröbner bases solve the *ideal membership problem* for polynomial ideals, i.e.,
given: $f, f_1, \ldots, f_m \in K[X]$,
decide: $f \in \langle f_1, \ldots, f_m\rangle$.

Let $G$ be a Gröbner basis for $I = \langle f_1, \ldots, f_m \rangle$. Then $f \in I$ if and only if the normal form of $f$ modulo $G$ is 0.

*Example 8.4.1* Suppose that we know the polynomial relations (axioms)

$$\begin{aligned} 4z - 4xy^2 - 16x^2 - 1 &= 0 , \\ 2y^2 z + 4x + 1 &= 0 , \\ 2x^2 z + 2y^2 + x &= 0 \end{aligned}$$

between the quantities $x, y, z$, and we want to decide whether the additional relation (hypothesis)

$$g(x, y) = 4xy^4 + 16x^2 y^2 + y^2 + 8x + 2 = 0$$

follows from them, i.e., whether we can write $g$ as a linear combination of the axioms or, in other words, whether $g$ is in the ideal $I$ generated by the axioms.

Trying to reduce the hypothesis $g$ w.r.t. the given axioms does not result in a reduction to 0. But we can compute a Gröbner basis for $I$ w.r.t. the lexicographic ordering with $x < y < z$, e.g., $G = \{g_1, g_2, g_3\}$ where

$$\begin{aligned} g_1 &= 32x^7 - 216x^6 + 34x^4 - 12x^3 - x^2 + 30x + 8 , \\ g_2 &= 2745y^2 - 112x^6 - 812x^5 + 10592x^4 - 61x^3 - 812x^2 + 988x + 2 , \\ g_3 &= 4z - 4xy^2 - 16x^2 - 1 . \end{aligned}$$

Now $g \longrightarrow_G^* 0$, i.e., $g(x, y) = 0$ follows from the axioms.

### Radical membership

Sometimes, especially in applications in geometry, we are not so much interested in the ideal membership problem but in the *radical membership problem*, i.e.,
given: $f, f_1, \ldots, f_m \in K[X]$,
decide: $f \in \text{radical}(\langle f_1, \ldots, f_m \rangle)$.

The radical of an ideal $I$ is the ideal containing all those polynomials $f$, some power of which is contained in $I$. So $f \in \text{radical}(I) \iff f^n \in I$ for some $n \in \mathbb{N}$. Geometrically $f \in \text{radical}(\langle f_1, \ldots, f_m \rangle)$ means that the hypersurface defined by $f$ contains all the points in the variety (algebraic set) defined by $f_1, \ldots, f_m$.

The following extremely important theorem relates the radical of an ideal $I$ to the set of common roots $V(I)$ of the polynomials contained in $I$.

**Theorem 8.4.2** (Hilbert's Nullstellensatz). Let $I$ be an ideal in $K[X]$, where $K$ is an algebraically closed field. Then radical$(I)$ consists of exactly those polynomials in $K[X]$ which vanish on all the common roots of $I$.

*Proof.* A proof of Hilbert's Nullstellensatz can be found in any introductory book on commutative algebra and algebraic geometry, e.g., in Cox et al. (1992). □

By an application of Hilbert's Nullstellensatz we get that $f \in \text{radical}(\langle f_1, \ldots, f_m \rangle)$ if and only if $f$ vanishes at every common root of $f_1, \ldots, f_m$ if and only if the system $f_1 = \ldots f_m = z \cdot f - 1 = 0$ has no solution, where $z$ is a new variable. That is,

$$f \in \text{radical}(\langle f_1, \ldots, f_m \rangle) \Longleftrightarrow 1 \in \langle f_1, \ldots, f_m, z \cdot f - 1 \rangle \ .$$

So the radical membership problem is reduced to the ideal membership problem.

Equality of ideals

We want to decide whether two given ideals are equal, i.e., we want to solve the *ideal equality problem*:
given: $f_1, \ldots, f_m, g_1, \ldots, g_k \in K[X]$,
decide: $\underbrace{\langle f_1, \ldots, f_m \rangle}_{I} = \underbrace{\langle g_1, \ldots, g_k \rangle}_{J}$.

Choose any admissible ordering. Let $G_I, G_J$ be the normed reduced Gröbner bases of $I$ and $J$, respectively. Then by Theorem 8.3.6 $I = J$ if and only if $G_I = G_J$.

Solution of algebraic equations

We consider a system of equations

$$\begin{aligned} f_1(x_1, \ldots, x_n) &= 0 \ , \\ &\vdots \\ f_m(x_1, \ldots, x_n) &= 0 \ , \end{aligned} \tag{8.4.1}$$

where $f_1, \ldots, f_m \in K[X]$. The system (8.4.1) is called a system of polynomial or algebraic equations. First let us decide whether (8.4.1) has any solutions in $\bar{K}^n$, $\bar{K}$ being the algebraic closure of $K$. Let $I = \langle f_1, \ldots, f_m \rangle$. The following theorem has first been proved in Buchberger (1970).

**Theorem 8.4.3.** Let $G$ be a normed Gröbner basis of $I$. (8.4.1) is unsolvable in $\bar{K}^n$ if and only if $1 \in G$.

*Proof.* If $1 \in G$ then $1 \in \langle G \rangle = I$, so every solution of (8.4.1) is also a solution of $1 = 0$. So there can be no solution.

On the other hand, assume that (8.4.1) is unsolvable. Then the polynomial 1 vanishes on every common root of (8.4.1). So by Hilbert's Nullstellensatz $1 \in$

radical($I$) and therefore also $1 \in I$. Since $G$ is a normed Gröbner basis of $I$, we must have $1 \longrightarrow_G 0$. This is only possible if $1 \in G$. □

Now suppose that (8.4.1) is solvable. We want to determine whether there are finitely or infinitely many solutions of (8.4.1) or, in other words, whether or not the ideal $I$ is 0-dimensional.

**Theorem 8.4.4.** Let $G$ be a Gröbner basis of $I$. Then (8.4.1) has finitely many solutions (i.e., $I$ is 0-dimensional) if and only if for every $i$, $1 \leq i \leq n$, there is a polynomial $g_i \in G$ such that $\text{lpp}(g_i)$ is a pure power of $x_i$. Moreover, if $I$ is 0-dimensional then the number of zeros of $I$ (counted with multiplicity) is equal to $\dim(K[X]_{/I})$.

*Proof.* $I$ is 0-dimensional if and only if $K[X]_{/I}$ has finite vector space dimension over $K$ and in this case the number of solutions and the vector space dimension agree (see, e.g., Gröbner 1949). By Theorem 8.4.1 that is the case if and only if the number of irreducible power products modulo $G$ is finite, i.e., for every variable $x_i$ there is a pure power of it in $\text{lpp}(G)$. □

The role of the Gröbner basis algorithm GRÖBNER_B in solving systems of algebraic equations is the same as that of Gaussian elimination in solving systems of linear equations, namely to triangularize the system, or carry out the elimination process. The crucial observation, first stated in Trinks (1978), is the elimination property of Gröbner bases. It states that if $G$ is a Gröbner basis of $I$ w.r.t. the lexicographic ordering with $x_1 < \ldots < x_n$, then the $i$-th elimination ideal of $I$, i.e., $I \cap K[x_1, \ldots, x_i]$, is generated by those polynomials in $G$ that depend only on the variables $x_1, \ldots, x_i$.

**Theorem 8.4.5** (Elimination property of Gröbner bases). Let $G$ be a Gröbner basis of $I$ w.r.t. the lexicographic ordering $x_1 < \ldots < x_n$. Then

$$I \cap K[x_1, \ldots, x_i] = \langle G \cap K[x_1, \ldots, x_i] \rangle ,$$

where the ideal on the right-hand side is generated over the ring $K[x_1, \ldots, x_i]$.

*Proof.* Obviously the right-hand side is contained in the left-hand side.

On the other hand, let $f \in I \cap K[x_1, \ldots, x_i]$. Then $f \longrightarrow_G^* 0$. All the polynomials in this reduction depend only on the variables $x_1, \ldots, x_i$. So we get a representation of $f$ as a linear combination $\sum h_j g_j$, where $g_j \in G \cap K[x_1, \ldots, x_i]$ and $h_j \in K[x_1, \ldots, x_i]$. □

Theorem 8.4.5 can clearly be generalized to product orderings, without changing anything in the proof.

*Example 8.4.2.* Consider the system of equations $f_1 = f_2 = f_3 = 0$, where

$$4xz - 4xy^2 - 16x^2 - 1 = 0 \;,$$
$$2y^2z + 4x + 1 = 0 \;,$$
$$2x^2z + 2y^2 + x = 0 \;,$$

are polynomials in $\mathbb{Q}[x, y, z]$. We are looking for solutions of this system of algebraic equations in $\bar{\mathbb{Q}}^3$, where $\bar{\mathbb{Q}}$ is the field of algebraic numbers.

Let $<$ be the lexicographic ordering with $x < y < z$. The algorithm GRÖBNER_B applied to $F = \{f_1, f_2, f_3\}$ yields (after reducing the result) the reduced Gröbner basis $G = \{g_1, g_2, g_3\}$, where

$$g_1 = 65z + 64x^4 - 432x^3 + 168x^2 - 354x + 104 \;,$$
$$g_2 = 26y^2 - 16x^4 + 108x^3 - 16x^2 + 17x \;,$$
$$g_3 = 32x^5 - 216x^4 + 64x^3 - 42x^2 + 32x + 5 \;.$$

By Theorem 8.4.3 the system is solvable. Furthermore, by Theorem 8.4.4, the system has finitely many solutions. The Gröbner basis $G$ yields an equivalent triangular system in which the variables are completely separated. So we can get solutions by solving the univariate polynomial $g_3$ and propagating the partial solutions upwards to solutions of the full system. The univariate polynomial $g_3$ is irreducible over $\mathbb{Q}$, and the solutions are

$$\Big(\alpha,\; \pm\frac{1}{\sqrt{26}}\sqrt{\alpha}\sqrt{16\alpha^3 - 108\alpha^2 + 16\alpha - 17}\;,$$
$$-\frac{1}{65}(64\alpha^4 - 432\alpha^3 + 168\alpha^2 - 354\alpha + 104)\Big) \;,$$

where $\alpha$ is a root of $g_3$. We can also determine a numerical approximation of a solution from $G$, e.g.,

$$(-0.1284722871,\; 0.3211444930,\; -2.356700326) \;.$$

*Example 8.4.3.* The same method can be applied to algebraic equations with symbolic coefficients. For example consider the system $f_1 = f_2 = f_3 = f_4 = 0$, where

$$f_1 = x_4 + b - d \;,$$
$$f_2 = x_4 + x_3 + x_2 + x_1 - a - c - d \;,$$
$$f_3 = x_3x_4 + x_1x_4 + x_2x_3 - ad - ac - cd \;,$$
$$f_4 = x_1x_3x_4 - acd$$

are polynomials in the variables $x_1, x_2, x_3, x_4$ containing the parameters $a, b, c, d$, i.e., $f_1, \ldots, f_4 \in \mathbb{Q}(a, b, c, d)[x_1, \ldots, x_4]$. Let $<$ be the lexicographic ordering with $x_1 < x_2 < x_3 < x_4$. The normed reduced Gröbner basis of $\langle f_1, \ldots, f_4\rangle$ is $G = \{g_1, \ldots, g_4\}$, where

$$
\begin{aligned}
g_1 &= x_4 + b - d\ , \\
g_2 &= x_3 - \frac{b^2 - 2bd + d^2}{acd} x_1^2 - \\
&\quad - \frac{abc + abd - acd - ad^2 + bcd - cd^2}{acd} x_1 - a - c - d\ , \\
g_3 &= x_2 + \frac{b^2 - 2bd + d^2}{acd} x_1^2 + \frac{abc + abd - ad^2 + bcd - cd^2}{acd} x_1 - b + d\ , \\
g_4 &= x_1^3 + \frac{ac + ad + cd}{b - d} x_1^2 + \frac{a^2cd + ac^2d + acd^2}{(b-d)^2} x_1 + \frac{a^2c^2d^2}{(b-d)^3}\ .
\end{aligned}
$$

Thus, the system has finitely many solutions. A particular root of $g_4$ is $-ad/(b-d)$, which can be continued to the solution

$$\left(-\frac{ad}{b-d},\ \frac{ab + b^2 - bd}{b-d},\ c,\ -b + d\right)\ .$$

Every minimal Gröbner basis $G$ of a 0-dimensional ideal $I$ w.r.t. the lexicographic ordering with $x_1 < \ldots < x_n$ has the form

$$
\begin{aligned}
g_{1,1}(x_1) &\in K[x_1]\ , \\
g_{2,1}(x_1, x_2), \ldots, g_{2,k_2}(x_1, x_2) &\in K[x_1, x_2]\ , \\
&\vdots \\
g_{n,1}(x_1, \ldots, x_n), \ldots, g_{n,k_n}(x_1, \ldots, x_n) &\in K[x_1, \ldots, x_n]\ .
\end{aligned}
$$

B. Roider (1986), M. Kalkbrener (1987) and P. Gianni (1987) have proved that if $b = (b_1, \ldots, b_{l-1}) \in \bar{K}^{l-1}$ is a solution of $g_{1,1} = \ldots = g_{l-1,k_{l-1}} = 0$, then for some index $j$, $1 \le j \le k_l$, we have

$$
\begin{aligned}
&\mathrm{lc}_{x_l}(g_{l,j})(b) \neq 0 \quad \text{(leading coefficient w.r.t. the variable } x_l\text{)} \quad \text{and} \\
&g_{l,j}(b, x_l) = \gcd(g_{l,1}(b, x_l), \ldots, g_{l,k_l}(b, x_l))\ .
\end{aligned}
$$

This implies that every partial solution $b$ can be continued to a solution of the full system.

In fact, for a 0-dimensional ideal $I$ in regular position a very strong structure theorem has been derived by Gianni and Mora (1987). $I$ is in *regular position* w.r.t. the variable $x_1$, if $a_1 \neq b_1$ for any two different zeros $(a_1, \ldots, a_n)$, $(b_1, \ldots, b_n)$ of $I$. Clearly it is very likely that an arbitrary 0-dimensional ideal is in regular position w.r.t. $x_1$. Otherwise, nearly every linear change of coordinates will make the ideal regular.

**Theorem 8.4.6** (Shape lemma). Let $I$ be a radical 0-dimensional ideal in $K[X]$, regular in $x_1$. Then there are $g_1(x_1), \ldots, g_n(x_1) \in K[x_1]$ such that $g_1$ is squarefree, $\deg(g_i) < \deg(g_1)$ for $i > 1$ and the normed reduced Gröbner basis $F$ for $I$ w.r.t. the lexicographic ordering $<$ with $x_1 < \cdots < x_n$ is of the form

$$\{g_1(x_1), x_2 - g_2(x_1), \ldots, x_n - g_n(x_1)\} .$$

On the other hand, if the normed reduced Gröbner basis for $I$ w.r.t. $<$ is of this form, then $I$ is a radical 0-dimensional ideal.

*Proof.* Since $I$ is in regular position, the first coordinates of zeros of $I$ are all different, say $a_{11}, \ldots, a_{1m}$. Then the squarefree polynomial $g_1(x_1) = \prod_{i=1}^{m}(x_1 - a_{1i})$ is in $I \cap K[x_1]$ and so it has to be in $F$. Since by Theorem 8.4.4 $m$ is the dimension of $K[X]_{/I}$, the normed reduced Gröbner basis for $I$ has to have the specified form.

To prove the converse, let $a_{11}, \ldots, a_{1m}$ be the zeros of $g_1(x_1)$. Then the zeros of $I$ are $\{(a_{1i}, g_2(a_{1i}), \ldots, g_n(a_{1i}) \mid i = 1, \ldots, m\}$. □

For a further discussion of the Shape lemma we refer to Becker et al. (1994).

Linear equations over $K[X]$

For given polynomials $f_1, \ldots, f_s, f$ in $K[X]$ we consider the linear equation

$$f_1 z_1 + \ldots + f_s z_s = f , \tag{8.4.2}$$

or the corresponding homogeneous equation

$$f_1 z_1 + \ldots + f_s z_s = 0 . \tag{8.4.3}$$

Let $F$ be the vector $(f_1, \ldots, f_s)$. The general solution of (8.4.2) and (8.4.3) is to be sought in $K[X]^s$. The solutions of (8.4.3) form a module over the ring $K[X]$, a submodule of $K[X]^s$ over $K[X]$.

*Definition 8.4.1.* Any solution of (8.4.3) is called a *syzygy* of the sequence of polynomials $f_1, \ldots, f_s$. The module of all solutions of (8.4.3) is the *module of syzygies* $\mathrm{Syz}(F)$ of $F = (f_1, \ldots, f_s)$.

It turns out that if the coefficients of this equation are a Gröbner basis, then we can immediately write down a generating set (basis) for the module $\mathrm{Syz}(F)$. The general case will be reduced to this one.

**Theorem 8.4.7.** If the elements of $F = (f_1, \ldots, f_s)$ are a Gröbner basis, then $S$ is a basis for $\mathrm{Syz}(F)$, where $S$ is defined as follows.

For $1 \le i \le s$ let $e_i = (0, \ldots, 0, 1, 0, \ldots, 0)$ be the $i$-th unit vector and for $1 \le i < j \le s$ let

$$t = \mathrm{lcm}(\mathrm{lpp}(f_i), \mathrm{lpp}(f_j)) ,$$
$$p_{ij} = \frac{1}{\mathrm{lc}(f_i)} \cdot \frac{t}{\mathrm{lpp}(f_i)}, \qquad q_{ij} = \frac{1}{\mathrm{lc}(f_j)} \cdot \frac{t}{\mathrm{lpp}(f_j)} ,$$

and $k_{ij}^1, \ldots, k_{ij}^s$ be the polynomials extracted from a reduction of $\mathrm{spol}(f_i, f_j)$ to 0, such that

$$\mathrm{spol}(f_i, f_j) = p_{ij} f_i - q_{ij} f_j = \sum_{l=1}^{s} k_{ij}^l f_l .$$

Then

$$S = \{\underbrace{p_{ij} \cdot e_i - q_{ij} \cdot e_j - (k_{ij}^1, \ldots, k_{ij}^s)}_{S_{ij}} \mid 1 \le i < j \le s\} .$$

*Proof.* Obviously every element of $S$ is a syzygy of $F$, since every S-polynomial reduces to 0.

On the other hand, let $z = (z_1, \ldots, z_s) \neq (0, \ldots, 0)$ be an arbitrary non-trivial syzygy of $F$. Let $p$ be the highest power product occurring in

$$f_1 z_1 + \ldots + f_s z_s = 0 , \tag{$*$}$$

i.e.,

$$p = \max_{<}\{t \in [X] \mid \mathrm{coeff}(f_i \cdot z_i, t) \neq 0 \ \text{ for some } \ i\}$$

and let $i_1 < \ldots < i_m$ be those indices such that $\mathrm{lpp}(f_{i_j} \cdot z_{i_j}) = p$. We have $m \ge 2$. Suppose that $m > 2$. By subtracting a suitable multiple of $S_{i_{k-1}, i_k}$ from $z$, we can reduce the number of positions in $z$ that contribute to the highest power product $p$ in $(*)$. Iterating this process $k - 2$ times, we finally reach a situation, where only two positions $i_1, i_2$ in the syzygy contribute to the power product $p$. Now the highest power product in $(*)$ can be decreased by subtracting a suitable multiple of $S_{i_1, i_2}$. Since $<$ is Noetherian, this process terminates, leading to an expression of $z$ as a linear combination of elements of $S$. □

Now that we are able to solve homogeneous linear equations in which the coefficients are a Gröbner basis, let us see how we can transform the general case to this one.

**Theorem 8.4.8.** Let $F = (f_1, \ldots, f_s)^T$ be a vector of polynomials in $K[X]$ and let the elements of $G = (g_1, \ldots, g_m)^T$ be a Gröbner basis for $\langle f_1, \ldots, f_s \rangle$. We view $F$ and $G$ as column vectors. Let the $r$ rows of the matrix $R$ be a basis for $\mathrm{Syz}(G)$ and let the matrices $A$, $B$ be such that $G = A \cdot F$ and $F = B \cdot G$.

Then the rows of $Q$ are a basis for Syz$(F)$, where

$$Q = \begin{pmatrix} I_s - B \cdot A \\ \cdots\cdots\cdots\cdots \\ R \cdot A \end{pmatrix} .$$

*Proof.* Let $b_1, \ldots, b_{s+r}$ be polynomials, $b = (b_1, \ldots, b_{s+r})$.

$$\begin{aligned}
&(b \cdot Q) \cdot F \\
&= ((b_1, \ldots, b_s) \cdot (I_s - B \cdot A) + (b_{s+1}, \ldots, b_{s+r}) \cdot R \cdot A) \cdot F \\
&= (b_1, \ldots, b_s) \cdot (F - \underbrace{B \cdot A \cdot F}_{=F}) + (b_{s+1}, \ldots, b_{s+r}) \cdot R \cdot \underbrace{A \cdot F}_{=G} = 0 .
\end{aligned}$$

So every linear combination of the rows of $Q$ is a syzygy of $F$.

On the other hand, let $H = (h_1, \ldots, h_s)$ be a syzygy of $F$. Then $H \cdot B$ is a syzygy of $G$. So for some $H'$ we can write $H \cdot B = H' \cdot R$, and therefore $H \cdot B \cdot A = H' \cdot R \cdot A$. Thus,

$$H = H \cdot (I_s - B \cdot A) + H' \cdot R \cdot A = (H, H') \cdot Q ,$$

i.e., $H$ is a linear combination of the rows of $Q$. □

What we still need is a particular solution of the inhomogeneous equation (8.4.2). Let $G = (g_1, \ldots, g_m)$ be a Gröbner basis for $\langle F \rangle$ and let $A$ be the transformation matrix such that $G = A \cdot F$ ($G$ and $F$ viewed as column vectors). Then a particular solution of (8.4.2) exists if and only if $f \in \langle F \rangle = \langle G \rangle$. If the reduction of $f$ to normal form modulo $G$ yields $f' \neq 0$, then (8.4.2) is unsolvable. Otherwise we can extract from this reduction polynomials $h'_1, \ldots, h'_m$ such that

$$g_1 h'_1 + \ldots + g_m h'_m = f .$$

So $H = (h'_1, \ldots, h'_m) \cdot A$ is a particular solution of (8.4.2).

Of course, once we are able to solve single linear equations over $K[X]$, we can also solve systems of linear equations by dealing with the equations recursively. An algorithm along these lines is presented in Winkler (1986). However, it is also possible to extend the concept of Gröbner bases from ideals to modules (see Furukawa et al. 1986, Mora and Möller 1986) and solve a whole system of linear equations by a single computation of a Gröbner basis for a submodule of $K[X]^s$.

*Example 8.4.4.* Consider the linear equation

$$\underbrace{\left( xz - xy^2 - 4x^2 - \frac{1}{4} \quad y^2 z + 2x + \frac{1}{2} \quad x^2 z + y^2 + \frac{1}{2} x \right)}_{F} \begin{pmatrix} z_1 \\ z_2 \\ z_3 \end{pmatrix} = 0 ,$$

where the coefficients are in $\mathbb{Q}[x, y, z]$. A basis for the syzygies can be computed as the rows of a matrix $Q$ according to Theorem 8.4.8. $Q^T$ may contain for instance the syzygy

$$\begin{pmatrix} z_1 \\ z_2 \\ z_3 \end{pmatrix} =$$
$$\begin{pmatrix} 2xy^2 + 4x^2y^4 + 2x^3y^2 + 4y^4 - 2x^4 - 8x^3 - 2x^2 - 8x^5 \\ -8x^3y^2 - 4x^5y^2 - 4xy^2 - 3x^2 - 19x^4 - 16x^6 \\ y^2 + 17x^2y^2 + 16x^4y^2 + 4x^3y^4 + 4xy^4 + 8x^4 + 2x^3 + 8x^2 + 2x \end{pmatrix}.$$

In fact, using the concept of Gröbner bases for modules, we get the following basis for $\mathrm{Syz}(F)$:

$$\begin{pmatrix} y^2z + 2x + \frac{1}{2} \\ -xz + xy^2 + 4x^2 + \frac{1}{4} \\ 0 \end{pmatrix}, \begin{pmatrix} x^2z + y^2 + \frac{1}{2}x \\ 0 \\ -xz + xy^2 + 4x^2 + \frac{1}{4} \end{pmatrix},$$
$$\begin{pmatrix} y^4 + \frac{1}{2}xy^2 - 2x^3 - \frac{1}{2}x^2 \\ -x^3y^2 - xy^2 - 4x^4 - \frac{3}{4}x^2 \\ xy^4 + 4x^2y^2 + \frac{1}{4}y^2 + 2x^2 + \frac{1}{2}x \end{pmatrix}, \begin{pmatrix} 0 \\ x^2z + y^2 + \frac{1}{2}x \\ -y^2z - 2x - \frac{1}{2} \end{pmatrix}.$$

### Effective ideal theoretic operations

In commutative algebra and algebraic geometry there is a strong correspondence between radical polynomial ideals and algebraic sets, the sets of zeros of such ideals over the algebraic closure of the field of coefficients. For any ideal $I$ in $K[x_1, \ldots, x_n]$ we denote by $V(I)$ the set of all points in $\mathbb{A}^n(\bar{K})$, the $n$-dimensional affine space over the algebraic closure of $K$, which are common zeros of all the polynomials in $I$. Such sets $V(I)$ are called *algebraic sets*. On the other hand, for any subset $V$ of $\mathbb{A}^n(\bar{K})$ we denote by $I(V)$ the ideal of all polynomials vanishing on $V$. Then for radical ideals $I$ and algebraic sets $V$ the functions $V(\cdot)$ and $I(\cdot)$ are inverses of each other, i.e.,

$$V(I(V)) = V \quad \text{and} \quad I(V(I)) = I .$$

This correspondence extends to operations on ideals and algebraic sets in the following way:

| ideal | algebraic set | |
|---|---|---|
| $I + J$ | $V(I) \cap V(J)$ | |
| $I \cdot J,\ I \cap J$ | $V(I) \cup V(J)$ | |
| $I : J$ | $V(I) - V(J) = \overline{V(I) - V(J)}$ | (Zariski closure of the difference) |

See, for instance, Cox et al. (1992: chap. 4). So we can effectively compute intersection, union, and difference of varieties if we can carry out the corresponding operations on ideals.

*Definition 8.4.2.* Let $I$, $J$ be ideals in $K[X]$.

- The *sum* $I + J$ of $I$ and $J$ is defined as $I + J = \{f + g \mid f \in I, g \in J\}$.
- The *product* $I \cdot J$ of $I$ and $J$ is defined as $I \cdot J = \langle\{f \cdot g \mid f \in I, g \in J\}\rangle$.
- The *quotient* $I : J$ of $I$ and $J$ is defined as $I : J = \{f \mid f \cdot g \in I$ for all $g \in J\}$.

**Theorem 8.4.9.** Let $I = \langle f_1, \ldots, f_r\rangle$ and $J = \langle g_1, \ldots, g_s\rangle$ be ideals in $K[X]$.

a. $I + J = \langle f_1, \ldots, f_r, g_1, \ldots, g_s\rangle$.
b. $I \cdot J = \langle f_i g_j \mid 1 \leq i \leq r, 1 \leq j \leq s\rangle$.
c. $I \cap J = (\langle t\rangle \cdot I + \langle 1 - t\rangle \cdot J) \cap K[X]$, where $t$ is a new variable.
d. $I : J = \bigcap_{j=1}^{s}(I : \langle g_j\rangle)$ and
$I : \langle g\rangle = \langle h_1/g, \ldots, h_m/g\rangle$, where $I \cap \langle g\rangle = \langle h_1, \ldots, h_m\rangle$.

*Proof.* (a) and (b) are easily seen.

c. Let $f \in I \cap J$. Then $tf \in \langle t\rangle \cdot I$ and $(1 - t)f \in \langle t - 1\rangle \cdot J$. Therefore $f = tf + (1 - t)f \in \langle t\rangle \cdot I + \langle 1 - t\rangle \cdot J$.

On the other hand, let $f \in (\langle t\rangle \cdot I + \langle 1 - t\rangle \cdot J) \cap K[X]$. So $f = g(X, t) + h(X, t)$, where $g \in \langle t\rangle I$ and $h \in \langle 1 - t\rangle J$. In particular, $h(X, t)$ is a linear combination of the basis elements $(1-t)g_1, \ldots, (1-t)g_s$ of $\langle 1-t\rangle J$. Evaluating $t$ at 0 we get

$$f = g(X, 0) + h(X, 0) = h(X, 0) \in J \ .$$

Similarly, by evaluating $t$ at 1 we get $f = g(X, 1) \in I$.

d. $h \in I : J$ if and only if $hg \in I$ for all $g \in J$ if and only if $hg_j \in I$ for all $1 \leq j \leq s$ if and only if $h \in I : \langle g_j\rangle$ for all $1 \leq j \leq s$.

If $f \in \langle h_1/g, \ldots, h_m/g\rangle$ and $a \in \langle g\rangle$ then $af \in \langle h_1, \ldots, h_m\rangle = I \cap \langle g\rangle \subset I$, i.e., $f \in I : \langle g\rangle$. Conversely, suppose $f \in I : \langle g\rangle$. Then $fg \in I \cap \langle g\rangle$. So $fg = \sum b_k h_k$ for some $b_k \in K[X]$. Thus,

$$f = \sum b_k \cdot (\underbrace{h_k/g}_{\text{polynomial}}) \in \langle h_1/g, \ldots, h_m/g\rangle \ . \qquad \square$$

So all these operations can be carried out effectively by operations on the bases of the ideals. In particular the intersection can be computed by Theorem 8.4.5.

We always have $I \cdot J \subset I \cap J$. However, $I \cap J$ could be strictly larger than $I \cdot J$. For example, if $I = J = \langle x, y\rangle$, then $I \cdot J = \langle x^2, xy, y^2\rangle$ and $I \cap J = I = J = \langle x, y\rangle$. Both $I \cdot J$ and $I \cap J$ correspond to the same variety. Since a basis for $I \cdot J$ is more easily computed, why should we bother with $I \cap J$? The reason is that the intersection behaves much better with respect to the operation of taking radicals (recall that it is really the radical ideals that uniquely correspond to algebraic sets). Whereas the product of radical ideals in general fails to be radical (consider $I \cdot I$), the intersection of radical ideals is always radical.

**Theorem 8.4.10.** Let $I$, $J$ be ideals in $K[X]$. Then $\sqrt{I \cap J} = \sqrt{I} \cap \sqrt{J}$ ($\sqrt{I}$ means the radical of $I$).

*Proof.* If $f \in \sqrt{I \cap J}$, then $f^m \in I \cap J$ for some integer $m > 0$. So $f \in \sqrt{I}$ and $f \in \sqrt{J}$.

On the other hand, if $f \in \sqrt{I} \cap \sqrt{J}$ then for some integers $m, p > 0$ we have $f^m \in I$ and $f^p \in J$. Thus, $f^{m+p} \in I \cap J$, i.e., $f \in \sqrt{I \cap J}$. $\square$

*Example 8.4.5.* Consider the ideals

$$\begin{aligned}
I_1 &= \langle 2x^4 - 3x^2y + y^2 - 2y^3 + y^4 \rangle \ , \\
I_2 &= \langle x, y^2 - 4 \rangle \ , \\
I_3 &= \langle x, y^2 - 2y \rangle \ , \\
I_4 &= \langle x, y^2 + 2y \rangle \ .
\end{aligned}$$

The coefficients are all integers, but we consider them as defining algebraic sets in the affine plane over $\mathbb{C}$. In fact, $V(I_1)$ is the tacnode curve (compare Sect. 1.1), $V(I_2) = \{(0, 2), (0, -2)\}$, $V(I_3) = \{(0, 2), (0, 0)\}$, $V(I_4) = \{(0, 0), (0, -2)\}$.

First, let us compute the ideal $I_5$ defining the union of the tacnode and the 2 points in $V(I_2)$. $I_5$ is the intersection of $I_1$ and $I_2$, i.e.,

$$\begin{aligned}
I_5 = I_1 \cap I_2 &= (\langle z \rangle I_1 + \langle 1 - z \rangle I_2) \cap \mathbb{Q}[\curvearrowleft, \curvearrowright] \\
&= \langle -4y^2 + 8y^3 - 3y^4 + 12x^2y - 8x^4 - 2y^5 + y^6 - 3x^2y^3 + 2y^2x^4, \\
&\qquad xy^2 - 2xy^3 + xy^4 - 3x^3y + 2x^5 \rangle \ .
\end{aligned}$$

Now let us compute the ideal $I_6$ defining $V(I_5) - V(I_3)$, i.e., the Zariski closure of $V(I_5) \setminus V(I_3)$, i.e., the smallest algebraic set containing $V(I_5) \setminus V(I_3)$.

$$\begin{aligned}
I_6 = I_5 : I_3 &= (I_5 : \langle x \rangle) \cap (I_5 : \langle y^2 - 2y \rangle) \\
&= \langle 2x^4 - 3x^2y + y^2 - 2y^3 + y^4 \rangle \cap \\
&\quad \langle y^5 - 3y^3 + 2y^2 - 3x^2y^2 + 2yx^4 - 6x^2y + 4x^4, \\
&\quad 2x^5 - 3x^3y + xy^2 - 2xy^3 + xy^4 \rangle \\
&= \langle y^5 - 3y^3 + 2y^2 - 3x^2y^2 + 2yx^4 - 6x^2y + 4x^4, \\
&\quad 2x^5 - 3x^3y + xy^2 - 2xy^3 + xy^4 \rangle \ .
\end{aligned}$$

$V(I_6)$ is the tacnode plus the point $(0, -2)$.

Finally, let us compute the ideal $I_7$ defining $V(I_6) - V(I_4)$, i.e., the Zariski closure of $V(I_6) \setminus V(I_4)$.

$$\begin{aligned} I_7 = I_6 : I_4 &= (I_6 : \langle x \rangle) \cap (I_6 : \langle y^2 + 2y \rangle) \\ &= \langle 2x^4 - 3x^2y + y^2 - 2y^3 + y^4 \rangle \cap \langle 2x^4 - 3x^2y + y^2 - 2y^3 + y^4 \rangle \\ &= I_1 \ . \end{aligned}$$

So we get back the ideal $I_1$ defining the tacnode curve.

### Exercises

1. Determine the singularities of the curve defined by $f(x, y) = x^6 + 3x^4y^2 - 4x^2y^2 + 3x^2y^4 + y^6 = 0$ in the projective plane $\mathbb{P}^2(\mathbb{C})$.
2. Let $I$ be a 0-dimensional prime ideal in $K[X]$. What is the shape of a Gröbner basis for $I$ w.r.t. a lexicographic ordering?
3. Let $P_1, \ldots, P_r \in \mathbb{A}^n(K)$, the affine space of dimension $n$ over the field $K$. Construct polynomials (separator polynomials) $f_1, \ldots, f_r \in K[x_1, \ldots, x_n]$ such that
$$f_i(P_j) = \begin{cases} 0 & \text{for } i \neq j, \\ 1 & \text{for } i = j. \end{cases}$$
4. Let $I$ be the ideal in $\mathbb{Q}[x, y, z]$ generated by $G = \{g_1, g_2, g_3\}$, where
$$\begin{aligned} g_1 &= z + x^4 - 2x + 1 \ , \\ g_2 &= y^2 + x^2 - 2 \ , \\ g_3 &= x^5 - 6x^3 + x^2 - 1 \ . \end{aligned}$$
   a. Is $G$ a Gröbner basis w.r.t. the lexicographic term ordering with $x < y < z$?
   b. How many solutions does the system of equations $g_1 = g_2 = g_3 = 0$ have?
   c. Give a basis for the vector space $\mathbb{Q}[x, y, z]_{/I}$ over $\mathbb{Q}$.

## 8.5 Speed-ups and complexity considerations

### Speeding up Buchberger's algorithm

In Example 8.3.1 there were 10 S-polynomials to be checked for reducibility to 0. But in order to arrive at the Gröbner basis we actually considered only 3 of them! This naturally leads to the question whether there are criteria for detecting such *unnecessary* S-polynomials which do not lead to new basis polynomials. Two of the best known criteria are the *product criterion* and the *chain criterion.*

**Theorem 8.5.1.** Let $G$ be a set of polynomials in $K[X]$, and $g_1, g_2 \in G$. If $\mathrm{lpp}(g_1) \cdot \mathrm{lpp}(g_2) = \mathrm{lcm}(\mathrm{lpp}(g_1), \mathrm{lpp}(g_2))$ then $\mathrm{spol}(g_1, g_2) \longrightarrow_G^* 0$.

The proof of this fact is easy and is left to the reader. From Theorem 8.5.1 one immediately gets the product criterion, which says that in step (2) of GRÖBNER_B the S-polynomial of $g_1, g_2$ can be discarded without reducing it to normal form if $\mathrm{lpp}(g_1) \cdot \mathrm{lpp}(g_2) = \mathrm{lcm}(\mathrm{lpp}(g_1), \mathrm{lpp}(g_2))$.

The following theorem has been derived in Buchberger (1979).

**Theorem 8.5.2.** Let $G$ be a set of polynomials in $K[X]$. $G$ is a Gröbner basis if and only if for all $g_1, g_2 \in G$ there are $h_1, \ldots, h_k \in G$ such that $g_1 = h_1$, $h_k = g_2$, $\mathrm{lcm}(\mathrm{lpp}(h_1), \ldots, \mathrm{lpp}(h_k))$ divides $\mathrm{lcm}(\mathrm{lpp}(g_1), \mathrm{lpp}(g_2))$, and $\mathrm{spol}(h_i, h_{i+1}) \longrightarrow_G^* 0$ for all $1 \leq i \leq k$.

From Theorem 8.5.2 one immediately gets the chain criterion, which says that in step (2) of GRÖBNER_B the S-polynomial of $g_1, g_2$ can be discarded without reducing it to normal form if there are $h_1, \ldots, h_k$ as in Theorem 8.5.2. Since testing for long chains might be costly, one often applies the chain criterion only for short chains, i.e., $k = 3$.

*Example 8.5.1* (Example 8.3.1 continued). After having added the polynomials $f_3, f_4, f_5$ to the basis, all the other S-polynomials can be discarded by applications of the product criterion and the chain criterion.

### Complexity of Gröbner basis computations

Already G. Hermann (1926) gives a double exponential upper bound for the degrees of polynomials in the basis of the module of syzygies for $f_1, \ldots, f_m$. For the special case of 2 variables the degrees of polynomials in a Gröbner basis can be bounded linearly in the degree of polynomials in the starting basis (Buchberger 1983a), and by a single exponential bound for the case of 3 variables (Winkler 1984b).

Let $D(n, d)$ be minimal with the property that for any finite set of polynomials $f_1, \ldots, f_m$ in $K[x_1, \ldots, x_n]$, where each $f_i$ has degree at most $d$, there is a Gröbner basis for $\langle f_1, \ldots, f_m \rangle$ such that each polynomial in the basis has degree bounded by $D(n, d)$. Yap (1991) proves a lower bound for $D(n, d)$, namely $D(n, d) \geq d^{2^k}$, where $k \sim n/2$ and $n, d$ sufficiently large.

Collecting various results on upper bounds for $D(n, d)$, Lazard (1992) shows that $D(n, d) \leq d^{2^l}$, where $l = (\log 3/\log 4)n + \mathcal{O}(\log n)$.

For 0-dimensional ideals Lakshman (1990) has been able to show that an upper bound for the complexity of constructing a Gröbner basis is polynomial in $d^n$.

## 8.6 Bibliographic notes

An excellent introduction to problems in polynomial ideal theory is contained in Hilbert (1890). G. Hermann's (1926) complexity analysis is based on this paper.

Hironaka (1964) introduced standard bases for ideals of power series, which have basically the same properties as Gröbner bases. Hironaka's notion of standard bases, however, is non-constructive. Becker (1990, 1993) is able to provide algorithms for some problems in the theory of standard bases.

There are various generalizations of the concept of Gröbner basis to different domains, e.g., to reduction rings (Buchberger 1983b, Stifter 1987, 1991, 1993), to polynomials over the integers (Pauer and Pfeifhofer 1988), to polynomials over Euclidean domains (Kandri-Rody and Kapur 1984), or to Gröbner bases for modules over $K[X]$ (Galligo 1979, Mora and Möller 1986, Furukawa et al. 1986).

$p$-Adic approximations of Gröbner bases over the rational numbers have been investigated by Winkler (1988a), Pauer (1992), Gräbe (1993).

The idea of computing syzygies by Gröbner bases was developed by Zacharias (1978). On the other hand, syzygies can be employed for computing Gröbner bases (Möller 1988).

Linear algebra methods for transforming a Gröbner basis for a 0-dimensional ideal to a Gröbner basis for the same ideal but a different ordering were developed by Faugère et al. (1993).

Ideas very similar to those in the theory of Gröbner bases appear also in the constructive approach to equational reasoning over first order terms, i.e., the Knuth–Bendix (1967) procedure for completing term-rewriting systems. The relation between these algorithms and procedures is investigated in Le Chenadec (1986), Winkler (1984a), Bündgen (1991). Improvements such as criteria for unnecessary S-polynomials can be adapted to the term-rewriting case (Winkler and Buchberger 1983).

Gröbner bases can be used for solving many more ideal theoretic problems, e.g., primary decomposition of polynomial ideals (Lazard 1985, Gianni et al. 1988), implicitization of parametric curves or surfaces (Arnon and Sederberg 1984, Kalkbrener 1991), computation of the dimension of algebraic sets (Kredel and Weispfenning 1988, Kalkbrener and Sturmfels 1992), computing in multiple algebraic extensions (Wall 1993), computation of the Hilbert function (Bayer and Stillman 1992). Another application of Gröbner bases is for geometry theorem proving (Kapur 1986; Kutzler and Stifter 1986a, b; Winkler 1988c, 1990, 1992). Further examples of solutions of systems of algebraic equations by Gröbner bases are given in Böge et al. (1986). Other applications are described in Winkler et al. (1985).

Results on the complexity of constructing Gröbner bases are reported in Mayr and Meyer (1982), Giusti (1984), Möller and Mora (1984), Bayer and Stillman (1988), Lakshman and Lazard (1990), Heintz and Morgenstern (1993).

Introductions to the theory of Gröbner bases are available in Buchberger (1985b), Cox et al. (1992), Becker and Weispfenning (1993), Mishra (1993), Adams and Loustaunau (1994).

# 9 Quantifier elimination in real closed fields

## 9.1 The problem of quantifier elimination

Many interesting problems of the geometry over the real numbers can be stated as systems of polynomial equations, inequations, and inequalities, usually with some structure of quantification. For instance, in Sect. 1.1 the piano movers problem in robotics has been mentioned. There are many other application areas, e.g., stability conditions for difference schemes. Quantifier elimination provides an approach to solving such polynomial problems over the real numbers.

In the sequel we collect a few definitions and facts about real fields as a basis for talking about Collins's quantifier elimination algorithm. These definitions and theorems can be found in van der Waerden (1970).

*Definition 9.1.1.* A field $K$ is *ordered* with ordering $>$, iff

a. for every $a \in K$ exactly one of the relations $a = 0$, $a > 0$, $-a > 0$ is valid, and
b. if $a > 0$ and $b > 0$ then $a + b > 0$ and $a \cdot b > 0$.

If we let $a > b :\Longleftrightarrow a - b > 0$, then the relation $> \in K^2$ satisfies the usual order axioms of transitivity and anti-symmetry. In an ordered field a sum of squares is always nonnegative,

$$\sum_{i=1}^{n} x_i^2 \geq 0 \ ,$$

and it can be zero only if all the summands are zero. In particular, $1 = 1^2$ is always positive, and for all $n \in \mathbb{N}$,

$$n = \underbrace{1 + \ldots + 1}_{n \text{ times}} = \underbrace{1^2 + \ldots + 1^2}_{n \text{ times}} > 0 \ .$$

So the characteristic of an ordered field is 0.

*Example 9.1.1.* Examples of ordered fields are $\mathbb{Q}$, $\mathbb{R}$, and the field of real algebraic numbers with the usual ordering.

Another example of a real field is $\mathbb{R}(x)$, the field of rational functions over $\mathbb{R}$. To see this we let $\mathbb{R}$ be ordered as usual and we let $x$ be positive but smaller than any positive real number. Now if $p(x) = a_n x^n + a_{n-1} x^{n-1} + \ldots + a_k x^k$

is a polynomial with $a_k \neq 0$, then we let $p(x) > 0$ if and only if $a_k > 0$. A rational function $p(x)/q(x)$ is positive if and only if $p(x)q(x) > 0$. This is indeed an ordering on $\mathbb{R}(x)$ satisfying the conditions (a) and (b). In fact, this ordering is a non-archimedean ordering, since $x$ is positive but strictly less than any $1/n, n \in \mathbb{N}$.

$\mathbb{C}$ cannot be ordered, since $1 = 1^2, -1 = i^2$, and not both of them can be positive.

*Definition 9.1.2.* A field $K$ is *(formally) real* iff $-1$ is not a sum of squares.

A real field always has characteristic 0, because in a field of characteristic $p > 0$ we have $-1 = (p-1) \cdot 1^2$.

*Definition 9.1.3.* A *real closed field* $K$ is a real field which does not admit any real algebraic extension, i.e., if $K'$ is an algebraic extension of $K$ for a real field $K'$, then $K = K'$.

*Example 9.1.2.* $\mathbb{R}$ and the field of real algebraic numbers are real closed fields. On the other hand, $\mathbb{Q}$ is not real closed, since $\mathbb{Q}(\sqrt{2})$ is a proper algebraic extension of $\mathbb{Q}$ to a real field.

Every real closed field can be ordered in one and only one way. Every positive element of a real closed field $K$ has a square root. Furthermore, over $K$ every polynomial of odd degree has at least one root. Conversely, if in an ordered field $K$ every positive element has a square root and every polynomial of odd degree has a root, then $K$ is a real closed field.

In the following we are interested in deciding formulas containing polynomial equations, inequations, and inequalities over real closed fields. For this purpose we introduce the formal setting of the elementary theory of real closed fields.

*Definition 9.1.4.* The *elementary theory of real closed fields* (ETRCF) is the first-order theory with the constants 0 and 1, function symbols $+, -, \cdot$, predicate symbols $=, >, \geq, <, \leq, \neq$ *(elementary algebra)*, and an axiom system consisting of the field axioms, the order axioms

$$(\forall a)(\forall b)[a > 0 \wedge b > 0 \Longrightarrow a + b > 0]$$
$$(\forall a)(\forall b)[a > 0 \wedge b > 0 \Longrightarrow a \cdot b > 0]$$

and axioms that guarantee roots of certain polynomials

$$(\forall a)(\exists b)[a = b^2 \vee -a = b^2]$$

$$\text{for every } n \geq 1: \quad (\forall a_0)(\forall a_1) \ldots (\forall a_{2n})(\exists b)[a_0 + a_1 b + \ldots + a_{2n} b^{2n} + b^{2n+1} = 0] \,.$$

Models of the elementary theory of real closed fields are exactly the real closed fields, as can be seen from the remark after Example 9.1.2. In this theory we can have polynomials as constituents of atomic formulas, and the coefficients of these polynomials have to be integers.

The principle of Tarski (see Seidenberg 1954) states:

> Every formula $\phi$ of elementary algebra that is valid in one model of the elementary theory of real closed fields is valid in every model of this theory.

So in order to decide such a formula $\phi$ for an arbitrary model, it suffices to decide it for $\mathbb{R}$.

*Definition 9.1.5.* A *standard atomic formula* is an expression of the form $p \sim 0$, where $p$ is a (multivariate) integral polynomial and $\sim$ is a predicate symbol in ETRCF. A *standard formula* is a formula in ETRCF in which the atomic formulas are all standard atomic formulas. A *standard prenex formula* is a standard formula in prenex form, i.e., a sequence of quantifiers followed by a quantifier free standard formula.

### The problem of quantifier elimination for ETRCF

The problem of quantifier elimination for ETRCF can be stated in the following way:

> For a given standard prenex formula $\phi$ find a standard quantifier-free formula $\psi$ such that $\phi$ is equivalent to $\psi$.

A. Tarski (1951) gave a quantifier elimination algorithm for transforming any formula $\phi$ of the theory of real closed fields into an equivalent formula that contains no quantifiers, i.e., for solving the problem of quantifier elimination for ETRCF. He also showed how to decide whether a formula $\phi'$, which does not contain quantifiers and variables, is true. Also other approaches have been suggested, e.g., in Seidenberg (1954) and Cohen (1969). Most of them, however, suffer from a prohibitively high complexity. In particular, even if one fixes the number of variables $r$ in the formula $\phi$, they have a computing time that is exponential in both $m$, the number of polynomials occurring in $\phi$, and $n$, the maximum degree of these polynomials. A real breakthrough was achieved in 1973 by G. E. Collins with his cad algorithm, published in Collins (1975).

## 9.2 Cylindrical algebraic decomposition

In describing Collins's cad algorithm we will not prove any results, rather we refer the reader to Collins (1975), Arnon et al. (1984a), and Arnon and Buchberger (1988) for background information.

The basic idea is to divide up the $r$-dimensional Euclidean space $\mathbb{R}^r$, where

$r$ is the number of variables occurring in the given formula $\phi$, into patches for which the validity of $\phi$ can be checked by simply inspecting particular points.

*Definition 9.2.1.* A nonempty connected subset of $\mathbb{R}^r$ is called a *region*. A *decomposition* $D$ of a subset $X$ of $\mathbb{R}^r$ is a finite collection $D = (D_1, \ldots, D_\mu)$ of disjoint regions whose union is $X$. An element of a decomposition is called a *cell* of the decomposition. A *sample point* for a cell of a decomposition is a point belonging to that cell. A *sample* of a decomposition $D = (D_1, \ldots, D_\mu)$ is a tuple $s = (s_1, \ldots, s_\mu)$ such that $s_i \in D_i$ for $1 \leq i \leq \mu$.

Let $A$ be a set of integral polynomials in $r$ variables. A decomposition $D$ of $\mathbb{R}^r$ is *$A$-invariant* iff every polynomial $p \in A$ is sign-invariant on every cell of $D$, i.e., is either positive, negative, or zero on the whole cell.

If $A$ is a set of integral polynomials, the decomposition $D$ is $A$-invariant, and $s$ is a sample for $D$, then the sign of a particular polynomial $p \in A$ on a particular cell $C$ in $D$ can be determined by evaluating $p$ at the sample point corresponding to $C$. Exact computation is essential.

*Definition 9.2.2.* A standard quantifier-free formula $\phi(x_1, \ldots, x_r)$ containing just the free variables $x_1, \ldots, x_r$ is a *defining formula* for the subset $X$ of $\mathbb{R}^r$ iff $X$ is the set of points in $\mathbb{R}^r$ satisfying $\phi$. A *standard definition* of the decomposition $D = (D_1, \ldots, D_\mu)$ is a sequence $(\phi_1, \ldots, \phi_\mu)$ such that, for $1 \leq i \leq \mu$, $\phi_i$ is a standard quantifier-free defining formula for $D_i$.

The validity of a standard formula $\phi$ can be effectively decided once we have a $\{\phi\}$-invariant decomposition of the appropriate space, which is algebraic, i.e., the cells of the decomposition are defined by polynomial equations or inequalities, and also the corresponding sample points are algebraic. Furthermore, it is also convenient to construct such a decomposition recursively, starting from a decomposition of the real line. For this purpose Collins introduced the notion of a cylindrical algebraic equation.

*Definition 9.2.3.* The *cylinder* over a region $R$, written $Z(R)$, is $R \times \mathbb{R}$. A *section* of $Z(R)$ is a set $S$ of points $(a_1, \ldots, a_r, f(a_1, \ldots, a_r))$, where $(a_1, \ldots, a_r)$ ranges over $R$, and $f$ is a continuous real-valued function on $R$. So $S$ is the graph of $f$ and it is also called the *$f$-section* of $Z(R)$. A *sector* of $Z(R)$ is a set $T$ of points $(a_1, \ldots, a_r, b)$, where $(a_1, \ldots, a_r)$ ranges over $R$ and $f_1(a_1, \ldots, a_r) < b < f_2(a_1, \ldots, a_r)$ for continuous, real-valued functions $f_1 < f_2$ on $R$. The constant functions $f_1 = -\infty$ and $f_2 = +\infty$ are allowed. $T$ is also called the *$(f_1, f_2)$-sector* of $Z(R)$.

Continuous, real-valued functions $f_1 < \ldots < f_k$, $k \geq 0$, defined on a region $R$, together with $f_0 = -\infty$ and $f_{k+1} = +\infty$, naturally determine a decomposition of $Z(R)$ consisting of the $(f_i, f_{i+1})$-sectors of $Z(R)$ for $0 \leq i \leq k$, and the $f_i$-sections of $Z(R)$ for $1 \leq i \leq k$. Such a decomposition is called the *stack* over $R$ determined by the functions $f_1, \ldots, f_k$.

A decomposition $D$ of $\mathbb{R}^r$ is *cylindrical* if either

a. $r = 1$, and $D = (D_1, \ldots, D_{2\nu+1})$, where $\nu = 0$ and $D_1 = \mathbb{R}$, or $\nu > 0$

and there exist real numbers $\alpha_1 < \alpha_2 < \ldots < \alpha_\nu$ such that $D_1 = (-\infty, \alpha_1)$, $D_{2i} = \{\alpha_i\}$ for $1 \leq i \leq \nu$, $D_{2i+1} = (\alpha_i, \alpha_{i+1})$ for $1 \leq i < \nu$, $D_{2\nu+1} = (\alpha_\nu, \infty)$; or

b. $r > 1$, and there is a cylindrical decomposition $D' = (D_1, \ldots, D_\mu)$ of $\mathbb{R}^{r-1}$ such that $D = (D_{1,1}, \ldots, D_{1,2\nu_1+1}, \ldots, D_{\mu,1}, \ldots, D_{\mu,2\nu_\mu+1})$ and for each $1 \leq i \leq \mu$ the decomposition $(D_{i,1}, \ldots, D_{i,2\nu_i+1})$ is a stack over $D_i$. $D'$ is unique and is called the *induced cylindrical decomposition* of $\mathbb{R}^{r-1}$. If $D$ is determined by algebraic functions $f_1, \ldots, f_k$, then it is a *cylindrical algebraic decomposition* (cad).

A sample $s = (s_1, \ldots, s_\mu)$ of a cad $D = (D_1, \ldots, D_\mu)$ is *algebraic* in case each $s_i$ is an algebraic point. The sample $s$ is *cylindrical* if either (1) $r = 1$ or (2) $r > 1$ and there is a cylindrical sample $s' = (s_1, \ldots, s_\mu)$ of a cad $D'$ of $\mathbb{R}^{r-1}$ such that $s = (s_{1,1}, \ldots, s_{1,2\nu_1+1}, \ldots, s_{\mu,1}, \ldots, s_{\mu,2\nu_\mu+1})$ and the first $r-1$ coordinates of $s_{i,j}$ are, respectively, the coordinates of $s_i$, for all $1 \leq i \leq \mu$, $1 \leq j \leq 2\nu_i + 1$. A sample that is both cylindrical and algebraic is called a *cylindrical algebraic sample* (cas).

*Example 9.2.1.* Let $\alpha_1 < \alpha_2 < \alpha_3$ be the three different real roots of the polynomial $f(x) = 10x^3 - 20x^2 + 10x - 1$. $\alpha_1 \approx 0.13$, $\alpha_2 \approx 0.59$, $\alpha_3 \approx 1.28$, see Fig. 9. So if we let $D_1 = (-\infty, \alpha_1)$, $D_2 = \{\alpha_1\}$, $D_3 = (\alpha_1, \alpha_2)$, $D_4 = \{\alpha_2\}$, $D_5 = (\alpha_2, \alpha_3)$, $D_6 = \{\alpha_3\}$, $D_7 = (\alpha_3, \infty)$, then $D = (D_1, \ldots, D_7)$ is a cylindrical algebraic decomposition of $\mathbb{R}^1$. Let $f_1(x)$ and $f_2(x)$ denote the greater and the smaller solutions, respectively, of the algebraic equation $x^2 + y^2 = (3/2)^2$ in the cylinder over $D_3$. Let $f_3(x)$ and $f_4(x)$ denote the greater and the smaller solutions, respectively, of the algebraic equation $x^2+y^2 = (3/2)^2$ in the cylinder over $D_5$. Then $f_1, f_2$ determine a stack $(D_{3,1}, D_{3,2}, D_{3,3}, D_{3,4}, D_{3,5})$ over $D_3$, and similarly $f_3, f_4$ determine a stack $(D_{5,1}, D_{5,2}, D_{5,3}, D_{5,4}, D_{5,5})$ over $D_5$.

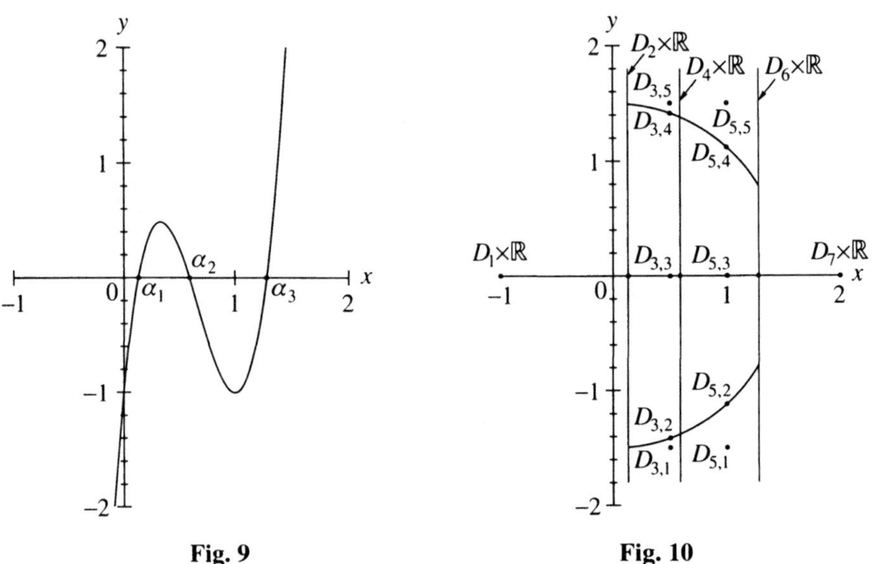


**Fig. 9** **Fig. 10**

So a possible cylindrical algebraic decomposition of $\mathbb{R}^2$ is

$$\begin{aligned}D = (&D_1 \times \mathbb{R}\ ,\\ &D_2 \times \mathbb{R}\ ,\\ &D_{3,1}, \ldots, D_{3,5}\ ,\\ &D_4 \times \mathbb{R}\ ,\\ &D_{5,1}, \ldots, D_{5,5}\ ,\\ &D_6 \times \mathbb{R}\ ,\\ &D_7 \times \mathbb{R})\ .\end{aligned}$$

As a cylindrical algebraic sample of $D$ we can take for instance

$$\begin{aligned}s = (&(-1, 0)\ ,\\ &(\alpha_1, 0)\ ,\\ &(1/2, -3/2), (1/2, -\sqrt{2}), (1/2, 0), (1/2, \sqrt{2}), (1/2, 3/2)\ ,\\ &(\alpha_2, 0)\ ,\\ &(1, -3/2), (1, -\sqrt{5}/2), (1, 0), (1, \sqrt{5}/2), (1, 3/2)\ ,\\ &(\alpha_3, 0)\ ,\\ &(2, 0))\ .\end{aligned}$$

See Fig. 10. For instance, $\phi_{5,3} \equiv f(x) < 0 \wedge x > 1/2 \wedge x^2 + y^2 < (3/2)^2$ is a defining formula for the cell $D_{5,3}$.

The top-level algorithm for solving the quantifier elimination problem by cylindrical algebraic decomposition is given in QE.

**Algorithm QE**(in: $\phi^*$; out: $\psi^*$);
[$\phi^* \equiv (Q_{k+1}x_{k+1}) \ldots (Q_r x_r)\phi(x_1, \ldots, x_r)$ is a standard prenex formula, where $0 \leq k \leq r$, $Q_i$ is either $\forall$ or $\exists$ for all $k+1 \leq i \leq r$, and $\phi$ is quantifier-free. $\psi^*$ is a standard quantifier-free formula equivalent to $\phi^*$.]

1. from $\phi^*$ extract $k$ and the set $A$ of distinct nonzero polynomials occurring in $\phi$;
2. apply the algorithm CAD for cylindrical algebraic decomposition to $A$ and $k$, obtaining $s$, a cas for an $A$-invariant cad $D$ of $\mathbb{R}^r$, and $\psi$, a standard definition of the cad $D'$ of $\mathbb{R}^k$ induced by $D$ if $k > 0$ or ( ) if $k = 0$;
3. construct $\psi^*$ from $\psi$ and $s$ by evaluating the polynomials in $A$ at the sample points in $s$;
   return.

Step (2) of QE needs some further description. The algorithm for computing a cylindrical algebraic decomposition, given the input $A = (a_1, \ldots, a_m)$ and $k$, proceeds in three phases: the projection phase, the base phase, and the extension phase. In the projection phase, if $r \geq 2$, a set $A^{(r-1)} = \text{proj}(A)$ (projection of $A$) of polynomials in $r - 1$ variables is computed, such that

(E) for every proj$(A)$-invariant cad $D'$ of $\mathbb{R}^{r-1}$ there is an $A$-invariant cad $D$ of $\mathbb{R}^r$ that induces $D'$, i.e., $D'$ can be extended to a cad of $\mathbb{R}^r$.

This projection process is applied to proj$(A)$ recursively until univariate polynomials are reached, i.e., we successively determine sets $A^{(r)} = A, A^{(r-1)}, \ldots, A^{(1)}$ of polynomials in $r, r-1, \ldots, 1$ variables, respectively. In the base phase we determine an $A^{(1)}$-invariant cad of $\mathbb{R}^1$. Finally, in the extension phase, an $A^{(i)}$-invariant cad $D^{(i)}$ is extended to an $A^{(i+1)}$-invariant cad $D^{(i+1)}$ for $1 \leq i < r$.

Let us take a closer look at the projection process. We treat only the case $r = 2$, i.e., we start out with polynomials in 2 variables, say in $\mathbb{Z}[x, y]$. Using the algorithms developed in previous chapters, we can assume that the elements of $A = A^{(2)}$ are all squarefree and relatively prime. Observe that an $\{\ldots, f_1 \cdot f_2 \cdots f_m, \ldots\}$-invariant cad is also $\{\ldots, f_1 \cdot f_2^2 \cdots f_m^m, \ldots\}$-invariant, and an $\{\ldots, f_1, f_2, g, \ldots\}$-invariant cad is also $\{\ldots, f_1 \cdot g, f_2 \cdot g, \ldots\}$-invariant.

**Theorem 9.2.1.** Let $A = \{a_1, \ldots, a_m\}$ be a set of squarefree and relatively prime polynomials in $\mathbb{Z}[x, y]$. Then the set

$$\begin{aligned} \text{proj}(A) = &\{\text{lc}_y(a_i) \mid 1 \leq i \leq m\} \cup \\ &\{\text{discr}_y(a_i) \mid 1 \leq i \leq m\} \cup \\ &\{\text{res}_y(a_i, a_j) \mid 1 \leq i < j \leq m\} \end{aligned}$$

of univariate polynomials in $\mathbb{Z}[x]$ satisfies the condition (E), i.e., is a suitable projection.

*Proof.* Let $B := \text{proj}(A)$. Let $D$ be a $B$-invariant cad of $\mathbb{R}^1$. (E) will be satisfied if we can show that for each 1-dimensional cell $C$ of $D$

a. each polynomial $a_i$ has a constant number of real roots over $C$, and
b. in the cylinder over $C$ the curves given by the real roots of the polynomials in $A$ do not intersect.

Let $C = (\alpha_j, \alpha_{j+1})$ be such a 1-dimensional cell of $D$ ($\alpha_j$ could be $-\infty$ and $\alpha_{j+1}$ could be $+\infty$). There are only two possibilities for the number of real roots of $a_i$ changing over $C$, namely a real root could go to infinity, i.e., $a_i$ has a pole at some $x_0$ in $(\alpha_i, \alpha_{i+1})$, or a pair of real roots could become complex. For $a_i(x, y)$ to have a pole at $x_0$, the leading coefficient of $a_i$ must vanish at $x_0$, i.e., $\text{lc}_y(a_i)(x_0) = 0$. But $\text{lc}_y(a_i) \in \text{proj}(A)$, so there can be no root of $\text{lc}_y(a_i)$ inside $C$. If a pair of real roots of $a_i(x, y)$ vanishes over $C$, there must be a point $x_0$ in $C$ over which $a_i(x_0, y)$ has a multiple root. But any such $x_0$ is a root of $\text{discr}_y(a_i)$, so there can be no such point inside $C$. Thus, (a) is satisfied.

If two curves given by real roots of polynomials $a_i, a_j \in A$ would intersect over $C$, then these curves could be given by different roots of the same polynomial, i.e., $i = j$, or by roots of different polynomials. In the first case there would be a root of $\text{discr}_y(a_i)$ inside $C$, which has been already excluded by the reasoning above. In the second case, $a_i(x_0, y)$ and $a_j(x_0, y)$ would have a nontrivial gcd. So there would be a root of $\mathrm{res}_y(a_i, a_j)$ inside $C$, which is impossible because all the resultants of polynomials in $A$ are in $B$. Thus, also (b) is satisfied. □

The size of the set of polynomials as well as the coefficients of these polynomials grow considerably in the projection step. In the general case ($n > 2$) the computation of projections is even more complicated, involving certain subresultant coefficients of the polynomials in $A$. For the details we refer to Collins (1975) and Arnon et al. (1984a). Improvements of the projection operation are described in McCallum (1988, 1993) and Hong (1990).

Now we can give an outline of the algorithm for computing cads.

**Algorithm CAD**(in: $A, k$; out: $s, \psi$);
[$A \subset \mathbb{Z}[x_1, \ldots, x_r]$ (finite), $0 \le k \le r$;
$s$ is a cas for some $A$-invariant cad $D$ of $\mathbb{R}^r$, and $\psi$ is a standard definition of the cad $D^*$ of $\mathbb{R}^k$ induced by $D$ if $k > 0$, and $\psi = (\ )$ if $k = 0$.]

1. [$r = 1$] If $r > 1$ then go to (2);
   isolate the real roots $\alpha_1, \ldots, \alpha_n$ of the irreducible factors of the nonzero elements of $A$;
   construct a cas $s$ for an $A$-invariant cad $D$;
   if $k = 0$ then set $\psi := (\ )$;
   otherwise, if $n = 0$, set $\psi :=$ "$0 = 0$";
   otherwise use the signs of the polynomials in $A$ in the cells of $D$ to construct a standard definition $\psi$ of $D$ (this might require an "augmented projection") and
   return;
2. [$r > 1$] if $k = r$ then set $k' := k - 1$, otherwise set $k' := k$;
   call CAD recursively with the input proj$(A)$ and $k'$, obtaining outputs $s'$ and $\psi'$;
   let $s' = (s'_1, \ldots, s'_p)$, $s'_j = (s'_{j,1}, \ldots, s'_{j,r-1})$;
   construct a cas $s$ for an $A$-invariant cad $D$ of $\mathbb{R}^r$ by isolating the real roots of $a(s'_{j,1}, \ldots, s'_{j,r-1}, x_r)$ for every $a \in A$ and $1 \le j \le p$;
   if $k < r$ then set $\psi := \psi'$, otherwise use the real roots of the derivations of the polynomials $a(s'_{j,1}, \ldots, s'_{j,r-1}, x_r)$ to extend $\psi'$ to a standard definition $\psi$ of $D$;
   return.

*Example 9.2.2.* We apply the algorithm QE to the formula

$$\phi^* \equiv (\exists y)\,(x^2 + y^2 - 4 < 0 \ \wedge \ y^2 - 2x + 2 < 0)\ ,$$

i.e., we determine a quantifier-free formula $\psi^*$ in $x$ which is equivalent to $\phi^*$.

In step (1) we set $k = 1$ and $A = \{y^2 + x^2 - 4,\ y^2 - 2x + 2\}$. In step (2) the algorithm CAD is called with the inputs

$$A = \{\underbrace{y^2 + x^2 - 4}_{a_1},\ \underbrace{y^2 - 2x + 2}_{a_2}\}, \qquad k = 1\ .$$

The number of variables in $A$ is 2, so step (2) of CAD is executed. Computing $\mathrm{proj}(A)$ according to Theorem 9.2.1 we get

$$\mathrm{proj}(A) = \{1,\ 4x^2 - 16,\ -8x + 8,\ (-x^2 - 2x + 6)^2\}\ .$$

W.l.o.g. we can make the polynomials in the projection squarefree and relatively prime. We can also scale them and eliminate constants. This results in the simplified projection set

$$B = \mathrm{proj}(A) = \{\underbrace{x^2 + 2x - 6}_{b_1},\ \underbrace{x^2 - 4}_{b_2},\ \underbrace{x - 1}_{b_3}\}\ .$$

So CAD is called recursively with the inputs $B$ and 1. Now $r = 1$, i.e., we are in the base phase, and the real roots of the polynomials in $B$ are

$$-1 - \sqrt{7}\ <\ -2\ <\ 1\ <\ -1 + \sqrt{7}\ <\ 2\ .$$

A cas for a $B$-invariant cad of $\mathbb{R}^1$ is

$$t = (-4, -1 - \sqrt{7}, -3, -2, 0, 1, 3/2, -1 + \sqrt{7}, 9/5, 2, 3)$$

and a standard definition of this cad is

$$\begin{aligned}
\psi \equiv (\ & b_1 > 0 \wedge b_2 > 0 \wedge b_3 < 0, && b_1 = 0 \wedge b_2 > 0 \wedge b_3 < 0,\\
& b_1 < 0 \wedge b_2 > 0 \wedge b_3 < 0, && b_1 < 0 \wedge b_2 = 0 \wedge b_3 < 0,\\
& b_1 < 0 \wedge b_2 < 0 \wedge b_3 < 0, && b_1 < 0 \wedge b_2 < 0 \wedge b_3 = 0,\\
& b_1 < 0 \wedge b_2 < 0 \wedge b_3 > 0, && b_1 = 0 \wedge b_2 < 0 \wedge b_3 > 0,\\
& b_1 > 0 \wedge b_2 < 0 \wedge b_3 > 0, && b_1 > 0 \wedge b_2 = 0 \wedge b_3 > 0,\\
& b_1 > 0 \wedge b_2 > 0 \wedge b_3 > 0\ )\ .
\end{aligned}$$

That finishes the recursive call of CAD.

The extension of $t$ to a cas for an $A$-invariant cad of $\mathbb{R}^2$ yields

$$
\begin{aligned}
s = (\ & (-4, 0)\ , \\
& (-1-\sqrt{7}, 0)\ , \\
& (-3, 0)\ , \\
& (-2, -1), (-2, 0), (-2, 1)\ , \\
& (0, -3), (0, -2), (0, 0), (0, 2), (0, 3)\ , \\
& (1, -2), (1, -\sqrt{3}), (1, -1), (1, 0), (1, 1), (1, \sqrt{3}), (1, 2)\ , \\
& (3/2, -2), (3/2, -\sqrt{7}/2), (3/2, -6/5), (3/2, -1), (3/2, 0)\ , \\
& \quad (3/2, 1), (3/2, 6/5), (3/2, \sqrt{7}/2), (3/2, 2)\ , \\
& (-1+\sqrt{7}, -2), (-1+\sqrt{7}, -\alpha), (-1+\sqrt{7}, 0), (-1+\sqrt{7}, \alpha)\ , \\
& \quad (-1+\sqrt{7}, 2)\ , \\
& (9/5, -2), (9/5, -\beta), (9/5, -1), (9/5, -\sqrt{19}/5), (9/5, 0)\ , \\
& \quad (9/5, \sqrt{19}/5), (9/5, 1), (9/5, \beta), (9/5, 2)\ , \\
& (2, -2), (2, -\sqrt{2}), (2, -1), (2, 0), (2, 1), (2, \sqrt{2}), (2, 2)\ , \\
& (3, -3), (3, -2), (3, 0), (3, 2), (3, 3)\ )\ ,
\end{aligned}
$$

where $\alpha = \sqrt{2}\sqrt{\sqrt{7}-2}$, $\beta = 2\sqrt{2}/\sqrt{5}$ (see Fig. 11). Further, $k < r$, so $\psi$ does not have to be extended. That finishes the call of CAD.

In step (3) of QE the polynomials in $A$ are evaluated at the sample points in $s$. Only the cylinders over the cells defined by $\psi_7$, $\psi_8$, and $\psi_9$ contain sample points that satisfy $a_1 < 0$ and $a_2 < 0$. Thus, a standard quantifier-free formula equivalent to $\phi^*$ is

$$\psi^* \equiv \psi_7 \vee \psi_8 \vee \psi_9\ .$$

In Fig. 11 the set of zeros of the polynomials in $A$ is shown, the sample

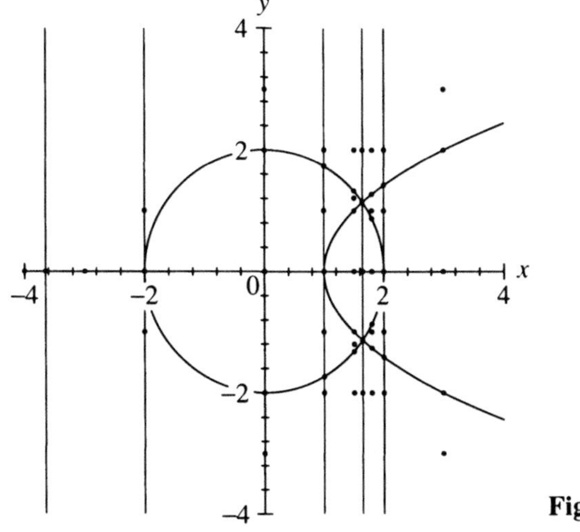


**Fig. 11**

points for $\mathbb{R}^1$ are indicated as lines parallel to the $y$-axis and the sample points for $\mathbb{R}^2$ as dots.

As is obvious from the above example, efficiency can be gained by combining neighboring cells, in which the polynomials involved have the same signs, into clusters of adjacent cells. Only one sample point is necessary for the whole cluster.

In Collins (1975) a complexity analysis of the quantifier elimination algorithm based on cylindrical algebraic decomposition is given. The complexity turns out to be

$$(2n)^{2^{2r+8}} m^{2^{r+6}} d^3 a \ ,$$

where $m$ is the number of polynomials occurring in the input formula $\phi^*$, $n$ is the maximum degree of these polynomials, $d$ is the maximum length of any integer coefficient of these polynomials, and $a$ is the number of occurrences of atomic formulas in $\phi^*$.

## 9.3 Bibliographic notes

For a thorough introduction to the theory of real algebraic geometry and semialgebraic sets we refer to Bochnak et al. (1987). An introduction to Collins's CAD algorithm can be found in Collins (1976), Arnon et al. (1984a), Hong (1993) and Mishra (1993). Collins (1996) includes an account of the historical development of cylindrical algebraic decomposition for quantifier elimination.

For adjacency and clustering of cells in decompositions we refer to Arnon et al. (1984b) and Arnon (1988). A further practical speed-up of the CAD algorithm is reported in Collins and Hong (1991). The complexity of deciding the theory of real closed fields is investigated in Davenport and Heintz (1988) and Renegar (1992a–c).

Some recent developments are contained in Caviness and Johnson (1996) and Jacob et al. (1994).

# 10 Indefinite summation

## 10.1 Gosper's algorithm

The problem of indefinite summation is very similar to the problem of indefinite integration, in fact, we can somehow think of it as a discrete analogon to the integration problem. Whereas in integration we start out with a continuous function $f(x)$ and want to determine another function $g(x)$ such that

$$\int f(x)\,\mathrm{d}x = g(x) \quad \text{and therefore} \quad \int_a^b f(x)\,\mathrm{d}x = g(b) - g(a)\ ,$$

in indefinite summation we are given a sequence $(a_n)_{n\in\mathbb{N}}$ and we want to determine another sequence $(s_n)_{n\in\mathbb{N}_0}$ (in which the function symbol $\sum$ is eliminated) such that any partial sum of the corresponding series can be expressed as

$$\sum_{n=m_1}^{m_2} a_n = s_{m_2} - s_{m_1-1}\ .$$

Of course we expect that the existence of algorithmic solutions for this indefinite summation problem will depend crucially on the class of functions that we take as input and possible output.

A *hypergeometric function* $f(z)$ is a function from $\mathbb{C}$ to $\mathbb{C}$ that can be written as

$$f(z) = \sum_{n=0}^{\infty} \underbrace{\frac{a_1^{\bar{n}} a_2^{\bar{n}} \cdots a_p^{\bar{n}}}{b_1^{\bar{n}} b_2^{\bar{n}} \cdots b_q^{\bar{n}}} \cdot \frac{z^n}{n!}}_{f_n}\ ,$$

for some $a_i, b_j \in \mathbb{C}$. $a^{\bar{n}}$ denotes the rising factorial of length $n$, i.e.,

$$a^{\bar{n}} = a(a+1)\cdots(a+n-1)\ .$$

The class of hypergeometric functions includes most of the commonly used special functions, e.g., exponentials, logarithms, trigonometric functions, Bessel functions, etc. Hypergeometric functions have the nice property that the quotient of successive terms $f_n/f_{n-1}$ is a rational function in the index $n$. Conversely, up to normalization, any rational function in $n$ can be written in this form. This fact gives rise to the notion of hypergeometric sequences.

*Definition 10.1.1.* Let $K$ be a field of characteristic 0. A sequence $(a_n)_{\mathbb{N}_0}$ of elements of $K$ is *hypergeometric* iff the quotient of successive elements of the sequence can be expressed as a rational function of the index $n$, i.e., there are polynomials $u(x), v(x) \in K[x]$ such that

$$\frac{a_n}{a_{n-1}} = \frac{u(n)}{v(n)} \quad \text{for all } n \in \mathbb{N} .$$

R. W. Gosper (1978) presented an algorithmic solution of the summation problem for the class of hypergeometric sequences, i.e., both $(a_n)_{n\in\mathbb{N}}$ and $(s_n)_{n\in\mathbb{N}_0}$ are hypergeometric in $n$. We will describe Gosper's algorithm.

So let us assume that we are given a hypergeometric sequence $(a_n)_{n\in\mathbb{N}}$ over the computable field $K$ of characteristic 0. We want to determine a hypergeometric sequence $(s_n)_{n\in\mathbb{N}_0}$ over $K$ such that

$$\sum_{n=1}^{m} a_n = s_m - s_0 \quad \text{for all } m \in \mathbb{N}_0 .$$

Clearly such an $(s_n)_{n\in\mathbb{N}_0}$ is determined only up to an additive constant. If $(s_n)_{n\in\mathbb{N}_0}$ exists, then it must have a very particular structure.

**Lemma 10.1.1.** Every rational function $u(n)/v(n)$ over $K$ can be written in the form

$$\frac{u(n)}{v(n)} = \frac{p(n) \cdot q(n)}{p(n-1) \cdot r(n)} ,$$

where $p, q, r$ are polynomials in $n$ satisfying the condition

$$\gcd(q(n), r(n+j)) = 1 \quad \text{for all } j \in \mathbb{N}_0 . \tag{10.1.1}$$

*Proof.* We determine $p, q, r$ by a recursive process of finitely many steps. Initially set

$$p(n) := 1, \quad q(n) := u(n), \quad r(n) := v(n) .$$

Let $R(j) = \mathrm{res}_n(q(n), r(n+j))$. The condition (10.1.1) is violated for $j^* \in \mathbb{N}_0$ if and only if $R(j^*) = 0$. If $R(j)$ has no roots in $\mathbb{N}_0$ then the process terminates and we have $p, q, r$ of the desired form. Otherwise let $j^*$ be a root of $R(j)$ in $\mathbb{N}_0$. We redefine $p, q, r$ according to the formula

$$g(n) := \gcd(q(n), r(n+j^*)) ,$$

$$p(n) := p(n) \prod_{k=0}^{j^*-1} g(n-k), \quad q(n) := \frac{q(n)}{g(n)}, \quad r(n) := \frac{r(n)}{g(n-j^*)} .$$

It is easy to see that the new $p, q, r$ are again a representation of the given rational function, i.e.,

$$\frac{u(n)}{v(n)} = \frac{p(n) \cdot q(n)}{p(n-1) \cdot r(n)} .$$

The process terminates because in every step the degree of $q$ is decreased. □

*Example 10.1.1.* Let us represent the rational function

$$a(n) = \frac{n^2 - 1}{n^2 + 2n}$$

as in the theorem. Initially we let

$$p(n) = 1, \quad q(n) = n^2 - 1, \quad r(n) = n^2 + 2n .$$

$\mathrm{res}_n(q(n), r(n+j)) = \mathrm{res}_n(n^2 - 1, n^2 + (2j+2)n + j^2 + 2j) = j^4 + 4j^3 + 2j^2 - 4j - 3$. The only non-negative integral root of the resultant is $j^* = 1$. So in the first iteration we get

$$\begin{aligned} g(n) &= \gcd(q(n), r(n+1)) = n + 1 , \\ p(n) &= 1 \cdot (n+1) = n + 1 , \\ q(n) &= (n^2 - 1)/(n+1) = n - 1 , \\ r(n) &= (n^2 + 2n)/n = n + 2 . \end{aligned}$$

Now $\mathrm{res}_n(q(n), r(n+j)) = j + 3$. This resultant has no non-negative integral root, so the process terminates.

*Definition 10.1.2.* If $a(n)/b(n)$ is a rational function over $K$ and $p(n), q(n), r(n)$ are as in Lemma 10.1.1, then we call $(p, q, r)$ a *regular representation* of $a/b$.

**Theorem 10.1.2.** Let $(a_n)_{n\in\mathbb{N}}$ be a hypergeometric sequence over $K$ and $(p, q, r)$ a regular representation of $a_n/a_{n-1}$. If

$$(s_n)_{n\in\mathbb{N}_0}, \quad \text{where } s_n = \sum_{i=1}^{n} a_m ,$$

is hypergeometric, then

$$s_n = \frac{q(n+1)}{p(n)} \cdot a_n \cdot f(n)$$

for a polynomial $f(n)$ satisfying the condition

$$p(n) = q(n+1) \cdot f(n) - r(n) \cdot f(n-1) . \tag{10.1.2}$$

*Proof.* Assume $(s_n)_{n\in\mathbb{N}_0}$ to be hypergeometric. Obviously we have

$$a_n = s_n - s_{n-1} \quad \text{for all} \quad n \in \mathbb{N} . \tag{10.1.3}$$

Let

$$f(n) := s_n \cdot \frac{p(n)}{q(n+1)\cdot a_n} . \tag{10.1.4}$$

Substituting (10.1.3) in (10.1.4) we get

$$f(n) = \frac{p(n)}{q(n+1)} \cdot \frac{s_n}{s_n - s_{n-1}} = \frac{p(n)}{q(n+1)} \cdot \frac{1}{1 - \frac{s_{n-1}}{s_n}} .$$

So we see that $f(n)$ is certainly a rational function in $n$. Substituting an appropriate version of (10.1.4) in (10.1.3) we derive

$$a_n = \frac{q(n+1)}{p(n)} \cdot a_n \cdot f(n) - \frac{q(n)}{p(n-1)} \cdot a_{n-1} \cdot f(n-1) .$$

Multiplying this by $p(n)/a_n$ and using the fact that $(p, q, r)$ is a regular representation of $a_n/a_{n-1}$, we get

$$p(n) = q(n+1) \cdot f(n) - r(n) \cdot f(n-1) ,$$

i.e., the rational function $f(n)$ satisfies the condition (10.1.2). What remains to be shown is that $f(x)$ is really a polynomial. This is proved in Lemma 10.1.3. □

**Lemma 10.1.3.** With the notation of Theorem 10.1.2, the rational function $f(n)$ is a polynomial.

*Proof.* Suppose

$$f(n) = \frac{c(n)}{d(n)} \quad \text{with} \quad \deg(d) > 0 \quad \text{and} \quad \gcd(c(n), d(n)) = 1 .$$

Then (10.1.2) can be written as

$$\begin{aligned} d(n)\cdot d(n-1)\cdot p(n) = c(n)\cdot d(n-1)\cdot q(n+1) \\ - d(n)\cdot c(n-1)\cdot r(n) . \end{aligned} \tag{10.1.5}$$

Let $j$ be the greatest integer such that

$$\gcd(d(n), d(n+j)) = g(n) \neq 1 . \tag{10.1.6}$$

Clearly $j$ exists and is non-negative. So

$$\gcd(d(n-1), d(n+j)) = 1 \tag{10.1.7}$$

and because of $g(n)|d(n+j)$

$$\gcd(d(n-1), g(n)) = 1\ . \tag{10.1.8}$$

Substitution of $n-j-1$ for $n$ in (10.1.6) yields

$$\gcd(d(n-j-1), d(n-1)) = g(n-j-1) \neq 1\ . \tag{10.1.9}$$

Substitution of $n-j$ for $n$ in (10.1.7) yields $\gcd(d(n-j-1), d(n)) = 1$, and since $g(n-j-1)|d(n-j-1)$ we get

$$\gcd(g(n-j-1), d(n)) = 1\ . \tag{10.1.10}$$

Now let us divide (10.1.5) by both $g(n)$ and $g(n-j-1)$. Because of (10.1.6) $g(n)$ divides $d(n)$, and because of (10.1.8) and the fact that $c$ and $d$ are relatively prime we have that $g(n)$ does not divide $d(n-1)$ and $c(n)$. Therefore, $g(n)|$ $q(n+1)$ and consequently

$$g(n-1)|q(n)\ . \tag{10.1.11}$$

Similarly, using (10.1.9) and (10.1.10) we derive $g(n-j-1)|r(n)$ and consequently

$$g(n-1)|r(n+j)\ . \tag{10.1.12}$$

So $j$ is a non-negative integer such that $g(n-1)|\gcd(q(n), r(n+j))$, and therefore

$$\gcd(q(n), r(n+j)) \neq 1\ ,$$

in contradiction to $(p, q, r)$ being a regular representation of $a_n/a_{n-1}$. Thus, the denominator $d(n)$ must be constant. □

The only remaining problem in the determination of $s_n$ is to find a polynomial solution $f(n)$ of Eq. (10.1.2). Obviously we could decide the existence of $f$ and also, provided $f$ exists, compute such an $f$ by a system of linear equations on its coefficients, if we had a degree bound for the solutions of (10.1.2).

**Theorem 10.1.4.** Let $(p, q, r)$ be a regular representation of $a_n/a_{n-1}$. Let $l^+ :=$ $\deg(q(n+1)+r(n))$, $l^- := \deg(q(n+1)-r(n))$. Then the degree of any solution of (10.1.2) is not greater than $\bar{k}$, where

a. for $l^+ \leq l^-$:
$\bar{k} = \deg(p(n)) - \deg(q(n+1) - r(n))$,

b. for $l^+ > l^-$:
$\bar{k} = \max\{k_0,\ \deg(p(n) - \deg(q(n)) + 1\}$, if

$k_0 = (-l^+\text{coeff}(q, l^+) - \text{coeff}(q, l^+ - 1) + \text{coeff}(r, l^+ - 1))/\text{coeff}(q, l^+)$
$\in \mathbb{Z}$, and
$\bar{k} = \deg(p(n)) - \deg(q(n)) + 1$, otherwise.

*Proof.* Let us replace (10.1.2) by the equivalent condition

$$p(n) = (q(n+1) - r(n)) \cdot \frac{f(n) + f(n-1)}{2} + (q(n+1) + r(n)) \cdot \frac{f(n) - f(n-1)}{2} . \tag{10.1.2'}$$

We have (assuming the notation $\deg(0) = -1$) the relation

$$\deg(f(n) + f(n-1)) = 1 + \deg(f(n) - f(n-1)) .$$

a. Assume $\deg(q(n+1) + r(n)) \le \deg(q(n+1) - r(n)) =: l$. Suppose $k$ is the degree of a solution $f$ of (10.1.2′), i.e., $f(n) = c_k n^k + \mathcal{O}(n^{k-1})$, and $f(n-1) = c_k n^k + \mathcal{O}(n^{k-1})$, where $c_k \ne 0$. From (10.1.2′) we get

$$p(n) = (q(n+1) - r(n)) \cdot (c_k n^k + \mathcal{O}(n^{k-1})) + (q(n+1) + r(n)) \cdot \mathcal{O}(n^{k-1}) ,$$

and therefore

$$p(n) = c \cdot c_k n^{k+l} + \mathcal{O}(n^{k+l-1}) ,$$

where $c = \text{lc}(q(n+1) - r(n))$. By comparison of the degrees of both sides of this equation we get $k = \deg(p(n)) - l$.

b. Assume $l := l^+ = \deg(q(n+1) + r(n)) > \deg(q(n+1) - r(n)) = l^-$. In this case $\deg(q(n)) = \deg(r(n)) = l$, so $q(n) = q_l n^l + \mathcal{O}(n^{l-1})$ and $r(n) = r_l n^l + \mathcal{O}(n^{l-1})$, where $q_l = r_l \ne 0$. We set

$$f(n) := c_k n^k + c_{k-1} n^{k-1} + \mathcal{O}(n^{k-2}) \quad \text{for } c_k \ne 0 ,$$

and therefore

$$f(n-1) = c_k n^k + (c_{k-1} - k \cdot c_k) n^{k-1} + \mathcal{O}(n^{k-2}) .$$

Substituting this into (10.1.2′) we get

$$\begin{aligned} p(n) &= \big(q(n+1) - r(n)\big) \cdot \big(c_k n^k + \mathcal{O}(n^{k-1})\big) \\ &\quad + \big(q(n+1) + r(n)\big) \cdot \big(\tfrac{k}{2} c_k n^{k-1} + \mathcal{O}(n^{k-2})\big) \\ &= \big(q_l (n+1)^l + q_{l-1}(n+1)^{l-1} - r_l n^l - r_{l-1} n^{l-1} + \mathcal{O}(n^{l-2})\big) \cdot \\ &\qquad \cdot \big(c_k n^k + \mathcal{O}(n^{k-1})\big) \\ &\quad + \big(q_l n^l + r_l n^l + \mathcal{O}(n^{l-1})\big) \cdot \big(\tfrac{k}{2} c_k n^{k-1} + \mathcal{O}(n^{k-2})\big) \end{aligned}$$

$$= \left(q_l n^l + q_l l n^{l-1} + q_{l-1} n^{l-1} - r_l n^l - r_{l-1} n^{l-1} + \mathcal{O}(n^{l-2})\right) \cdot$$
$$\cdot \left(c_k n^k + \mathcal{O}(n^{k-1})\right)$$
$$+ \left(q_l n^l + r_l n^l + \mathcal{O}(n^{l-1})\right) \cdot \left(\tfrac{k}{2} c_k n^{k-1} + \mathcal{O}(n^{k-2})\right)$$
$$= \underbrace{\left(q_l l + q_{l-1} - r_{l-1} + \tfrac{k}{2}(q_l + r_l)\right)}_{L(k)} \cdot c_k n^{k+l-1} + \mathcal{O}(n^{k+l-2}) \, .$$

Let $k_0$ be the root of the linear equation $L(k) = 0$, i.e.,

$$k_0 = \frac{-l q_l - q_{l-1} + r_{l-1}}{q_l} \, .$$

So we get the bound

$$\bar{k} = \begin{cases} \max\{k_0, \deg(p(n)) - l + 1\} & \text{if } k_0 \in \mathbb{Z}, \\ \deg(p(n)) - l + 1 & \text{otherwise.} \end{cases}$$ □

Combining all these facts we have an algorithm for computing $(s_n)_{n\in\mathbb{N}_0}$ such that

$$\sum_{i=1}^{n} a_i = s_n - s_0 \quad \text{for all } n \in \mathbb{N} \, .$$

We get the sequence of partial sums after normalizing to $(s_n - s_0)_{n\in\mathbb{N}_0}$. Now we are ready for stating Gosper's algorithm for summation of hypergeometric sequences.

**Algorithm SUM_G**(in: $(a_n)_{n\in\mathbb{N}}$; out: $(s_n)_{n\in\mathbb{N}_0}$, FLAG);
[$(a_n)_{n\in\mathbb{N}}$ is a hypergeometric sequence over $K$;
if the sequence of partial sums is hypergeometric then FLAG = "hypergeometric" and $s_n = \sum_{i=1}^{n} a_i$ for $n \in \mathbb{N}_0$, and if the sequence of partial sums is not hypergeometric then FLAG = "not hypergeometric".]

1. FLAG := "hypergeometric";
2. if $a_n \equiv 0$ then $\{s_n := 0;$ return$\}$;
3. [regular representation]
   $p(n) := 1$; $q(n) :=$ numerator of $a_n/a_{n-1}$; $r(n) :=$ denominator of $a_n/a_{n-1}$;
   while $\mathrm{res}_n(q(n), r(n+j))$ has a non-negative integral root do
   $\{j^* :=$ a non-negative integral root of $\mathrm{res}_n(q(n), r(n+j))$;
   $g(n) := \gcd(q(n), r(n+j^*))$;
   $p(n) := p(n)\prod_{i=1}^{j^*-1} g(n-k)$; $q(n) := q(n)/g(n)$; $r(n) := r(n)/g(n - j^*)\}$;
4. [degree bound for $f$]
   $l^+ := \deg(q(n+1) + r(n))$; $l^- := \deg(q(n+1) - r(n))$;
   if $l^+ \le l^-$
   then $k := \deg(p(n)) - l^-$
   else $\{k_0 := (-l^+ q_{l^+} - q_{l^+-1} + r_{l^+-1})/q_{l^+}$;

if $k_0 \in \mathbb{Z}$
then $k := \max\{k_0, \deg(p(n)) - l^+ + 1\}$
else $k := \deg(p(n)) - l^+ + 1$ };
if $k < 0$ then {FLAG := "not hypergeometric"; return};

5. [determination of $f$]
determine a polynomial $f(n)$ satisfying $p(n) = q(n+1)f(n) - r(n)f(n-1)$ and of degree $\leq k$ by solving a system of linear equations over $K$ for the indeterminate coefficients of $f(n) = c_k n^k + \ldots + c_0$;
6. [combination of partial results]
$s'_n := q(n+1) \cdot a_n \cdot f(n)/p(n)$;
$s_n := s'_n - s'_0$;
return.

*Example 10.1.1* (continued). We want to solve the summation problem for the series

$$\sum_{n=1}^{\infty} \underbrace{\frac{1}{n^2 + 2n}}_{a_n}$$

over $\mathbb{Q}$. $(n+1, n-1, n+2)$ is a regular representation of $a_n/a_{n-1}$. $l^+ = 1, l^- = 0$, $k_0 = 2$. So as a bound for $f(n) = c_k n^k + \ldots + c_0$ we get $k = 2$. Now we have to determine $c_2, c_1, c_0$ such that

$$n + 1 = n \cdot (c_2 n^2 + c_1 n + c_0) - (n+2) \cdot (c_2(n-1)^2 + c_1(n-1) + c_0) .$$

The corresponding linear system in $c_2, c_1, c_0$ has the solutions

$$c_0 = \lambda, \qquad c_1 = \frac{5 + 6\lambda}{4}, \qquad c_2 = \frac{3 + 2\lambda}{4}$$

for any $\lambda \neq 0$. So we get (for $\lambda = 0$)

$$f(n) = \frac{3n^2 + 5n}{4} .$$

Therefore the sequence of partial sums is hypergeometric and in fact

$$s_n = \sum_{i=1}^{n} \frac{1}{n^2 + 2n} = \frac{3n^2 + 5n}{4(n^2 + 3n + 2)} .$$

*Example 10.1.2.* We want to solve the summation problem for the series

$$\sum_{n=1}^{\infty} n \cdot x^n \quad \text{(i.e., } a_n = n \cdot x^n\text{)}$$

over $\mathbb{Q}(x)$. $a_n/a_{n-1} = n \cdot x/(n-1)$ is a rational function in $n$, so the sequence

$(a_n)_{n\in\mathbb{N}}$ is hypergeometric. In step (3) of SUM_G we start with the representation $p(n) = 1, q(n) = n \cdot x, r(n) = n-1$. $j^* = 1$ is a root of $\text{res}_n(q(n), r(n+j))$. The updating process yields the regular representation $p(n) = n, q(n) = x, r(n) = 1$. The degrees $l^+$ and $l^-$ in step (4) are equal, so we get the degree bound $k = 1$. Comparing coefficients of like powers in

$$n = x \cdot (c_1 n + c_0) - 1 \cdot (c_1(n-1) + c_0)$$

we get

$$c_1 = \frac{1}{x-1}, \quad c_2 = \frac{1}{(x-1)^2}, \qquad \text{i.e., } f(n) = \frac{1}{x-1}n - \frac{1}{(x-1)^2} .$$

Setting $s_n' := x \cdot (n \cdot x^n) \cdot f(n)/n = (1/(x-1)^2) \cdot (n \cdot x^{n+2} - (n+1) \cdot x^{n+1})$ and normalizing to $s_n := s_n' - s_0'$, we finally arrive at the formula

$$s_n = \sum_{i=1}^{n} n \cdot x^n = \frac{n \cdot x^{n+2} - (n+1) \cdot x^{n+1} + x}{(x-1)^2}$$

for the partial sum.

### Exercises

1. Apply Gosper's algorithm for solving the summation problem for the series

$$\sum_{n=1}^{\infty} \underbrace{\frac{\prod_{j=1}^{n-1} bj^2 + cj + d}{\prod_{j=1}^{n} bj^2 + cj + c}}_{a_n} .$$

2. Are there hypergeometric solutions for the following summation problems?
   a. $\sum_{i=1}^{n} i^2 2^i$
   b. $\sum_{i=1}^{n} i \cdot i!$
   c. $\sum_{i=1}^{n} (1/i!)$
   d. $\sum_{i=1}^{n} i!$
3. What is a good way of determining whether the resultant in step (3) of SUM_G has an integral root, and in fact finding one if it exists?

## 10.2 Bibliographic notes

Approaches to the summation problem are discussed in Lafon (1983) and in Graham et al. (1994). An extension of Gosper's approach is described in Karr (1985). Recently Zeilberger (1990, 1991; Wilf and Zeilberger 1992) has greatly advanced the field with his method of creative telescoping based on holonomic systems. Results in this direction are also reported in Paule (1993, 1994). For an overview of these recent developments we refer to Paule and Strehl (1994).

# 11 Parametrization of algebraic curves

## 11.1 Plane algebraic curves

Algebraic geometry is the study of geometric objects defined as the zeros of polynomial equations. So it is not surprising that many of the techniques in algebraic geometry become computationally feasible once we have algorithmic solutions for the relevant problems in commutative algebra, i.e., the algebra of polynomial rings. We take a look at one particular problem in algebraic geometry, the rational parametrization of algebraic curves. For an introduction to algebraic curves we refer to Walker (1950) or Fulton (1969).

Throughout this chapter let $K$ be an algebraically closed field of characteristic 0.

*Definition 11.1.1.* For any field $F$, the *$n$-dimensional affine space* over $F$ is defined as

$$\mathbb{A}^n(F) := F^n = \{(a_1, \ldots, a_n) \mid a_i \in F\} .$$

$\mathbb{A}^2(F)$ is the *affine plane* over $F$.

An *affine plane algebraic curve* $\mathcal{C}$ in $\mathbb{A}^2(F)$ is the set of zeros of a polynomial $f(x, y) \in F[x, y]$, i.e.,

$$\mathcal{C} = \{(a_1, a_2) \mid f(a_1, a_2) = 0,\ (a_1, a_2) \in \mathbb{A}^2(F)\} .$$

$f$ is called a *defining polynomial* of the curve $\mathcal{C}$. The curve $\mathcal{C}$ has *degree $d$*, iff $d$ is the degree of a defining polynomial of $\mathcal{C}$ with smallest degree. The curve $\mathcal{C}$ is *irreducible* iff it has an absolutely irreducible polynomial $f$ defining it.

Obviously a particular curve $\mathcal{C}$ can be defined by many different polynomials, e.g., the circle is defined by $x^2 + y^2 - 1$, but also by $(x^2 + y^2 - 1)^2$. Two polynomials define the same curve if and only if they have the same squarefree factors. So the circle is, of course, a curve of degree 2. In fact, w.l.o.g. we can always assume that a defining polynomial is squarefree.

The problem with affine space is that, for instance, although curves of degrees $m$ and $n$, respectively, generally intersect in $m \cdot n$ points (unless they have a common component), this might not be true for particular examples. For example, two parallel lines do not intersect, a hyperbola does not intersect its asymptotic lines. This problem is resolved by considering the curves in projective space. We consider the homogenization $f^*(x, y, z)$ of the polynomial

$f(x, y)$ of degree $d$, i.e., if

$$f(x, y) = f_d(x, y) + f_{d-1}(x, y) + \ldots + f_0(x, y) ,$$

where the $f_i$'s are forms of degree $i$, respectively (i.e., all the terms occurring in $f_i$ are of the same degree and $\deg(f_i) = i$), then

$$f^*(x, y, z) = f_d(x, y) + f_{d-1}(x, y) \cdot z + \ldots + f_0(x, y) \cdot z^d .$$

$f^*$ is a homogeneous polynomial. For all $\alpha \in K^*$ we have $f^*(a, b, c) = 0 \Longleftrightarrow f^*(\alpha a, \alpha b, \alpha c) = 0$, and $f(a, b) = 0 \Longleftrightarrow f^*(a, b, 1) = 0$. $f(a, b, 0) = f_d(a, b) = 0$ means that there is a zero of $f$ "at infinity" in the direction $(a, b)$. By adding these "points at infinity" to affine space we get the corresponding projective space.

*Definition 11.1.2.* For any field $F$, the *n-dimensional projective space* over $F$ is defined as

$$\mathbb{P}^n(F) := \{(a_1 : \ldots : a_{n+1}) \mid (a_1, \ldots, a_{n+1}) \in F^{n+1} \setminus \{(0, \ldots, 0)\}\} ,$$

where $(a_1 : \ldots : a_{n+1}) = \{(\alpha a_1, \ldots, \alpha a_{n+1}) \mid \alpha \in F^*\}$. So a point in $\mathbb{P}^n(F)$ has many representations as an $(n + 1)$-tuple, since $(a_1 : \ldots : a_{n+1})$ and $(\alpha a_1 : \ldots : \alpha a_{n+1})$, for any $\alpha \in F^*$, denote the same point $P$. $(a_1 : \ldots : a_{n+1})$ are *homogeneous coordinates* for $P$. $\mathbb{P}^2(F)$ is the *projective plane* over $F$.

A *projective plane algebraic curve* $\mathcal{C}$ in $\mathbb{P}^2(F)$ is the set of zeros of a homogeneous polynomial $f(x, y, z) \in F[x, y, z]$, i.e.,

$$\mathcal{C} = \{(a_1 : a_2 : a_3) \mid f(a_1, a_2, a_3) = 0, \ (a_1 : a_2 : a_3) \in \mathbb{P}^2(F)\} .$$

$f$ is called a *defining polynomial* of the curve $\mathcal{C}$.

We write simply $\mathbb{A}^2$ or $\mathbb{P}^2$ for $\mathbb{A}^2(K)$ or $\mathbb{P}^2(K)$, respectively. Whenever we have a curve $\mathcal{C}$ in $\mathbb{A}^2$ defined by a polynomial $f(x, y)$, we can associate with it the curve $\mathcal{C}^*$ in $\mathbb{P}^2$ defined by $f^*(x, y, z)$. Any affine point $(a, b)$ of $\mathcal{C}$ corresponds to a point $(a : b : 1)$ of $\mathcal{C}^*$, and in addition to these points $\mathcal{C}^*$ contains only finitely many points "at infinity," namely with coordinates $(a : b : 0)$. These are the zeros of $f_d$, the form of highest degree in $f$.

In $\mathbb{P}^2$ Bezout's theorem holds, which states that if $f, g \in K[x, y, z]$ are relatively prime homogeneous polynomials, i.e., the projective curves $\mathcal{C}$ and $\mathcal{D}$ defined by $f$ and $g$, respectively, do not have a common component, then $\mathcal{C}$ and $\mathcal{D}$ have exactly $\deg(f) \cdot \deg(g)$ projective points in common, counting multiplicities.

*Definition 11.1.3.* a. Let $\mathcal{C}$ be a curve in $\mathbb{A}^2$ defined by the polynomial $f(x, y)$. Let $P = (a, b)$ be a point on $\mathcal{C}$. $P$ is a *simple point* on $\mathcal{C}$ iff $\frac{\partial f}{\partial x}(P) \neq 0$ or

$\frac{\partial f}{\partial y}(P) \neq 0$. In this case the *tangent* to $C$ at $P$ is uniquely determined as

$$\frac{\partial f}{\partial x}(P) \cdot (x-a) + \frac{\partial f}{\partial y}(P) \cdot (y-b) = 0 .$$

If $P$ is not simple, i.e., both partial derivatives vanish at $P$, then $P$ is called a *multiple point* or *singularity* on $C$. Let $m$ be such that for all $i + j < m$ the partial derivative $\frac{\partial^{i+j} f}{\partial x^i \partial y^j}$ vanishes at $P$, but at least one of the partial derivatives of order $m$ does not vanish at $P$. Then $m$ is called the *multiplicity* of $P$ on $C$, or, in other words, $P$ is an $m$-fold point on $C$. In this case the polynomial

$$\sum_{i+j=m} \frac{1}{i!j!} \cdot \frac{\partial^m f}{\partial x^i \partial y^j}(P) \cdot (x-a)^i \cdot (y-b)^j \qquad (11.1.1)$$

factors completely into linear factors, the *tangents* of $C$ at $P$. An $m$-fold point $P$ on $C$ is *ordinary* iff all the $m$ linear factors of (11.1.1) are different, i.e., all the tangents are different.

b. Let $C$ be a curve in $\mathbb{P}^2$ defined by the homogeneous polynomial $f(x, y, z)$. Let $P = (a : b : c)$ be a point on $C$. W.l.o.g. let $c = 1$ (for the other coordinates proceed analogously). $P$ is a simple or multiple point on $C$ depending on whether $Q = (a, b)$ is a simple or multiple point on the affine curve defined by $f(x, y, 1)$.

An irreducible curve has only finitely many singularities. The multiplicity of the origin $(0, 0)$ on an affine curve $C$ defined by the polynomial $f(x, y)$ is particularly easy to deduce: the multiplicity is the least degree of any term occurring in $f$. The tangents of $C$ at $(0, 0)$ are the factors of the form of least degree in $f$.

*Example 11.1.1.* Some plane algebraic curves can be rationally parametrized, e.g., the tacnode curve (see Fig. 3) defined by the polynomial

$$f(x, y) = 2x^4 - 3x^2y + y^2 - 2y^3 + y^4 = 0$$

in $\mathbb{A}^2(\mathbb{C})$ can be parametrized as

$$x(t) = \frac{t^3 - 6t^2 + 9t - 2}{2t^4 - 16t^3 + 40t^2 - 32t + 9}, \qquad y(t) = \frac{t^2 - 4t + 4}{2t^4 - 16t^3 + 40t^2 - 32t + 9} .$$

That is, the points on the tacnode curve are exactly the values of $(x(t), y(t))$ for $t \in \mathbb{C}$, except for finitely many exceptions. We will see in Sect. 11.2 how such a parametrization can be computed.

*Definition 11.1.4.* The irreducible affine curve $C$ in $\mathbb{A}^2(K)$ defined by the irreducible polynomial $f(x, y)$ is called *rational* (or *parametrizable*) iff there are rational functions $\phi(t), \chi(t) \in K(t)$ such that

a. for almost all (i.e., for all but a finite number of exceptions) $t_0 \in K$, $(\phi(t_0), \chi(t_0))$ is a point on $\mathcal{C}$, and
b. for almost every point $(x_0, y_0)$ on $C$ there is a $t_0 \in K$ such that $(x_0, y_0) = (\phi(t_0), \chi(t_0))$.

In this case $(\phi, \chi)$ are called a *(rational) parametrization* of $\mathcal{C}$.

The irreducible projective curve $\mathcal{C}$ in $\mathbb{P}^2(K)$ defined by the irreducible polynomial $f(x, y, z)$ is called *rational* (or *parametrizable*) iff there are rational functions $\phi(t), \chi(t), \psi(t) \in K(t)$ such that

a$'$. for almost all $t_0 \in K$, $(\phi(t_0) : \chi(t_0) : \psi(t_0))$ is a point on $\mathcal{C}$, and
b$'$. for almost every point $(x_0 : y_0 : z_0)$ on $\mathcal{C}$ there is a $t_0 \in K$ such that $(x_0 : y_0 : z_0) = (\phi(t_0) : \chi(t_0) : \psi(t_0))$.

In this case $(\phi, \chi, \psi)$ are called a *(rational) parametrization* of $\mathcal{C}$.

*Example 11.1.2.* An example of a curve which is not rational over $\mathbb{C}$ is the curve $\mathcal{C}_1$, defined by $x^3 + y^3 = 1$. Suppose $\phi = p(t)/r(t)$, $\chi = q(t)/r(t)$ is a parametrization of $\mathcal{C}_1$, where $\gcd(p, q, r) = 1$. Then

$$p^3 + q^3 - r^3 = 0 .$$

Differentiating this equation by $t$ we get

$$3 \cdot (p'p^2 + q'q^2 - r'r^2) = 0 .$$

So $p^2, q^2, r^2$ are a solution of the system of linear equations with coefficient matrix

$$\begin{pmatrix} p & q & -r \\ p' & q' & -r' \end{pmatrix} .$$

Gaussian elimination reduces this coefficient matrix to

$$\begin{pmatrix} qp' - q'p & 0 & q'r - qr' \\ 0 & qp' - q'p & r'p - rp' \end{pmatrix} .$$

So

$$p^2 : q^2 : r^2 = qr' - rq' : rp' - pr' : pq' - qp' .$$

Since $p, q, r$ are relatively prime, this proportionality implies

$$p^2 | (qr' - rq'), \quad q^2 | (rp' - pr'), \quad r^2 | (pq' - qp') .$$

Suppose $\deg(p) \geq \deg(q), \deg(r)$. Then the first divisibility implies $2\deg(p) \leq \deg(q) + \deg(r) - 1$, a contradiction. Similarly we see that $\deg(q) \geq \deg(p), \deg(r)$ and $\deg(r) \geq \deg(p), \deg(q)$ are impossible. Thus, there can be no parametrization of $\mathcal{C}_1$.

The rationality problem for an affine curve is equivalent to the rationality problem for the associated projective curve.

**Lemma 11.1.1.** Let $\mathcal{C}$ be an irreducible affine curve and $\mathcal{C}^*$ its corresponding projective curve. Then $\mathcal{C}$ is rational if and only if $\mathcal{C}^*$ is rational. A parametrization of $\mathcal{C}$ can be computed from a parametrization of $\mathcal{C}^*$ and vice versa.

*Proof.* Let

$$\phi^*(t) = \frac{u_1(t)}{u_2(t)}, \quad \chi^*(t) = \frac{v_1(t)}{v_2(t)}, \quad \psi^*(t) = \frac{w_1(t)}{w_2(t)}$$

be a parametrization of $\mathcal{C}^*$, i.e., $f^*(\phi^*(t), \chi^*(t), \psi^*(t)) = 0$, where $f(x, y)$ is a defining polynomial of $\mathcal{C}$. Observe that $w_1(t) \neq 0$, since the curve $\mathcal{C}^*$ can have only finitely many points at infinity. Hence,

$$\phi(t) = \frac{u_1(t)w_2(t)}{u_2(t)w_1(t)}, \quad \chi(t) = \frac{v_1(t)w_2(t)}{v_2(t)w_1(t)}$$

is a parametrization of the affine curve $\mathcal{C}$.

Conversely, a rational parametrization of $\mathcal{C}$ can always be extended to a parametrization of $\mathcal{C}^*$ by setting the $z$-coordinate to 1. □

A good measure for the complexity of an algebraic curve is the genus of the curve. It turns out that exactly the curves of genus 0 are rational. In defining the genus and also later in computing a parametrization, we treat only the case in which all the singular points of the curve are ordinary. For an algebraic treatment of the non-ordinary case we refer to Sendra and Winkler (1991).

*Definition 11.1.5.* Let $\mathcal{C}$ be an irreducible curve of degree $d$ in $\mathbb{P}^2(K)$ with singular points $P_1, \ldots, P_n$, having the multiplicities $r_1, \ldots, r_n$, respectively. Let all these singular points be ordinary points. Then the *genus* of $\mathcal{C}$ is defined as

$$\text{genus}(\mathcal{C}) = \tfrac{1}{2}\left[(d-1)(d-2) - \sum_{i=1}^{n} r_i(r_i - 1)\right].$$

The genus of an irreducible affine curve is the genus of the associated projective curve (if all the singularities are ordinary).

The genus of $\mathcal{C}$, genus($\mathcal{C}$), is a nonnegative integer for any irreducible curve $\mathcal{C}$. In fact, $(d-1)(d-2)$ is a bound for $\sum_{i=1}^{n} r_i(r_i - 1)$ for any irreducible curve. The curves of genus 0 are those curves, which achieve the theoretically highest possible count of singularities. These, moreover, are exactly the curves which can be parametrized. A proof of this fact can be found in Walker (1950: theorem 5.1). So, irreducible conics (curves of degree 2) are rational, and an irreducible cubic (curve of degree 3) is rational if and only if it has a double point. This, again, shows that the curve in Example 11.1.2 is not rational.

**Theorem 11.1.2.** An algebraic curve $\mathcal{C}$ (having only ordinary singularities) is rationally parametrizable if and only if genus($\mathcal{C}$) = 0.

A projective curve $\mathcal{C}$ of degree $d$ is defined by a polynomial of the form

$$f(x, y, z) = \sum_{\substack{i,j,k \in \mathbb{N}_0 \\ i+j+k=d}} a_{ijk} x^i y^j z^k \; . \tag{11.1.2}$$

$\mathcal{C}$ is uniquely determined by the coefficients $a_{ijk}$, and on the other hand the coefficients, up to a common factor, determine uniquely the curve $\mathcal{C}$, if we distinguish curves having different multiplicities of their respective components, i.e., if we assume that, e.g., $xy^2$ and $x^2y$ define different curves. We can identify a curve with its defining polynomial (up to a constant factor), i.e., we can view the $(d+1)(d+2)/2$ coefficients in (11.1.2) as projective coordinates of the curve $f(x, y, z)$. Collecting all curves of degree $d$ satisfying an equation of the form (11.1.2) we get a projective space $\mathbb{P}^{N_d}$, where $N_d = \frac{(d+1)(d+2)}{2} - 1 = \frac{d(d+3)}{2}$.

*Definition 11.1.6.* Let $\mathbb{P}^{N_d}$ be the projective space of curves of degree $d$ as introduced above. A linear subspace $\mathbb{P}^m$ of $\mathbb{P}^{N_d}$ is called a *linear system (of curves) of degree d*. Such a linear system is determined by $m + 1$ linearly independent curves $f_0, f_1, \ldots, f_m$ in the system, i.e., every other curve in the system is of the form

$$\sum_{i=0}^{m} \lambda_i f_i(x, y, z) \; .$$

Let $P \in \mathbb{P}^2$ be such that all partial derivations of order $< r$ vanish at $P$, for any curve in the linear system $\mathbb{P}^m$. Then we call $P$ a *base point of multiplicity r* of the linear system $\mathbb{P}^m$.

Linear systems of curves are often created by specifying base points for them. So we might consider the linear subsystem $L$ of $\mathbb{P}^{N_d}$ having the points $P_1, \ldots, P_s$ as base points of multiplicities $r_1, \ldots, r_s$, respectively. A base point of multiplicity $r$ implies $\frac{r(r+1)}{2}$ linear conditions on the coefficients in the system.

*Example 11.1.3.* Let us determine the linear system $\hat{L}$ of curves of degree 2, having $O = (0:0:1)$ as a base point of multiplicity 2, and $P = (0:1:1)$ as a base point of multiplicity 1.

The full space of curves of degree 2 is

$$\mathbb{P}^{N_2} = \{(a_1 : \ldots : a_6) \mid a_1, \ldots, a_6 \in \mathbb{C}\} \; ,$$

i.e., an arbitrary such curve has the equation

$$f(x, y, z) = a_1x^2 + a_2y^2 + a_3xy + a_4xz + a_5yz + a_6z^2 \; .$$

The base point requirements lead to the following linear equations

$$\begin{aligned}
a_6 &= 0 && \left(\text{from } f(O) = 0\right), \\
a_4 &= 0 && \left(\text{from } \tfrac{\partial f}{\partial x}(O) = 0\right), \\
a_5 &= 0 && \left(\text{from } \tfrac{\partial f}{\partial y}(O) = 0\right), \\
a_2 + a_5 + a_6 &= 0 && \left(\text{from } f(P) = 0\right).
\end{aligned}$$

So $\hat{L}$ consists of all quadratic curves of the form

$$\hat{f}(x, y, z) = a_1 x^2 + a_3 xy = x(a_1 x + a_3 y) .$$

i.e., every curve in $\hat{L}$ decomposes into the line $x = 0$ and an arbitrary line through the origin.

### Exercises

1. Can the result of Example 11.1.2 be generalized to $x^n + y^n = 1$ and an arbitrary algebraically closed field $K$? Or, if not, for which $n$ and $K$ is the corresponding curve irrational?
2. Can a projective curve have infinitely many points at infinity? What does this mean for a defining polynomial?
3. What is the genus of the affine curve defined by

$$f(x, y) = (x^2 + 4y + y^2)^2 - 16(x^2 + y^2) \ ?$$

4. Give a linearly independent basis for the system $\hat{L}$ of Example 11.1.3.

## 11.2 A parametrization algorithm

For this section we assume that $K$ is a computable field of characteristic 0, we call it the *field of definition*. $\bar{K}$ denotes the algebraic closure of $K$. The defining polynomial of the curve $C$ that we want to parametrize will have coefficients in $K$. The curve $C$ itself, however, is considered to be a curve over $\bar{K}$. In the parametrization process we might have to extend $K$ algebraically. We will ultimately construct a parametrization (if one exists) over some field $K(\gamma) \subseteq \bar{K}$, where $\gamma$ is algebraic over $K$. Of course, we will be interested in keeping the degree of $\gamma$ as low as possible.

Points on algebraic curves occur only in full conjugacy classes. If we choose all the points in such a conjugacy class as base points of a linear system, then we need no algebraic extension of $K$ for expressing the equation of the linear system.

**Lemma 11.2.1.** Let $f(x, y, z)$ be a homogeneous polynomial in $K[x, y, z]$ defining an algebraic curve $C$ in $\mathbb{P}^2(\bar{K})$. Let $(a_1(\alpha) : a_2(\alpha) : a_3(\alpha))$ be a point of multiplicity $r$ on $C$, $\alpha$ algebraic over $K$ with minimal polynomial $p(t)$, and $a_1, a_2, a_3 \in K[t]$. Then for every conjugate $\beta$ of $\alpha$, also the point $(a_1(\beta) : a_2(\beta) : a_3(\beta))$ is a point of multiplicity $r$ of $C$.

*Proof.* Let $g(x, y, z)$ be a derivative of $f(x, y, z)$ of order $i$, $0 \leq i < r$. Then $g(a_1(\alpha), a_2(\alpha), a_3(\alpha)) = 0$ if and only if $g(a_1(\beta), a_2(\beta), a_3(\beta)) = 0$. So also $(a_1(\beta) : a_2(\beta) : a_3(\beta))$ is a point of multiplicity $r$ on $C$. □

*Definition 11.2.1.* If $p(t) \in K[t]$ is irreducible and $a_1, a_2, a_3 \in K[t]$ with $p \not| \gcd(a_1, a_2, a_3)$, then we call

$$\{ (a_1(\alpha) : a_2(\alpha) : a_3(\alpha)) \mid p(\alpha) = 0 \}$$

a *family of conjugate algebraic points.*

**Lemma 11.2.2.** Let $L$ be a linear system of curves of degree $d$, defined over $K$. Let $P_\alpha = \{(a_1(\alpha) : a_2(\alpha) : a_3(\alpha) \mid p(\alpha) = 0\}$ be a family of conjugate algebraic points. Then also the subsystem $\hat{L}$ of $L$, having all the points in $P_\alpha$ as base points of multiplicity $r$, is defined over $K$.

*Proof.* The linear system $L$ is defined by a polynomial $h$ with some indetermined coefficients $t_i$,

$$h(x, y, z) = \sum_i c_i t_i x^{m_{i,1}} y^{m_{i,2}} z^{d-m_{i,1}-m_{i,2}} ,$$

where the $c_i$'s are in $K$. We restrict $L$ to the subsystem $\hat{L}$ having all the points in $P_\alpha$ as base points of multiplicity 1 by requiring

$$\tilde{h}(\alpha_i) = 0 \quad \text{for all } \beta \text{ conjugate to } \alpha ,$$

where $\tilde{h}(t) = h(a_1(t), a_2(t), a_3(t))$. This, however, means that $\tilde{h}$ must be divisible by the minimal polynomial $p(t)$, i.e.,

$$\mathrm{rem}(\tilde{h}(t), p(t)) = 0 ,$$

leading to linear conditions on the indeterminate coefficients $t_i$ of $L$. So the resulting subsystem $\hat{L}$ will be defined over $K$.

The same idea applies to base points of higher multiplicity. We only have to use derivatives of $h$ instead of $h$. □

A parametrization of a curve $C$ is a generic point of $C$, i.e., of the ideal $I = \langle f \rangle$, where $f$ is a defining polynomial of least degree for $C$ (see van der

Waerden 1970). Only prime ideals have a generic point, so only irreducible curves can be parametrizable.

The simplest case of a parametrizable curve $C$ is an irreducible curve of degree $d$ having a point $P$ of multiplicity $d-1$. Of course, this must be a rational point and there can be no other singularity of $C$. W.l.o.g. we can assume that $P$ is the origin $O=(0,0)$. Otherwise $P$ can be moved to the origin by a linear change of coordinates. Now we consider lines through the origin, i.e., the linear system $L^1$ of curves of degree 1 having $O$ as a point of multiplicity 1. The equation for $L^1$ is

$$y = tx ,$$

for an undetermined coefficient $t$. Intersecting $C$ with an element of $L^1$ we get, of course, the origin as an intersection point of multiplicity $d-1$. By Bezout's theorem, we must (in general) get exactly one more intersection point $P$ depending rationally on the value of $t$. On the other hand, every point on the curve $C$ can be constructed in this way. So the coordinates of $P$ are rational functions in $t$,

$$P = \left(\frac{u_1(t)}{u_2(t)}, \frac{v_1(t)}{v_2(t)}\right) .$$

These rational functions are a parametrization of the curve $C$. We give a more precise statement in the following lemma.

**Lemma 11.2.3.** Let $C$ be an irreducible affine curve of degree $d$ defined by the polynomial $f(x,y) = f_d(x,y) + \ldots + f_0(x,y)$, having a $(d-1)$-fold point at the origin. Then $C$ is rational and a rational parametrization is

$$x(t) = -\frac{f_{d-1}(1,t)}{f_d(1,t)}, \qquad y(t) = -t \cdot \frac{f_{d-1}(1,t)}{f_d(1,t)} .$$

*Proof.* Since the origin is a $(d-1)$-fold point of $C$, the defining polynomial $f$ is of the form $f(x,y) = f_d(x,y) + f_{d-1}(x,y)$, where $f_d$ and $f_{d-1}$ are non-vanishing forms of degree $d$ and $d-1$, respectively. Thus, for every $t_0 \in \bar{K}$, with the exception of the roots of $f_d(1,t_0)$, the polynomial $f$ vanishes on the point

$$P_{t_0} = \left(-\frac{f_{d-1}(1,t_0)}{f_d(1,t_0)}, -t_0 \cdot \frac{f_{d-1}(1,t_0)}{f_d(1,t_0)}\right) .$$

On the other hand, for every point $(x_0, y_0)$ on $C$ such that $x_0 \neq 0$ and $y_0 \neq 0$ one has

$$x_0^{d-1}\big(x_0 \cdot f_d(1, \frac{y_0}{x_0}) + f_{d-1}(1, \frac{y_0}{x_0})\big) = 0,$$

$$y_0^{d-1}\big(y_0 \cdot f_d(\frac{x_0}{y_0}, 1) + f_{d-1}(\frac{x_0}{y_0}, 1)\big) = 0 .$$

Thus, since $x_0 \neq 0$, $y_0 \neq 0$, we get

$$x_0 = -\frac{f_{d-1}(1, \frac{y_0}{x_0})}{f_d(1, \frac{y_0}{x_0})} ,$$

$$y_0 = -\frac{f_{d-1}(\frac{x_0}{y_0}, 1)}{f_d(\frac{x_0}{y_0}, 1)} = -\frac{y_0}{x_0} \cdot \frac{f_{d-1}(1, \frac{y_0}{x_0})}{f_d(1, \frac{y_0}{x_0})} .$$

Therefore the point $(x_0, y_0)$ is generated by the value $y_0/x_0$ of the parameter $t$. The number of intersections of $C$ and the axes is finite. That concludes the proof. □

*Example 11.2.1.* Let $C_1$ be the affine curve defined by $f_1(x, y)$. See Fig. 12.

$$f_1(x, y) = x^3 + x^2 - y^2 .$$

$C_1$ has a double point at the origin $O = (0, 0)$. Intersecting $C_1$ by the line $y = tx$, we get the additional intersection point $(t^2 - 1, t^3 - t)$. So

$$x(t) = t^2 - 1, \quad y(t) = t^3 - t$$

is a parametrization of $C_1$.

Rational cubics can always be parametrized in this way. In general, a parametrizable curve will not have this nice property of being parametrizable by lines. What we can do in the general situation is to determine a linear system of curves $L^{d-2}$ of degree $d - 2$ (or some other suitable degree), having every $r$-fold singularity of $C$ as a base point of multiplicity $r - 1$. This linear system is called the system of *adjoint curves* or *adjoints* of $C$. Then we know that the number of intersections of $C$ and a general element of $L^{d-2}$ will be

$$d(d-2) = \underbrace{\sum_{P \in C} m_P(m_P - 1)}_{=(d-1)(d-2)} + (d-2) ,$$

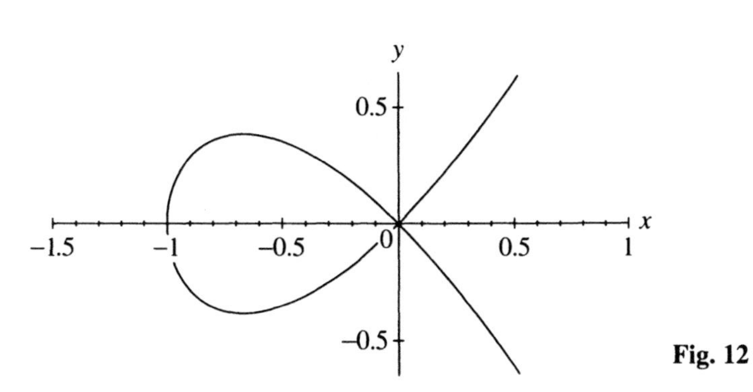


**Fig. 12**

where $m_P$ denotes the multiplicity of $P$ on the curve $\mathcal{C}$. If we fix $d-3$ simple points on $\mathcal{C}$ and make them base points of multiplicity 1 on $L^{d-2}$, then there will be exactly 1 other intersection point of $\mathcal{C}$ and a general element of $L^{d-2}$ depending rationally on $t$. This "free" intersection point will give us a parametrization of $\mathcal{C}$ as in the case of parametrization by lines.

*Example 11.2.2.* Let $\mathcal{C}_2$ be the affine curve defined by $f_2(x, y)$. See Fig. 13.

$$f_2(x, y) = (x^2 + 4y + y^2)^2 - 16(x^2 + y^2) .$$

$\mathcal{C}_2$ has a double point at the origin $(0, 0)$ as the only affine singularity. But if we move to the associated projective curve $\mathcal{C}_2^*$ defined by the homogeneous polynomial

$$f_2^*(x, y, z) = (x^2 + 4yz + y^2)^2 - 16(x^2 + y^2)z^2 ,$$

we see that the singularities of $\mathcal{C}_2^*$ are

$$O = (0 : 0 : 1), \qquad P_{1,2} = (1 : \pm i : 0) .$$

$P_{1,2}$ is a family of conjugate algebraic points on $\mathcal{C}_2^*$. All of these singularities have multiplicity 2, so the genus of $\mathcal{C}_2^*$ is 0, i.e., it can be parametrized. We also know that the affine curve $\mathcal{C}_2$ is parametrizable, from Lemma 11.1.1.

In order to achieve a parametrization, we need a simple point on $\mathcal{C}_2^*$. Intersecting $\mathcal{C}_2^*$ by the line $x = 0$, we get of course the origin as a multiple intersection point. The other intersection point is

$$Q = (0 : -8 : 1) .$$

So now we construct the system $L^2$ of curves of degree 2, having $O$, $P_{1,2}$ and $Q$ as base points of multiplicity 1. The full system of curves of degree 2 is of

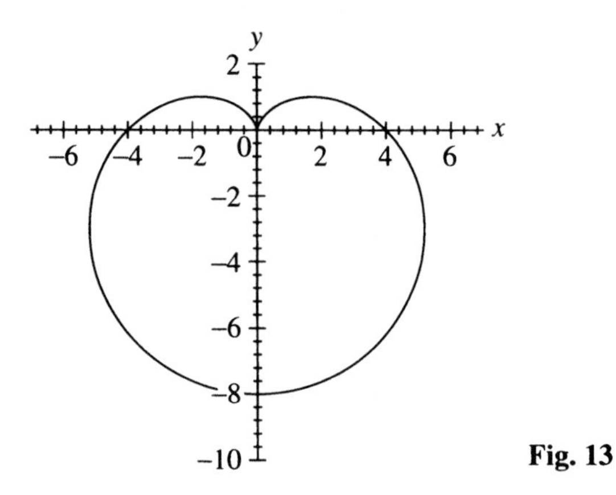


**Fig. 13**

the form

$$a_1x^2 + a_2y^2 + a_3z^2 + a_4xy + a_5xz + a_6yz$$

for arbitrary coefficients $a_1, \ldots, a_6$. Requiring that $O$ be a base point leads to the linear equation

$$a_3 = 0 \ .$$

We apply Lemma 11.2.2 for making $P_{1,2}$ base points of $L_2$. This leads to the equations

$$a_4 = 0 \ ,$$
$$a_1 - a_2 = 0 \ .$$

Finally, to make $Q$ a base point we have to satisfy

$$64a_2 + a_3 - 8a_6 = 0 \ .$$

This leaves exactly 2 parameters unspecified, say $a_1$ and $a_5$. Since curves are defined uniquely by polynomials only up to a nonzero constant factor, we can set one of these parameters to 1. Thus, the system $L^2$ depends on 1 free parameter $a_1 = t$, and its defining equation is

$$h(x, y, z, t) = tx^2 + ty^2 + xz + 8tyz \ .$$

The affine version $L_a^2$ of $L^2$ is defined by

$$h_a(x, y, t) = tx^2 + ty^2 + x + 8ty \ .$$

Now we determine the free intersection point of $L_a^2$ and $C_2$. The non-constant factors of $\mathrm{res}_x(f_2(x, y), h_a(x, y, t))$ are

$$y^2 \ ,$$
$$y + 8 \ ,$$
$$(256t^4 + 32t^2 + 1)y + (2048t^4 - 128t^2) \ .$$

The first two factors correspond to the affine base points of the linear system $L^2$, and the third one determines the $y$-coordinate of the free intersection point depending rationally on $t$.

Similarly, the non-constant factors of $\mathrm{res}_y(f_2(x, y), h_a(x, y, t))$ are

$$x^3 \ ,$$
$$(256t^4 + 32t^2 + 1)x + 1024t^3.$$

The first factor corresponds to the affine base points of the linear system $L^2$, and the second one determines the $x$-coordinate of the free intersection point depending rationally on $t$.

So we have found a rational parametrization of $C_2$, namely

$$x(t) = \frac{-1024t^3}{256t^4 + 32t^2 + 1},$$

$$y(t) = \frac{-2048t^4 + 128t^2}{256t^4 + 32t^2 + 1}.$$

In the previous example we were lucky enough to find a rational simple point on the curve, allowing us to determine a rational parametrization over the field of definition $\mathbb{Q}$. In fact, there are methods for determining whether a curve of genus 0 has rational simple points, and if so find one. We cannot go into more details here, but we refer the reader to Sendra and Winkler (1994). Of course, we can easily find a simple point on the curve having coordinates in a quadratic algebraic extension of $\mathbb{Q}$. We simply intersect $C_2$ by an arbitrary line through the origin. Using such a point, we would ultimately get a parametrization of $C_2$ having coefficients in the respective quadratic extension of $\mathbb{Q}$.

From the work of Noether (1884) and Hilbert and Hurwitz (1890) we know that it is possible to parametrize any curve $\mathcal{C}$ of genus 0 over the field of definition $K$, if $\deg(\mathcal{C})$ is odd, and over some quadratic extension of $K$, if $\deg(\mathcal{C})$ is even. An algorithm which actually achieves this optimal field of parametrization is presented in Sendra and Winkler (1994). Moreover, if the field of definition is $\mathbb{Q}$, we can also decide if the curve can be parametrized over $\mathbb{R}$, and if so, compute a parametrization over $\mathbb{R}$. Space curves can be handled by projecting them to a plane along a suitable axis (Garrity and Warren 1989), parametrizing the plane curve, and inverting the projection.

### Exercises

1. Let $P_\alpha$ be a family of conjugate algebraic points on the projective curve $\mathcal{C}$. Prove that all the points in the family have the same multiplicity on $\mathcal{C}$.
2. Which parametrization of $C_2$ do you get if you use the point $Q_2$ as a base point for $L^2$ in Example 11.2.2?

## 11.3 Bibliographic notes

A computational method for curve parametrization is described in Abhyankar and Bajaj (1988). However, the question of algebraic extension of the field of definition is not addressed. Schicho (1992) investigates more general linear systems for parametrization.

Sederberg (1986) gives an algorithm for transforming any rational parametrization into a proper one, traversing the curve only once, roughly speaking.

An alternative method for parametrization, based on computing a generating element $p$ of $\bar{K}(x)[y]/\langle f\rangle$, the function field of the curve, is presented in van Hoeij (1994).

# Solutions of selected exercises

### Section 1.3

1. By inspection. $\deg(p \cdot q) = \deg(p) + \deg(q)$, for non-zero polynomials.
2. a. Suppose $(a + b\sqrt{-5})(c + d\sqrt{-5}) = (ac - 5bd) + (cb + ad)\sqrt{-5} = 0$ and both factors are non-zero. Clearly $c = 0$ is impossible. From the integral part of the equation we get $a = 5bd/c$. Substituting this into the coefficient of $\sqrt{-5}$, we get $c^2 = -5d^2$, which is impossible.

   b. For $\alpha = a + b\sqrt{-5}$ define the norm $N(\alpha)$ of $\alpha$ by $N(\alpha) = |\alpha|^2 = a^2 + 5b^2$. $N$ is multiplicative and non-negative for all $\alpha \in R$. The units $\pm 1$ of $R$ are exactly the elements of norm 1. Now

$$9 = 3 \cdot 3 = (2 + \sqrt{-5}) \cdot (2 - \sqrt{-5}),$$
$$6 = 2 \cdot 3 = (2 + \sqrt{-5}) \cdot (1 - \sqrt{-5}).$$

All these factors are irreducible, since there are no elements of norm 2 or 3.

### Section 2.1

1. a. Obviously $L_\beta(a) = L_\beta(-a)$. So let us assume that $a > 0$. $L_\beta(a) = n$ if and only if $\beta^{n-1} \leq a < \beta^n$. The statement follows immediately from this relation.

   b. For $a > 0$ we have $\log_\beta a = \log_\gamma a \cdot \log_\beta \gamma$. Let $c$ be a positive constant such that for all integers $a$, $1 \leq a < \gamma$, $L_\beta(a) \leq c \cdot L_\gamma(a)$. There are only finitely many such $a$, so such a constant obviously exists. Let $d := 2\max(\log_\beta \gamma, 1)$. Then for $\gamma \leq a$

$$L_\beta(a) = \lfloor \log_\gamma a \cdot \log_\beta \gamma \rfloor + 1 \leq \tfrac{d}{2} \cdot (\log_\gamma a + 1)$$
$$\leq d \cdot \log_\gamma a \leq d \cdot (\lfloor \log_\gamma a \rfloor + 1) = d \cdot L_\gamma(a)\ .$$

So for $e := \max(c, d)$ we have $L_\beta(a) \leq e \cdot L_\gamma(a)$ for all $a$.

2. We assume an algorithm SIGN for determining the sign of a single digit. We consider the following algorithm:

**Algorithm INT_SIGN**(in: $a$ (in modified representation); out: sign($a$));
1. if $a = [\ ]$ then $\{s := 0$; return$\}$;
2. $d := 0$; $a' := a$;
   while $d = 0$ do $\{d :=$ FIRST($a'$); $a' :=$ REST($a'$)$\}$;
   $s :=$ SIGN($d$); return.

For $1 \leq k < n$ there are $2(\beta - 1)^2\beta^{n-k-1}$ integers in $\mathbb{Z}_n$, i.e., of length $n$, for which exactly $k$ digits have to be examined. There are $2(\beta - 1)$ integers in $\mathbb{Z}_n$ for which all $n$ digits have to be examined. $\mathbb{Z}_n$ has $2(\beta - 1)\beta^{n-1}$ elements.

Let $A$ denote the constant time for executing step (1), and $B$ the constant time for executing one loop in step (2). Then

$t^*_{\text{INT\_SIGN}}(n) = \sum_{a\in\mathbb{Z}_n} t_{\text{INT\_SIGN}}(a)/|\mathbb{Z}_n| = A+B(n\beta^{-n+1}+\sum_{k=1}^{n-1} k(\beta-1)\beta^{-k}) < A+B(\sum_{k=1}^{\infty} k\beta^{-k+1})$. By the quotient criterion this series converges to some $c \in \mathbb{R}^+$, so we have $t^*_{\text{INT\_SIGN}}(n) < A + cB \sim 1$.

3. Determine whether $|a| \leq |b|$. This can be done in average time proportional to $\min(L(a), L(b))$. If $|a| = |b|$ then return 0. Otherwise suppose, w.l.o.g., that $|a| < |b|$. So $\text{sign}(a + b) = \text{sign}(b)$. Subtract the digits of $a$ from the digits of $b$, as in the classical subtraction algorithm until $a$ is exhausted. The carry propagation can be analyzed analogously to INT_SUM1.
5. Assume that $\beta = 2$. A similar reasoning applies for $\beta \geq 2$. $\frac{1}{2} f(n) \leq \sum_{i=1}^{n} \log a_i = \log(\prod_{i=1}^{n} a_i) \leq g(n) \leq 2(\lfloor\prod_{i=1}^{n} a_i\rfloor) \leq 2\sum_{i=1}^{n}(\lfloor\log a_i\rfloor + 1) = 2 \cdot f(n)$.
6. **Algorithm INT_DIVPS**(in: $a, b$; out: $q, r$);
   [$a, b$ are positive integers; $q = \text{quot}(a, b), r = \text{rem}(a, b)$. We neglect signs in $\beta$-representations.]
   1. $m := L(a)$; $n := L(b)$; $q := [\ ]$; $r := a$;
      if $m < n$ then return;
   2. [normalization]
      $d := \lfloor\beta/(b_{n-1} + 1)\rfloor$;
      $a' := a \cdot d$; $b' := b \cdot d$;
   3. [decompose $a'$ into leading part $a''$ and rest $a'''$]
      $a'' := a'$; $a''' := [\ ]$;
      for $i = 1$ to $m - n$ do {$c :=$ FIRST($a''$); $a'' :=$ REST($a''$); $a''' :=$ CONS($c, a'''$)};
   4. [determination of digits $q_{m-n}, \ldots, q_0$]
      for $j = 0$ to $m - n$ do {[set $a''_i = 0$ for $L(a'') < i + 1$]
      4.1. [guess] $q^* := \min(\beta - 1, \lfloor(a''_n\beta + a''_{n-1})/b'_{n-1}\rfloor)$;
      4.2. [correction] while INT_DIFF($a'', q^*b'$) $< 0$ do $q^* := q^* - 1$;
      4.3. [record digit and continue] $q :=$ CONS($q^*, q$); $a'' :=$ INT_DIFF($a'', q^*b'$);
      if $j < m - n$
      then {$c :=$ FIRST($a'''$); $a''' :=$ REST($a'''$); $a'' :=$ CONS($c, a''$)}};
   5. [prepare output] $q' :=$ INV($q$);
      if FIRST($q'$) $= 0$ then $q :=$ INV(REST($q'$));
      $r := a''/d$; return.

9. An application of the Euclidean algorithm is characterized by the sequence of quotients it produces. Let $a, b$ be integers with $a_{(\beta)} = [+, a_0, \ldots, a_{m-1}]$, $b_{(\beta)} = [+, b_0, \ldots, b_{n-1}]$, with $m \geq n$. Set $a'_{(\beta)} = [+, a_{n-1}, \ldots, a_{m-1}]$, $b'_{(\beta)} = [+, b_{n-1}+1]$, $a''_{(\beta)} = [+, a_{n-1} + 1, \ldots, a_{m-1}]$, $b''_{(\beta)} = [+, b_{n-1}]$. According to Exercise 8, apply INT_GCDEE for computing the sequence of quotients both for $a', b'$ and for $a'', b''$, as long as they agree. When a difference is noted, the exact numbers in the remainder sequences are reconstructed from the sequence of quotients and the cofactors, and the process starts anew.

**Section 2.2**

1. $\mathcal{O}(l \cdot (d + 1)^n)$.
2. $\mathcal{O}(l^{\log_2 3} \cdot (d + 1)^{2n})$, if we use the Karatsuba algorithm for the multiplication of integer coefficients.

5. The precomputation in step (2) takes $m-n$ ring operations. The computation of $a'$ in the $i$-th iteration takes $\mathcal{O}(\deg(m+1-i))$ operations. So the overall complexity is $\mathcal{O}((m-n+1)(m+n))$.
6. $p_n x^n + p_{n-1}x^{n-1} + p_{n-2}x^{n-2} + \ldots + p_0 : x - a = p_n x^{n-1} + (p_n a + p_{n-1})x^{n-2} + \ldots$

$$\begin{array}{l} \underline{-p_n x^n + p_n a x^{n-1}} \\ \quad (p_n a + p_{n-1})x^{n-1} + p_{n-2}x^{n-2} + \ldots \\ \quad \underline{-(p_n a + p_{n-1})x^{n-1} + (p_n a + p_{n-1})a x^{n-2}} \\ \quad\quad ((p_n a + p_{n-1})a + p_{n-2})x^{n-2} + \ldots \\ \quad\quad\quad \ddots \\ \quad\quad\quad\quad (\ldots(p_n a + p_{n-1})a + \ldots)a + p_0 . \end{array}$$

Similarly for division by $x^2 - a^2$.

**Section 2.3**

4. In roughly 25% of the cases the result of $(ad+bd)/bd$ can be simplified. In general, the length almost doubles.

**Section 2.5**

3. $\beta = 2\alpha^3 + \alpha$, $\beta^3 = 2\alpha^2 + 2\alpha$, $\beta^9 = 2\alpha^2$, $b^{27} = \alpha^3 + 2\alpha^2$, $\beta^{81} = 2\alpha^3 + \alpha = \beta$. So $m_\beta(x) = \prod_{i=0}^{3}(x - \beta^{p^i}) = x^4 + x^2 + x + 1$.

**Section 3.1**

2. Let $I$ be an ideal in $D$. If $I = \{0\}$ we are done. Otherwise let $g \in I$ be such that $\deg(g)$ is minimal in $I^*$. Now let $a$ be an arbitrary element of $I^*$. Write $a = qg + r$ according to the division property. Because of the minimality of $g$, $r = 0$. So every element of $I$ is a multiple of $g$, i.e., $I = \langle g \rangle$.
3. The existence of factorization into finitely many irreducibles of every non-unit element $a$ in $D$ follows from Exercise 1. If $a = b \cdot c$ for $b, c$ non-units, we have $\deg(b), \deg(c) < \deg(a)$. So by induction on the degree we get the existence of a factorization of $a$.

   We still have to prove uniqueness. Suppose $a = a_1 \cdots a_n = b_1 \cdots b_m$ for non-units $a_i, b_j$. We proceed by induction on $n$. If $n = 1$ then $a_1$ is irreducible, so $m$ must be 1 and $a_1 = b_1$. For $n > 1$ we use the fact that (*) in a principal ideal domain every irreducible element is prime, i.e., if an irreducible element $p$ divides a product $q \cdot r$ it must divide either $q$ or $r$. By Exercise 2 the Euclidean domain $D$ is a principal ideal domain. So $a_1$ must divide one of the $b_j$'s, w.l.o.g. $b_1 = \epsilon \cdot a_1$ for a unit $\epsilon$. Substituting this factorization for $b_1$ and cancelling $a_1$ we get $a_2 \cdots a_n = (\epsilon b_2) \cdot b_3 \cdots b_m$. These factorizations must be identical (up to reordering and associates) by the induction hypothesis.

   *Proof of (*):* In a principal ideal domain $\langle p \rangle$ is maximal, for if $\langle p \rangle \subset \langle q \rangle$ and $\langle p \rangle \neq \langle q \rangle$, then $p = q \cdot r$ for a non-unit $r$, which is impossible. So $F := D_{/\langle p \rangle}$ is a field. Now if $p | q \cdot r$, then $q \cdot r = 0$ in $F$, therefore $q = 0$ or $r = 0$ in $F$, and therefore $p | q$ or $p | r$.
4. Let the norm function $N$ on $\mathbb{Z}[i]$ be defined as $N(a + bi) = a^2 + b^2$. We let $N$ be the degree function on $\mathbb{Z}[i]$. Observe that $N(\alpha\beta) = N(\alpha)N(\beta)$, so that for $\alpha, \beta \neq 0$ we have $N(\alpha\beta) \geq N(\alpha)$. Now consider $\alpha, \beta \in \mathbb{Z}[i]$ with $\beta \neq 0$. Let $\lambda, \mu \in \mathbb{Q}$ such that $\alpha\beta^{-1} = \lambda + \mu i$. Each rational number is within $\frac{1}{2}$ of an

integer, so for some $\lambda_1, \mu_1 \in \mathbb{Z}$ and $\lambda_2, \mu_2 \in \mathbb{Q}$ with $|\lambda_2|, |\mu_2| \leq \frac{1}{2}$ we can write $\lambda = \lambda_1 + \lambda_2, \mu = \mu_1 + \mu_2$. Now $\alpha = (\lambda_1 + \mu_1 i)\beta + (\lambda_2 + \mu_2 i)\beta$. Setting $\lambda_1 + \mu_1 i = \gamma$, $(\lambda_2 + \mu_2 i)\beta = \eta$, we get $\alpha = \gamma\beta + \eta$. $\eta = \alpha - \gamma\beta \in \mathbb{Z}[i]$, and $N(\eta) \leq (\frac{1}{4} + \frac{1}{4})N(\beta) < N(\beta)$. So $\gamma, \eta$ are the quotient and remainder of $\alpha, \beta$.

We apply the technique from above.

$$\frac{5-8i}{7+3i} = \frac{(5-8i)(7-3i)}{(7+3i)(7-3i)} = \frac{11}{58} - \frac{71}{58}i \ .$$

The Gaussian integer nearest to this is $-i$, so $\mathrm{quot}(5-8i, 7+3i) = -i$, and we get $5 - 8i = -i(7+3i) + (2-i)$ as the result of the first division. Next we divide $7+3i$ by $2-i$ in $\mathbb{C}$ to get $\frac{11}{5} + \frac{13}{5}i$. The Gaussian integer nearest to this is $2+3i$, so the result of the second division is $7+3i = (2+3i)(2-i) - i$. Here we can stop, because the last remainder $-i$ is a unit, i.e., the gcd of $5-8i$ and $7+3i$ is trivial.

5. Applying E_EUCLID to $f$ and $g$, we get

$$1 = \underbrace{(1/13)(-x+3)}_{u} \cdot f + \underbrace{(1/13)(x^2-5x+7)}_{v} \cdot g \ .$$

So $h = u' \cdot f + v' \cdot g$, where $u' = \mathrm{rem}(h \cdot u, g) = (1/13)(12x+3)$ and $v' = h \cdot v + \mathrm{quot}(h \cdot u, g) \cdot f = (1/13)(x^2 + 8x - 6)$.

6. $r(x) = 5x^4 + 10x^3 + 7x^2 - 5x - 4$.

8. For simplicity let us assume that $n$ is a power of 2, i.e., $n = 2^k$. Then there will be $k = \log_2 n$ levels of recursion. Working our way up from the bottom, at level $k+1-i$ we have to solve $2^{k-i}$ CRP$_2$'s on integers of size $i$, so the complexity of this level is proportional to $i^2 \cdot 2^{k-i}$. So we get the complexity bound

$$\mathcal{O}\left(\sum_{i=1}^{k} i^2 \cdot 2^{k-i}\right) \preceq \mathcal{O}\left(k^2 \sum_{i=1}^{k} 2^{k-i}\right) = \mathcal{O}(k^2 \cdot 2^k) = \mathcal{O}(n \log^2 n) \ .$$

9. We only have to demonstrate that $1 \in I + I_1 \cap \ldots \cap I_n$. Choose $u_k \in I$, $v_k \in I_k$ such that $u_k + v_k = 1$, for all $1 \leq k \leq n$. Then

$$1 = \prod_{k=1}^{n}(u_k + v_k) = \sum_{k=1}^{n} u_k \prod_{i \neq k}(u_i + v_i) \ + v_1 \ldots v_n \in I + I_1 \cap \ldots \cap I_n \ .$$

10. We consider the canonical homomorphism $\phi$ from $R$ to $\prod_{j=1}^{n} R_{/I_j}$ with kernel $I_1 \cap \ldots \cap I_n$. It remains to show that $\phi$ is surjective, i.e., for arbitrary elements $r_1, \ldots, r_n \in R$ we have to show that there exists an $r \in R$ satisfying

$$r \equiv r_i \bmod I_i \quad \text{for } 1 \leq i \leq n \ .$$

Everything is trivial if $n = 1$. For $n \geq 2$ let us assume that we have an element $r' \in R$ satisfying

$$r' \equiv r_i \bmod I_i \quad \text{for } 1 \leq i \leq n-1 \ .$$

By Exercise 9 we have $I_n + \prod_{j=1}^{n-1} I_j = R$. So there is $c \in \bigcap_{j=1}^{n-1} I_j$ such that $1 - c \in I_n$. Hence, $r = r' + c(r_n - r')$ solves our problem.

11. By relative primeness, there are $\tilde{u}_1, \ldots, \tilde{u}_r$ such that $1 = \sum_{i=1}^{r}(\tilde{u}_i \prod_{j=1,j\neq i}^{r} a_j)$. Let $u_i := (c\tilde{u}_i \bmod a_i)$. Obviously $d = \sum_{i=1}^{r}(u_i \prod_{j=1,j\neq i}^{r} a_j)$ is of degree less than $n$ and satisfies $d \equiv c \bmod a_i$ for $1 \leq i \leq r$. So $d \equiv c \bmod \prod_{i=1}^{r} a_i$ and therefore $c = d$.

**Section 3.3**

4. Analogous to convolution theorem in Aho et al. (1974).

**Section 4.1**

1. By application of Gauss's lemma.
2. Maple program for PRS_SR:

```
prs_sr := proc(f1,f2,x)
      local F,g,h,fp,i,delta,result;
# (1)
      result:=[f1,f2];
      F[1]:=f1; F[2]:=f2;
      g:=1; h:=1; fp:=f2; i:= 3;
# (2)
      while  (fp<>0  and degree(fp,x)>0)
            do
            delta:=degree(F[i-2],x) - degree(F[i-1],x);
            fp := prem(F[i-2],F[i-1],x);
            if fp<>0
            then     F[i]:=simplify(fp/(g*h^delta));
                     result:=[op(result),F[i]];
                     g:= lcoeff(F[i-1],x);
                     h:=h^(1-delta)*g^delta;
                     i:=i+1
             fi;
             od;
      RETURN(result)
      end;
```

4. Obviously $\prod_{i=1}^{r} f_i \in I$. So we have to show that every $g \in I$ is divisible by each $f_i$, $1 \leq i \leq r$. Suppose for some $i$ the factor $f_i$ does not divide $g$. W.l.o.g. (perhaps after renaming the variables) we can assume that $f_i$ is primitive w.r.t. $x_n$ and we can write $f_i$ as
$$f_i = a_t x_n^t + \ldots + a_1 x_n + a_0 ,$$
where $a_j \in K[x_1, \ldots, x_{n-1}]$, $t > 0$, and $a_t \neq 0$. Let $g = g_1(x_1, \ldots, x_{n-1}) \cdot g_2(x_1, \ldots, x_n)$, where $g_2$ is primitive w.r.t. $x_n$. By Gauss's lemma $f_i$ and $g_2$ are also relatively prime in $K(x_1, \ldots, x_{n-1})[x_n]$. So there are polynomials $h_1, h_2 \in K[x_1, \ldots, x_n]$ and $d \in K[x_1, \ldots, x_{n-1}]$, $d \neq 0$, such that
$$d = h_1 \cdot f_i + h_2 \cdot g_2 .$$
Let $(c_1, \ldots, c_{n-1})$ be a point in $\mathbb{A}^{n-1}(K)$ such that
$$(d \cdot a_t \cdot g_1)(c_1, \ldots, c_{n-1}) \neq 0 .$$

Now let $c_n \in K$ such that $f_i(c_1, \ldots, c_n) = 0$. Then $(c_1, \ldots, c_n) \in H$ and therefore $g_2(c_1, \ldots, c_n) = 0$. But this is a contradiction to $d(c_1, \ldots, c_{n-1}) \neq 0$.

5. By Gauss's lemma $f$ and $g$ are also relatively prime in $K(x)[y]$. So there are $a, b \in K[x, y]$ and $d \in K[x]$ such that $d = af + bg$. Let $(a_1, a_2)$ be such that $f(a_1, a_2) = g(a_1, a_2) = 0$. So $d(a_1) = 0$. But $d$ has only finitely many roots, i.e., there are only finitely many possible values for the $x$-coordinate of common roots of $f$ and $g$. By the same argument there are only finitely many possible values for the $y$-coordinates of common roots of $f$ and $g$.

**Section 4.2**

2. Calling PRS_SR with main variable $y$ we get the result

$$
\begin{aligned}
f_1 &= f, \quad f_2 = g\,,\\
f_3 &= (-x^2+2x)y^3 + (-x^3+2x^2)y^2 + (x^5-4x^4+3x^3+4x^2-4x)y +\\
&\quad + x^6 - 4x^5 + 3x^4 + 4x^3 - 4x^2\,,\\
f_4 &= -x^4y^2 + (-x^{10}+6x^9-10x^8-4x^7+23x^6-11x^5-12x^4+8x^3)y -\\
&\quad - x^{11} + 6x^{10} - 10x^9 - 4x^8 + 23x^7 - 10x^6 - 12x^5 + 8x^4\,,\\
f_5 &= (x^{17}-10x^{16}+36x^{15}-40x^{14}-74x^{13}+228x^{12}-84x^{11}-312x^{10}+\\
&\quad + 321x^9 + 117x^8 - 280x^7 + 49x^6 + 80x^5 - 32x^4)y +\\
&\quad + x^{18} - 10x^{17} + 36x^{16} - 40x^{15} - 74x^{14} + 228x^{13} - 84x^{12} - 312x^{11} +\\
&\quad + 321x^{10} + 117x^9 - 280x^8 + 49x^7 + 80x^6 - 32x^5\,.
\end{aligned}
$$

The primitive part of $f_5$ is $y + x$ and this is the gcd.

Now let us apply GCD_MODm. Choosing $x = 0$ as the evaluation point, we get the gcd $y^5$. $x = 1$ yields the gcd $y^2 + y$. So 0 was a bad evaluation point, and it is discarded. $x = 2$ yields the gcd $y + 2$. So 1 was a bad evaluation point, and it is discarded. $x = 3$ yields the gcd $y + 3$. Interpolation gives us the gcd candidate $y + x$, which is correct.

3. $h = x^2 - 2x + 1$. $\mathrm{res}(f, g) = -(37)(1619)$. The gcd modulo both these factors has degree 3, so both of them are unlucky.

**Section 4.3**

1. $\mathrm{res}_y(a, b) = 4x^8(2x^4 - 4x^3 + 4x + 2)$.
2. $\mathrm{res}_y(\mathrm{res}_z(f_1, f_2), \mathrm{res}_z(f_1, f_3)) = -4x^2(2x^5 - 8x^4 - 16x^3 - 4x^2 + 14x - 4)$, but there is no solution with $x$-coordinate 0.
3. $\mathrm{res}_y(\mathrm{res}_z(f_1, f_2), \mathrm{res}_z(f_1, f_3)) = \frac{1}{256}x^6 \cdot r(x)$, and $\mathrm{res}_y(\mathrm{res}_z(f_1, f_2), \mathrm{res}_z(f_2, f_3)) = \frac{1}{1024}x^4(4x+1)^2 \cdot r(x)$, where $r(x) = 32x^5 - 216x^4 + 64x^3 - 42x^2 + 32x + 5$. But there is no solution with $x$-coordinate 0. The other roots of $r(x)$ actually lead to solutions of the whole system.
4. Let $t = \deg(a)$. So $a = a_t + \ldots + a_0$, where $a_i$ is a homogeneous polynomial of degree $i$, i.e., every term in $a_i$ is of degree $i$. Now there exist $c_1, \ldots, c_r \in K$ such that $c_r \neq 0$ and $a_t(c_1, \ldots, c_r) \neq 0$ (if $K$ is infinite). Now we let $\hat{a} := a(x_1 + c_1x_r, \ldots, x_{r-1} + c_{r-1}x_r, c_rx_r)$, and $\hat{b} := b(x_1 + c_1x_r, \ldots, x_{r-1} + c_{r-1}x_r, c_rx_r)$. Then for every $(\alpha_1, \ldots, \alpha_r) \in K^r$ we have that $(\alpha_1, \ldots, \alpha_r)$ is a common root of

$\hat{a}$ and $\hat{b}$ iff $(\alpha_1 + c_1\alpha_r, \ldots, \alpha_{r-1} + c_{r-1}\alpha_r, c_r\alpha_r)$ is a common root of $a$ and $b$, and $(\alpha_1, \ldots, \alpha_r)$ is a common root of $a$ and $b$ iff $(\alpha_1 - \frac{c_1}{c_r}\alpha_r, \ldots, \alpha_{r-1} - \frac{c_{r-1}}{c_r}\alpha_r, \frac{1}{c_r}\alpha_r)$ is a common root of $\hat{a}$ and $\hat{b}$. Furthermore, it can be easily seen that $\deg(\hat{a}) = t$ and $\mathrm{lc}(\hat{a})_{x_r} = a_t(c_1, \ldots, c_r)$.

**Section 4.4**

1. $a(x) = (x^2 + 1)(x - 1)^2(x + 1)^3$.
2. See Cox et al. (1992: p. 180).
3. See Akritas (1989: p. 295).
4. $p(x) \equiv (x^2 + 2x)(x + 1)^2 \bmod 3$, and $p(x)$ is squarefree modulo 11.

**Section 4.5**

1. See Horowitz (1969).
2. $q_1 = x + 1, q_2 = 1, q_3 = x - 1, a_1 = -1/8, a_2 = 1, a_3 = (1/8)(x^2 - 4x + 7)$.

**Section 4.6**

1. $q^{*\prime}(\alpha_i) = \prod_{j \neq i}(\alpha_i - \alpha_j)$. The partial fraction decomposition of $h/q^*$ is

$$\frac{h}{q^*} = \sum_{i=1}^{n} \frac{h(\alpha_i)/q^{*\prime}(\alpha_i)}{x - \alpha_i} ,$$

as can be seen by putting the right-hand side over the common denominator $q^*$ and comparing the numerators at the $n$ different values $\alpha_1, \ldots, \alpha_n$.

2. $x^4 - 2 = (x - \sqrt[4]{2})(x + \sqrt[4]{2})(x - i\sqrt[4]{2})(x + i\sqrt[4]{2})$, and $c_i = (8x/(x^4 - 2)')|_{\alpha_i}$, i.e., $c_1 = c_2 = \sqrt{2}, c_3 = c_4 = -\sqrt{2}$. So

$$\begin{aligned}\int \frac{8x}{x^4 - 2}\,dx &= \sqrt{2}(\log(x - \sqrt[4]{2}) + \log(x + \sqrt[4]{2})) - \\ &\quad - \sqrt{2}(\log(x - i\sqrt[4]{2}) + \log(x + i\sqrt[4]{2})) \\ &= \sqrt{2}\log(x^2 - \sqrt{2}) - \sqrt{2}\log(x^2 + \sqrt{2}) .\end{aligned}$$

Now let us compute the integral according to Theorem 4.6.4. $r(c) = \mathrm{res}_x(8x - c(x^4 - 2)', x^4 - 2) = -2048(c^2 - 2)^2$. Hence, $c_1 = \sqrt{2}, c_2 = -\sqrt{2}$. Furthermore $v_1 = \gcd(8x - \sqrt{2}(4x^3), x^4 - 2) = x^2 - \sqrt{2}$ in $\mathbb{Q}(\sqrt{2})[x]$, and $v_2 = x^2 + \sqrt{2}$.

3. Yes. Let $c_i = c_{i_1}$. The conjugates $c_{i_2}, \ldots, c_{i_k}$ of $c_i$ are also roots of $r(c)$, in fact $\prod_{j=1}^{k}(c - c_{i_j}) \in \mathbb{Q}[c]$ is an irreducible factor of $r(c)$. The corresponding $v_{i_j}$ is obtained from $v_i$ by applying the automorphism $c_i \mapsto c_{i_j}$.
4. The complete result is

$$\begin{aligned}\int \frac{p(x)}{q(x)}\,dx &= \tfrac{2}{3}\log(3x^2 - 2x + 1) - \tfrac{1}{2}\log(x^2 + 1) + \frac{2\sqrt{2}}{3}\arctan\Big(\frac{6x - 2}{2\sqrt{2}}\Big) + \\ &\quad + \log(x) + \arctan(x) + \frac{4x^3 - 4x^2 + 2x - 1}{2x^4 + 2x^2} .\end{aligned}$$

**Section 5.1**

2. $x^2 + x + 1$ occurs with multiplicity 2 in $a$. The complete factorization of $a$ is $(x^2+x+1)^2(x^2+3x+3)(x+4)$.
3. 4 factors modulo 2 and modulo $8k+1$, 2 factors in the other cases. $u(x)$ is irreducible over the integers.

**Section 5.2**

1. Yes, it is. $a(x) = (5x+4)(x-5)(x+6)$.
3. By the Berlekamp algorithm

$$a(x) \equiv \underbrace{(2x^2+4x+3)}_{u_1} \cdot \underbrace{(x^2+2)}_{u_2} \cdot \underbrace{(x+2)}_{u_3} \cdot \underbrace{(x+3)}_{u_4} \bmod 5 .$$

By an application of LIFT_FACTORS we get

$$a(x) \equiv \underbrace{(2x^2-11x+3)}_{v_1} \cdot \underbrace{(x^2-10x-8)}_{v_2} \cdot \underbrace{(x+7)}_{v_3} \cdot \underbrace{(x-7)}_{v_4} \bmod 25 .$$

Factor combinations yield

$$a(x) = (2x^4 - 6x^3 - 3x^2 + 8x + 1)(x+7)(x-7) .$$

**Section 5.3**

2. $\|b_i^*\|^2 + \mu_{i\,i-1}^2 \|b_{i-1}^*\|^2 \geq \frac{3}{4}\|b_{i-1}^*\|^2$, so
$\|b_i^*\|^2 \geq \frac{3}{4}\|b_{i-1}^*\| - \mu_{i\,i-1}^2\|b_{i-1}^*\|^2 \geq \frac{3}{4}\|b_{i-1}^*\|^2 - \frac{1}{4}\|b_{i-1}^*\|^2 = \frac{1}{2}\|b_{i-1}^*\|^2$.
3. By induction on $i$. Clearly $d_1 = \|b_1^*\|^2$. Let $M^{(i)} = (\langle b_j, b_l\rangle)_{1\leq j,l\leq i}$. By elementary column operations (adding columns to the last one) $M^{(i)}$ can be transformed into

$$\tilde{M}^{(i)} = \begin{pmatrix} & & & \vdots & 0 \\ & M^{(i-1)} & & \vdots & \vdots \\ & & & \vdots & 0 \\ \cdots & \cdots & \cdots & \cdots & \cdots \\ M_{i1}^{(i)} & \cdots & M_{i\,i-1}^{(i)} & \vdots & \|b_i^*\|^2 \end{pmatrix} .$$

So by the induction hypothesis and expansion of the determinant of $\tilde{M}^{(i)}$ w.r.t. the $i$-th row we get $d_i = |M^{(i)}| = |\tilde{M}^{(i)}| = |M^{(i-1)}| \cdot \|b_i^*\|^2 = \prod_{j=1}^{i} \|b_j^*\|^2$.

We show how to transform $M^{(i)}$ into $\tilde{M}^{(i)}$ just for the case $i = 3$. $b_1 = b_1^*$, $b_2 = b_2^* + \mu_{21}b_1^*$, $b_3 = b_3^* + \mu_{31}b_1^* + \mu_{32}b_2^*$. Multiplying the first column in $M^{(3)}$ by $(-\mu_{31} + \mu_{21}\mu_{32})$ and the second column by $-\mu_{32}$ and adding the results to the third column, we get $\tilde{M}^{(3)}$.
4. See Lenstra et al. (1982).
5. See Lenstra et al. (1982).
6. BASIS_REDUCTION is called with $a_1 = (5^{12}, 0)$, $a_2 = (46966736, 1)$. $n = 2$. The "while"-loop is entered with $k = 2$, $l = 1$. After execution of (2.1) we have $b_1 = (46966736, 1)$, $b_2 = (5^{12}, 0)$. REDUCE changes $b_2$ to $(9306945, -5)$ and after

execution of (2.1) we have $b_1 = (9306945, -5)$, $b_2 = (46966736, 1)$. REDUCE changes $b_2$ to $(432011, 26)$ and after execution of (2.1) we have $b_1 = (432011, 26)$, $b_2 = (9306945, -5)$. REDUCE changes $b_2$ to $(-197297, -577)$ and after execution of (2.1) we have $b_1 = (-197297, -577)$, $b_2 = (432011, 26)$. REDUCE changes $b_2$ to $(37417, -1128)$ and after execution of (2.1) we have $b_1 = (37417, -1128)$, $b_2 = (-197297, -577)$. REDUCE changes $b_2$ to $(-10212, -6217)$ and after execution of (2.1) we have $b_1 = (-10212, -6217)$, $b_2 = (37417, -1128)$. REDUCE changes $b_2$ to $(6781, -19779)$ and now (2.2) is executed. $k$ is set to 3, the computation terminates and the output is $b_1 = (-10212, -6217)$, $b_2 = (6781, -19779)$.

**Section 5.4**

1. Let $\alpha_1, \ldots, \alpha_n$ be the roots of $p$, and $\beta_1, \ldots, \beta_m$ the roots of $h$. Then

$$p(x) = b_n \prod_{i=1}^{n} (x - \alpha_i), \qquad h(x, \alpha) = h_m \prod_{i=1}^{m} (x - \beta_i) .$$

According to van der Waerden (1970: sect. 5.9)

$$\mathrm{res}_y(h(x, y), p(y)) = \underbrace{(-1)^{nm} b_n^m}_{\text{constant}} \cdot \underbrace{\prod_{i=1}^{n} h(x, \alpha_i)}_{\text{norm}} .$$

So in fact, if $p$ is monic, then the norm and the resultant agree up to the sign.

2. We compute the factorization with the help of Maple V.2. In fact, the call `evala(Factor(f))` would produce the factorization. But we want to see the partial results.

```
> p:=x^3-x+1;
                              3
                        p := x  - x + 1
> alias(al=RootOf(p)):
> f := x^5+al^2*x^4+(al+1)*x^3+(al^2+al-1)*x^2+al^2 x+al^2;
          5    3         2   2     2  4    2       2     2        3     2
   f := x  + x  al + x  al  + al  x  + x  al - x  + al  x + x  + al
> evala(Factor(f));
                     2     2          3             2
                   (x  + al  x + 1) (x  + al x + al )
> py:=subs(x=y,p):
> op(factor(resultant(subs(al=y,f),py,y)));
 9      6    5      4           6      5      4      3      2
x  + 2 x  - x  + 3 x  + x + 1, x  + 2 x  + 4 x  + 5 x  + 4 x  + 2 x + 1
> n1:="[1]:   n2:=""[2]:
> evala(Gcd(n1,f));
                            3             2
                           x  + al x + al
> evala(Gcd(n2,f));
                            2     2
                           x  + al  x + 1
```

**Section 5.5**

1. $f$ is not monic as a polynomial in $y$. So we replace $f$ by

$$\hat{f} = (x^2 - 4x + 4) \cdot f\left(x, \frac{y}{x^2 - 4x + 4}\right) = y^2 + x^4 - 4x^3 + 4x^2 .$$

In fact, $\hat{g} = iy + x^2 - 2x$ is a factor of $\hat{f}$, so

$$\mathrm{pp}_y(\hat{g}(x, \mathrm{lc}_y(f)y)) = \frac{i(x^2 - 4x + 4)y + x^2 - 2x}{x - 2} = ixy - 2iy + x$$

is a factor of $f$.

$\hat{f}(0, y)$ is not squarefree, but $\hat{f}(1, y)$ is.

2. $f(x, y) = (y^2 - ix^2y - i)(y^2 + ix^2y + i)$.

**Section 6.1**

1. Let

$$f_1 = c_k x^{n+k} + \ldots + c_1 x^{n+1} + a_n x^n + \ldots + a_0 ,$$
$$f_2 = c_k x^{m+k} + \ldots + c_1 x^{m+1} + b_m x^m + \ldots + b_0 ,$$
$$g = d_l x^l + \ldots + d_0 .$$

Let $c_i = 0$ for $i > k$. For $1 \le j \le k$ (i.e., the highest exponents in $f_1 g$) the coefficient of $x^{n+l+j}$ in $f_1 g$ is

$$c_j d_l + c_{j+1} d_{l-1} + \ldots + c_{j+l} d_0 ,$$

which is also the coefficient of $x^{m+l+j}$ in $f_2 g$.

2. $g(x) = x^4 + 2x^2 + x + 2$ and $h(x) = x^2 - x$.

**Section 7.1**

2. The extended coefficient matrix $A^{[4]}$ turns out to be

$$A^{[4]} = \begin{pmatrix} 57 & 0 & 0 & 0 & \vdots & 57 \\ 0 & 57 & 0 & 0 & \vdots & 57 \\ 0 & 0 & 57 & 0 & \vdots & 57 \\ 0 & 0 & 0 & 57 & \vdots & 57 \end{pmatrix} ,$$

so the solution is $x = (1, 1, 1, 1)$.

**Section 7.3**

1. Let $g(x) = g_m x^m + \ldots + g_0$, and let $M_n^{(k)}$ be the $n \times n$ principal submatrix of $\varphi(x^k/f)$. Then we see that $M_n^{-1} \cdot M_n^{(k)} = C^k$ for $0 \le k \le m$. Furthermore, $H_n = g_m M_n^{(m)} + \ldots + g_0 M_n^{(0)}$. Therefore, $M_n^{-1} \cdot H_n = g(C)$.
2. $R(x) = g(x)/f(x)$ if and only if $f \cdot (x^{2n-1} R) = x^{2n-1} \cdot g$.

**Section 8.1**

1. No. Let $M = \mathbb{N} \times \mathbb{N}$, $(m, n) \longrightarrow (m', n') \Longleftrightarrow m > m'$ or $m = m'$ and $n > n'$.

2.

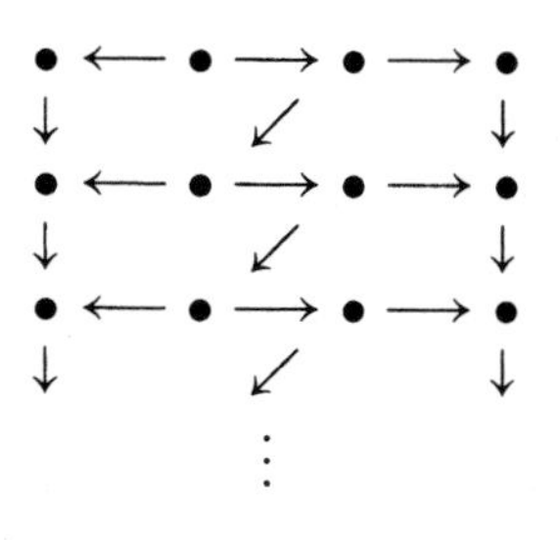

Or see Huet (1980).

**Section 8.2**

1. We have to show that every sequence $f_1 \gg f_2 \gg \ldots$ is finite. This is achieved by Noetherian induction on $\mathrm{lpp}(f_1)$ w.r.t. $>$.
2. (a), (b), (c) are obvious. d. Let $s$ be the power product in $g_1$ that is reduced, i.e., $s = u \cdot \mathrm{lpp}(f)$ for some $u \in [X]$, $f \in F$. If $\mathrm{coeff}(s, h) = 0$ then $g_1 + h \longrightarrow_F g_2 + h$ by the same polynomial $f$. If $\mathrm{coeff}(s, h) = -\mathrm{coeff}(s, g_1)$ then $g_2 + h \longrightarrow_F g_1 + h$ by $f$. Otherwise $g_1 + h \longrightarrow_F \tilde{g} \longleftarrow_F g_2 + h$ by $f$, where $\tilde{g} = g_1 + h - (\mathrm{coeff}(s, g_1 + h)/\mathrm{lc}(f)) \cdot u \cdot f = g_2 + h - (\mathrm{coeff}(s, g_2 + h)/\mathrm{lc}(f)) \cdot u \cdot f$.
3. $\longleftrightarrow_F^*$ is the smallest equivalence relation containing $\longrightarrow_F$. If $g \longrightarrow_F h$ then by the definition of the reduction relation, $g - h \in \langle F \rangle$, i.e., $g \equiv_{\langle F \rangle} h$. Because $\longrightarrow_F \subseteq \longleftrightarrow_F^*$ and $\equiv_{\langle F \rangle}$ is an equivalence relation we have $\longleftrightarrow_F^* \subseteq \equiv_{\langle F \rangle}$.

   On the other hand, let $g \equiv_{\langle F \rangle} h$, i.e., $g = h + \sum_{j=1}^{m} c_j \cdot u_j \cdot f_j$ for $c_j \in K$, $u_j \in [X]$, $f_j \in F$. If we can show $g \longleftrightarrow_F^* h$ for the case $m = 1$, then the statement follows by induction on $m$. $f_1 \longrightarrow_F 0$. So by Lemma 8.2.5 $g = h + c_1 \cdot u_1 \cdot f_1 \downarrow_F^* h$ and therefore $g \longleftrightarrow_F^* h$.

**Section 8.3**

1. (a) $\Longrightarrow$ (b): $f \equiv_{\langle F \rangle} 0$, so by Theorem 8.2.6 $f \longleftrightarrow_F^* 0$, and by the Church–Rosser property of $\longrightarrow_F$ we get $f \longrightarrow_F^* 0$.

   (b) $\Longrightarrow$ (c): Obvious.

   (c) $\Longrightarrow$ (a): Let $g \longleftrightarrow_F^* h$ and $\tilde{g}, \tilde{h}$ normal forms of $g, h$ w.r.t. $\longrightarrow_F$, respectively. By Theorem 8.2.6 $\tilde{g} - \tilde{h} \in I$. So, since $\tilde{g} - \tilde{h}$ is irreducible, $\tilde{g} = \tilde{h}$.

   Clearly (a) implies (d) and (d) implies (b).
2. $\mathrm{spol}(f_1, f_2) = 2xy + x^3 + x^2$ is irreducible and leads to a new basis element $f_3 = xy + 2x^3 + 2x^2$. $\mathrm{spol}(f_2, f_3) = x^4 + x^3 + x$ is irreducible and is added to the basis as $f_4$. All the other S-polynomials are reducible to 0. So $\{f_1, f_2, f_3, f_4\}$ is a Gröbner basis for the ideal. $f_1$ can be reduced to 0 w.r.t. the other basis polynomials, so it is cancelled from the basis. Also $f_2$ is cancelled for the same reason. $\{f_3, f_4\}$ is the normed reduced Gröbner basis for the ideal.
3. $G = \{z + x^4 - 2x + 1, y^2 + x^2 - 2, x^5 - 6x^3 + x^2 - 1\}$. $\dim \mathbb{Q}[x, y, z]_{/I} = 10$.

**Section 8.4**

1. The projective curve is defined by the homogenization of $f(x, y)$, i.e., $\tilde{f}(x, y, z) = x^6 + 3x^4y^2 - 4x^2y^2z^2 + 3x^2y^4 + y^6 = 0$. The singularities $(a : b : c)$ with $c = 1$ are the solutions of the system of equations $f(x, y) = (\partial f/\partial x)(x, y) = (\partial f/\partial y)(x, y) = 0$. A Gröbner basis for the corresponding ideal w.r.t. the lexicographic ordering is

$\{x, y\}$, so the only solution is $(0:0:1)$. Similarly we dehomogenize $\tilde{f}$ w.r.t. to $y$ and $x$, getting the singularities $(\pm i : 1 : 0)$ at infinity.

2. From the Shape lemma we know that the normed reduced Gröbner basis w.r.t. the lexicographic ordering will have the shape $\{g_1(x_1), x_2 - g_2(x_1), \ldots, x_n - g_n(x_1)\}$ if $I$ is regular in $x_1$. Since $I$ is prime, the polynomial $g_1$ must be irreducible.
3. For $1 \le i \le r$ let $I_i = \langle x_1 - a_{i1}, \ldots, x_n - a_{in}\rangle$ where $P_i = (a_{i1}, \ldots, a_{in})$, i.e., $I_i$ is the ideal of all polynomials vanishing on $P_i$. Now we set $J_i = \bigcap_{j \neq i} I_j$, i.e., $J_i$ is the ideal of all polynomials vanishing on $P_1, \ldots, P_{i-1}, P_{i+1}, \ldots, P_r$. There must be an element $f_i$ in the basis of $J_i$ which does not vanish on $P_i$. After an appropriate scaling, this is the $i$-th separator polynomial.

**Section 10.1**

1. $p(n) = 1$, $q(n) = b(n-1)^2 + c(n-1) + d$, $r(n) = bn^2 + cn + c$ is a regular representation of $a_n/a_{n-1}$. The bound for the degree of $f$ turns out to be 0. The solution of the corresponding linear system is $f = 1/(d-c)$. Thus,

$$s_n = s_n' - s_0' = \frac{1 - \prod_{j=1}^{n} \frac{bj^2+cj+d}{bj^2+cj+c}}{c-d}$$

**Section 11.1**

1. The result holds for $n > 2$ and $n$ not a divisor of the characteristic of $K$.
2. Yes, if the line at infinity $z = 0$ is a component of the curve, i.e., $z | f(x, y, z)$.
3. The corresponding projective curve is defined by $f^*(x, y, z) = (x^2 + 4yz + y^2)^2 - 16(x^2 + y^2)z^2$, having double points $(0:0:1), (1 : \pm i : 0)$ in $\mathbb{P}^2(\mathbb{C})$. So the genus is 0.
4. $x^2, xy$.

# References


Abhyankar, S. S., Bajaj, C. L. (1988): Automatic parametrization of rational curves and surfaces III: algebraic plane curves. Comput. Aided Geom. Des. 5: 309–321.

Adams, W. W., Loustaunau, P. (1994): An introduction to Gröbner bases. AMS, Providence, RI (Graduate studies in mathematics, vol. 3)

Aho, A. V., Hopcroft, J. E., Ullman, J. D. (1974): The design and analysis of computer algorithms. Addison-Wesley, Reading, MA.

Akritas, A. G. (1989): Elements of computer algebra with applications. Wiley, New York.

Alagar, V. S., Thanh, M. (1985): Fast polynomial decomposition algorithms. In: Caviness, B. F. (ed.): EUROCAL '85. Proceedings, vol. 2. Springer, Berlin Heidelberg New York Tokyo, pp. 150–153 (Lecture notes in computer science, vol. 204).

Arnon, D. S. (1988): A cluster-based cylindrical algebraic decomposition algorithm. J. Symb. Comput. 5: 189–212.

Arnon, D. S., Buchberger, B. (eds.) (1988): Algorithms in real algebraic geometry. J. Symb. Comput. 5 no. 1–2.

Arnon, D. S., Sederberg, T. W. (1984): Implicit equations for a parametric surface by Groebner basis. In: Golden, V. E. (ed.): Proceedings of the 1984 MACSYMA Users' Conference. General Electric, Schenectady, NY, pp. 431–436.

Arnon, D. S., Collins, G. E., McCallum, S. (1984a): Cylindrical algebraic decomposition I: the basic algorithm. SIAM J. Comput. 13: 865–877.

Arnon, D. S., Collins, G. E., McCallum, S. (1984b): Cylindrical algebraic decomposition II: an adjacency algorithm for the plane. SIAM J. Comput. 13: 878–889.

Bajaj, C., Canny, J., Garrity, T., Warren, J. (1993): Factoring rational polynomials over the complex numbers. SIAM J. Comput. 22: 318–331.

Bareiss, E. H. (1968): Sylvester's identity and multistep integer-preserving Gaussian elimination. Math. Comput. 22: 565–578.

Barton, D. R., Zippel, R. E. (1985): Polynomial decomposition algorithms. J. Symb. Comput. 1: 159–168.

Bayer, D., Stillman, M. (1988): On the complexity of computing syzygies. J. Symb. Comput. 6: 135–147.

Bayer, D., Stillman, M. (1992): Computation of Hilbert functions. J. Symb. Comput. 14: 31–50.

Becker, E., Marinari, M. G., Mora, T., Traverso, C. (1994): The shape of the Shape lemma. In: von zur Gathen, J., Giesbrecht, M. (eds.): Proceedings of the International Symposium on Symbolic and Algebraic Computation. Association for Computing Machinery, New York, pp. 129–133.

Becker, T. (1990): Standard bases and some computations in rings of power series. J. Symb. Comput. 10: 165–178.

Becker, T. (1993): Standard bases in power series rings: uniqueness and superfluous critical pairs. J. Symb. Comput. 15: 251–265.

Becker, T., Weispfenning, V. (1993): Gröbner bases – a computational approach to commutative algebra. Springer, Berlin New York Heidelberg Tokyo (Graduate texts in mathematics, vol. 141).

Berlekamp, E. R. (1968): Algebraic coding theory. McGraw-Hill, New York.

Bini, D., Pan, V. (1993): Parallel computations with Toeplitz-like and Hankel-like matrices. In: Bronstein, M. (ed.): Proceedings of the International Symposium on Symbolic and Algebraic Computation. Association for Computing Machinery, New York, pp. 193–200.

Bobrow, D. G. (ed.) (1968): IFIP Working Conference on Symbol Manipulation Languages and Techniques. North-Holland, Amsterdam.

Bochnak, J., Coste, M., Roy, M.-F. (1987): Géométrie algébrique réelle. Springer, Berlin Heidelberg New York Tokyo.

Böge, W., Gebauer, R., Kredel, H. (1986): Some examples of solving systems of algebraic equations by calculating Groebner bases. J. Symb. Comput. 2: 83–98.

Book, R. V. (ed.) (1986): Studies in complexity theory. Pitman, London.

Boyle, A., Caviness, B. F. (1990): Future directions for research in symbolic computation. SIAM, Philadelphia.

Brackx, F., Constales, D., Ronveaux, A., Serras, H. (1989): On the harmonic and monogenic decomposition of polynomials. J. Symb. Comput. 8: 297–304.

Brent, R. P., Gustavson, F. G., Yun, D. Y. Y. (1980): Fast solution of Toeplitz system of equations and computation of Padé approximants. J. Algorithms 1: 259–295.

Bronstein, M. (ed.) (1993): Proceedings of the International Symposium on Symbolic and Algebraic Computation. Association for Computing Machinery, New York.

Brown, W. S. (1971): On Euclid's algorithm and the computation of polynomial greatest common divisors. J. ACM 18: 478–504.

Brown, W. S. (1978): The subresultant PRS algorithm. ACM Trans. Math. Software 4: 237–249.

Brown, W. S., Traub, J. F. (1971): On Euclid's algorithm and the theory of subresultants. J. ACM 18: 505–514.

Buchberger, B. (1965): Ein Algorithmus zum Auffinden der Basiselemente des Restklassenringes nach einem nulldimensionalen Polynomideal. Ph. D. dissertation, University of Innsbruck, Innsbruck, Austria.

Buchberger, B. (1970): Ein algorithmisches Kriterium für die Lösbarkeit eines algebraischen Gleichungssystems. Aequ. Math. 4: 374–383.

Buchberger, B. (1979): A criterion for detecting unnecessary reductions in the construction of Gröbner-bases. In: Ng, W. (ed.): Symbolic and algebraic computation. Springer, Berlin Heidelberg New York, pp. 3–21 (Lecture notes in computer science, vol. 72).

Buchberger, B. (1983a): A note on the complexity of constructing Gröbner-bases. In: van Hulzen, J. A. (ed.): Computer algebra. EUROCAL '83. Springer, Berlin Heidelberg New York Tokyo, pp. 137–145

Buchberger, B. (1983b): A critical-pair/completion algorithm in reduction rings. In: Börger, E., Hasenjäger, G., Rödding, D. (eds.): Proceedings Logic and Machines: Decision Problems and Complexity. Springer, Berlin Heidelberg New York Tokyo, pp. 137–161 (Lecture notes in computer science, vol. 171).

Buchberger, B. (ed.) (1985a): EUROCAL '85, proceedings, vol. 1. Springer, Berlin Heidelberg New York Tokyo (Lecture notes in computer science, vol. 203).
Buchberger, B. (1985b): Gröbner-bases: an algorithmic method in polynomial ideal theory. In: Bose, N. K. (ed.): Multidimensional systems theory. Reidel, Dordrecht, pp. 184–232.
Buchberger, B., Collins, G. E., Loos, R. (eds.) (1983): Computer algebra, symbolic and algebraic computation, 2nd edn. Springer, Wien New York
Buchberger, B., Collins, G. E., Encarnación, M., Hong, H., Johnson, J., Krandick, W., Loos, R., Mandache, A., Neubacher, A., Vielhaber, H. (1993): SACLIB 1.1 user's guide. Techn. Rep. RISC Linz 93-19.
Bündgen, R. (1991): Term completion versus algebraic completion. Ph. D. dissertation, University of Tübingen, Tübingen, Federal Republic of Germany.
Calmet, J. (ed.) (1982): Computer algebra. EUROCAM '82. Springer, Berlin Heidelberg New York (Lecture notes in computer science, vol. 144).
Calmet, J. (ed.) (1986): Algebraic algorithms and error-correcting codes. Springer, Berlin Heidelberg New York Tokyo (Lecture notes in computer science, vol. 229).
Cannon, J. J. (1984): An introduction to the group theory language Cayley. In: Atkinson, M. D. (ed.). Computational group theory. Academic Press, London, pp. 145–183.
Cantor, D. G., Zassenhaus, H. (1981): A new algorithm for factoring polynomials over finite fields. Math. Comput. 36: 587–592.
Cassels, J. W. S. (1971): An introduction to the geometry of numbers. Springer, Berlin Heidelberg New York.
Caviness, B. F. (1970): On canonical forms and simplification. J. ACM 17: 385–396.
Caviness, B. F. (ed.) (1985): EUROCAL '85, proceedings, vol. 2. Springer, Berlin Heidelberg New York Tokyo (Lecture notes in computer science, vol. 204).
Caviness, B. F. (1986): Computer algebra: past and future. J. Symb. Comput. 2: 217–236.
Caviness, B. F., Collins, G. E. (1976): Algorithms for Gaussian integer arithmetic. In: Jenks, R. D. (ed.): Proceedings of the ACM Symposium on Symbolic and Algebraic Computation. Association for Computing Machinery, New York, pp. 36–45.
Caviness, B. F., Fateman, R. J. (1976): Simplification of radical expressions. In: Jenks, R. D. (ed.): Proceedings of the ACM Symposium on Symbolic and Algebraic Computation. Association for Computing Machinery, New York, pp. 329–338.
Caviness, B. F., Johnson, J. R. (eds.) (1996): Quantifier elimination and cylindrical algebraic decomposition. Springer, Wien New York (forthcoming).
Char, B. W. (ed.) (1986): Proceedings of the 1986 Symposium on Symbolic and Algebraic Computation (SYMSAC '86). Association for Computing Machinery, New York.
Char, B. W., Geddes, K. O., Gonnet, G. H. (1989): GCDHEU: heuristic polynomial GCD algorithm based on integer GCD computation. J. Symb. Comput. 7: 31–48.
Char, B. W., Geddes, K. O., Gonnet, G. H., Leong, B. L., Monagan, M. B., Watt, S. M. (1991a): Maple V library reference manual. Springer, Berlin Heidelberg New York Tokyo.
Char, B. W., Geddes, K. O., Gonnet, G. H., Leong, B. L., Monagan, M. B., Watt, S. M. (1991b): Maple V language reference manual. Springer, Berlin Heidelberg New York Tokyo.
Char, B. W., Geddes, K. O., Gonnet, G. H., Leong, B. L., Monagan, M. B., Watt, S. M.

(1992): First leaves: a tutorial introduction to Maple V. Springer, Berlin Heidelberg New York Tokyo.

Chistov, A. L., Grigoryev, D. Y. (1983): Subexponential-time solving systems of algebraic equations I. LOMI preprint E-9-83, Steklov Institute, Leningrad (St. Petersburg).

Chudnovsky, D. V., Jenks, R. D. (eds.) (1990): Computers in mathematics. Marcel Dekker, New York (Lecture notes in pure and applied mathematics, vol. 125).

Cohen, A. M. (ed.) (1991): The SCAFI papers – proceedings of the 1991 Seminar on Studies in Computer Algebra for Industry. CAN Expertise Centre, Amsterdam.

Cohen, A. M., van Gastel, L. J. (1995): Computer algebra in industry 2. Proceedings of SCAFI II, CAN Expertise Centre, Amsterdam, November 1992. Wiley, Chichester.

Cohen, G., Mora, T., Moreno, O. (eds.) (1993): Applied algebra, algebraic algorithms and error-correcting codes. Springer, Berlin Heidelberg New York Tokyo (Lecture notes in computer science, vol. 637).

Cohen, H. (1993): A course in computational algebraic number theory. Springer, Berlin Heidelberg New York Tokyo (Graduate texts in mathematics, vol. 138).

Cohen, P. J. (1969): Decision procedures for real and p-adic fields. Comm. Pure Appl. Math. 22: 131–151.

Cohn, P. M. (1974): Algebra. Wiley, New York.

Collins, G. E. (1967): Subresultants and reduced polynomial remainder sequences. J. ACM 14: 128–142.

Collins, G. E. (1971): The calculation of multivariate polynomial resultants. J. ACM 19: 515–532.

Collins, G. E. (1973): Computer algebra of polynomials and rational functions. Am. Math. M. 80: 725–755.

Collins, G. E. (1975): Quantifier elimination for real closed fields by cylindrical algebraic decomposition. In: Brakhage, H. (ed.): Automata theory and formal languages. Springer, Berlin Heidelberg New York Tokyo, pp. 134–183 (Lecture notes in computer science, vol. 33).

Collins, G. E. (1976): Quantifier elimination for real closed fields by cylindrical algebraic decomposition – a synopsis. ACM SIGSAM Bull. 10/1: 10–12.

Collins, G. E. (1980): ALDES/SAC-2 now available. ACM SIGSAM Bull. 14/2 (2): 19.

Collins, G. E. (1996): Quantifier elimination by cylindrical algebraic decomposition. In: Caviness, B. F., Johnson, J. R. (eds.): Quantifier elimination and cylindrical algebraic decomposition. Springer, Wien New York (forthcoming).

Collins, G. E., Hong, H. (1991): Partial cylindrical algebraic decomposition for quantifier elimination. J. Symb. Comput. 12: 299–328.

Collins, G. E., Musser, D. R. (1977): Analysis of the Pope–Stein division algorithm. Inf. Process. Lett. 6: 151–155.

Collins, G. E., Mignotte, M., Winkler, F. (1983): Arithmetic in basic algebraic domains. In: Buchberger, B., Collins, G. E., Loos, R. (eds.): Computer algebra, symbolic and algebraic computation, 2nd edn. Springer, Wien New York, pp. 189–220.

Cooley, J. M., Tukey, J. W. (1965): An algorithm for the machine calculation of complex Fourier series. Math. Comput. 19: 297–301.

Cooley, J. M., Lewis, P. A. W., Welch, P. D. (1967): History of the fast Fourier transform. Proc. IEEE 55: 1675–1677.

Cooley, J. M., Lewis, P. A. W., Welch, P. D. (1969): The fast Fourier transform and its applications. IEEE Trans. Educ. 12: 27–34.

Cox, D., Little, J., O'Shea, D. (1992): Ideals, varieties, and algorithms. Springer, Berlin Heidelberg New York Tokyo.

Davenport, J. H. (ed.) (1989): EUROCAL '87. Springer, Berlin Heidelberg New York Tokyo (Lecture notes in computer science, vol. 378).

Davenport, J. H., Heintz, J. (1988): Real quantifier elimination is doubly exponential. J. Symb. Comput. 5: 29–35.

Davenport, J. H., Siret, Y., Tournier, E. (1988): Computer algebra, systems and algorithms for algebraic computation. Academic Press, London.

Davis, P. J., Hersh, R. (1981): The mathematical experience. Birkhäuser, Boston.

Della Dora, J., Fitch, J. (1989): Computer algebra and parallelism. Academic Press, London.

Dickson, L. E. (1913): Finiteness of the odd perfect and primitive abundant numbers with $n$ distinct prime factors. Am. J. Math. 35: 413–426.

Duval, D. (1991): Absolute factorization of polynomials: a geometric approach. SIAM J. Comput. 20: 1–21.

Dvornicich, R., Traverso, C. (1987): Newton symmetric functions and the arithmetic of algebraically closed fields. In: Huguet, L., Poli, A. (eds.): Applied algebra, algebraic algorithms and error-correcting codes. Springer, Berlin Heidelberg New York Tokyo, pp. 216–224 (Lecture notes in computer science, vol. 356).

Encarnación, M. J. (1994): On a modular algorithm for computing GCDs of polynomials over algebraic number fields. In: von zur Gathen, J., Giesbrecht, M. (eds.): Proceedings of the International Symposium on Symbolic and Algebraic Computation. Association for Computing Machinery, New York, pp. 58–65.

Fateman, R. J. (1972): Essays in algebraic simplification. Ph. D. thesis, Project MAC, MIT, Cambridge, MA.

Fateman, R. J. (ed.) (1977): MACSYMA Users' Conference. MIT Press, Cambridge, MA.

Faugère, J. C., Gianni, P., Lazard, D., Mora, T. (1993): Efficient computation of zero-dimensional Gröbner bases by change of ordering. J. Symb. Comput. 16: 377–399.

Fitch, J. (ed.) (1984): EUROSAM 84. Springer, Berlin Heidelberg New York Tokyo (Lecture notes in computer science, vol. 174).

Fitch, J. P. (1985): Solving algebraic problems with REDUCE. J. Symb. Comput. 1: 211–227.

Floyd, R. W. (ed.) (1966): Proceedings Association for Computing Machinery Symposium on Symbolic and Algebraic Manipulation (SYMSAM '66). Commun. ACM 9: 547–643.

Freeman, T. S., Imirzian, G. M., Kaltofen, E. (1986): A system for manipulating polynomials given by straight-line programs. Techn. Rep. 86-15, Department of Computer Science, Rensselaer Polytechnic Institute, Troy, New York.

Fulton, W. (1969): Algebraic curves. Benjamin/Cummings, Menlo Park, CA.

Furukawa, A., Sasaki, T., Kobayashi, H. (1986): Gröbner basis of a module over $K[x_1, \ldots, x_n]$ and polynomial solutions of a system of linear equations. In: Char, B. W. (ed.): Proceedings of the 1986 Symposium on Symbolic and Algebraic Computation (SYMSAC '86). Association for Computing Machinery, New York, pp. 222–224.

Galligo, A. (1979): Théorème de division et stabilité en géométrie analytique locale. Ann. Inst. Fourier 29: 107–184.
Gantmacher, F. R. (1977): Theory of matrices, vol. I, II. Chelsea, New York.
Garrity, T., Warren, J. (1989): On computing the intersection of a pair of algebraic surfaces. Comput. Aided Geom. Des. 6: 137–153.
Geddes, K. O., Czapor, S. R., Labahn, G. (1992): Algorithms for computer algebra. Kluwer, Boston.
Gemignani, L. (1994): Solving Hankel systems over the integers. J. Symb. Comput. 18: 573–584.
Gianni, P. (1987): Properties of Gröbner bases under specializations. In: Davenport, J. H. (ed.): EUROCAL '87. Springer, Berlin Heidelberg New York Tokyo, pp. 293–297 (Lecture notes in computer science, vol. 378).
Gianni, P. (ed.) (1989): Symbolic and algebraic computation. Springer, Berlin Heidelberg New York Tokyo (Lecture notes in computer science, vol. 358).
Gianni, P., Mora, T. (1987): Algebraic solution of systems of polynomial equations using Groebner bases. In: Huguet, L., Poli, A. (eds.): Applied algebra, algebraic algorithms and error-correcting codes. Springer, Berlin Heidelberg New York Tokyo, pp. 247–257 (Lecture notes in computer science, vol. 356).
Gianni, P., Trager, B., Zacharias, G. (1988): Gröbner bases and primary decomposition of polynomial ideals. J. Symb. Comput. 6: 149–167.
Giusti, M. (1984): Some effectivity problems in polynomial ideal theory. In: Fitch, J. (ed.): EUROSAM 84. Springer, Berlin Heidelberg New York Tokyo, pp. 159–171 (Lecture notes in computer science, vol. 174).
Gonnet, G. H. (ed.) (1989): Proceedings of the ACM-SIGSAM 1989 International Symposium on Symbolic and Algebraic Computation (ISSAC '89). Association for Computing Machinery, New York.
Gonnet, G. H., Gruntz, D. W. (1993): Algebraic manipulation: systems. In: Ralston, A., Reilly, E. D. (eds): Encyclopedia of computer science, 3rd edn. Van Nostrand Reinhold, New York.
Gosper, R. W. (1978): Decision procedure for indefinite hypergeometric summation. Proc. Natl. Acad. Sci. USA 75: 40–42.
Gräbe, H.-G. (1993): On lucky primes. J. Symb. Comput. 15: 199–209.
Graham, R. L., Knuth, D. E., Patashnik, O. (1994): Concrete mathematics, a foundation for computer science, 2nd edn. Addison-Wesley, Reading.
Gröbner, W. (1949): Moderne algebraische Geometrie. Springer, Wien Innsbruck.
Grossman, R. (1989): Symbolic computation, applications to scientific computing. SIAM, Philadelphia.
Gutierrez, J., Recio, T. (1992): A practical implementation of two rational function decomposition algorithms. In: Wang, P. S. (ed.): Proceedings of the International Symposium on Symbolic and Algebraic Computation. Association for Computing Machinery, New York, pp. 152–157.
Gutierrez, J., Recio, T., Ruiz de Velasco, C. (1988): A polynomial decomposition algorithm of almost quadratic complexity. In: Mora, T. (ed.): Applied algebra, algebraic algorithms and error-correcting codes. Springer, Berlin Heidelberg New York Tokyo, pp. 471–476 (Lecture notes in computer science, vol. 357).
Hardy, G. H. (1916): The integration of functions of a single variable, 2nd edn. Cambridge University Press, Cambridge.

Hardy, G. H., Wright, E. M. (1979): An introduction to the theory of numbers, 5th edn. Oxford University Press, Oxford.

Harper, D., Wooff, C., Hodgkinson, D. (1991): A guide to computer algebra systems. Wiley, Chichester.

Heck, A. (1993): Introduction to Maple. Springer, Berlin Heidelberg New York Tokyo.

Hehl, F. W., Winkelmann, V., Meyer, H. (1992): Computer-Algebra, ein Kompaktkurs über die Anwendung von REDUCE. Springer, New York Berlin Heidelberg Tokyo.

Heindel, L. E. (1971): Integer arithmetic algorithms for polynomial real zero determination. J. ACM 18: 533–548.

Heinig, G., Rost, K. (1984): Algebraic methods for Toeplitz-like matrices and operators. Birkhäuser, Basel.

Heintz, J., Morgenstern, J. (1993): On the intrinsic complexity of elimination theory. J. Complex. 9: 471–498.

Heintz, J., Sieveking, M. (1981): Absolute primality of polynomials is decidable in random polynomial time in the number of variables. In: Even, S., Kariv, O. (eds.): Automata, languages and programming. Springer, Berlin Heidelberg New York, pp. 16–28 (Lecture notes in computer science, vol. 115).

Henrici, P. (1956): A subroutine for computations with rational numbers. J. ACM 3: 6–9.

Hermann, G. (1926): Die Frage der endlich vielen Schritte in der Theorie der Polynomideale. Math. Ann. 95: 736–788.

Hilbert, D. (1890): Über die Theorie der algebraischen Formen. Math. Ann. 36: 473–534.

Hilbert, D., Hurwitz, A. (1890): Über die Diophantischen Gleichungen vom Geschlecht Null. Acta Math. 14: 217–224.

Hironaka, H. (1964): Resolution of singularities on an algebraic variety over a field of characteristic zero: I, II. Ann. Math. 79: 109–326.

Hong, H. (1990): An improvement of the projection operator in cylindrical algebraic decomposition. In: Watanabe, S., Nagata, M. (eds.): Proceedings of the International Symposium on Symbolic and Algebraic Computation (ISSAC '90). Association for Computing Machinery, New York, pp. 261–264.

Hong, H. (ed.) (1993): Computational quantifier elimination. Comput. J. 36 no. 5.

Hong, H., Sendra, J. R. (1996): Computation of variant resultants. In: Caviness, B. F., Johnson, J. R. (eds.): Quantifier elimination and cylindrical algebraic decomposition. Springer, Wien New York (forthcoming).

Horowitz, E. (1969): Algorithms for symbolic integration of rational functions. Ph. D. thesis, University of Wisconsin, Madison, Wisconsin.

Horowitz, E. (1971): Algorithms for partial fraction decomposition and rational function integration. In: Petrick, S. R. (ed.): Proceedings of the ACM Symposium on Symbolic and Algebraic Manipulation (SYMSAM '71). Association for Computing Machinery, New York, pp. 441–457.

Horowitz, E., Sahni, S. (1976): Fundamentals of data structures. Computer Science Press, Potomac, MD.

Huet, G. (1980): Confluent reductions: abstract properties and applications to term rewriting systems. J. ACM 27: 797–821.

Huguet, L., Poli, A. (eds.) (1989): Applied algebra, algebraic algorithms and error-correcting codes. Springer, Berlin Heidelberg New York Tokyo (Lecture notes in computer science, vol. 356).

Jacob, W. B., Lam, T.-Y., Robson, R. O. (eds.) (1994): Recent advances in real algebraic geometry and quadratic forms. AMS, Providence, RI (Contemporary mathematics, vol. 155).

Janßen, R. (ed.) (1987): Trends in computer algebra. Springer, Berlin Heidelberg New York Tokyo (Lecture notes in computer science, vol. 296).

Jebelean, T. (1993): An algorithm for exact division. J. Symb. Comput. 15: 169–180.

Jenks, R. D. (ed.) (1974): Proceedings of EUROSAM '74. ACM SIGSAM Bull. 8 no. 3.

Jenks, R. D. (ed.) (1976): Proceedings of the ACM Symposium on Symbolic and Algebraic Computation. Association for Computing Machinery, New York.

Jenks, R. D., Sutor, R. S. (1992): Axiom – the scientific computation system. Springer, Berlin Heidelberg New York Tokyo.

Kalkbrener, M. (1987): Solving systems of algebraic equations by using Gröbner bases. In: Davenport, J. H. (ed.): EUROCAL '87. Springer, Berlin Heidelberg New York Tokyo, pp. 282–292 (Lecture notes in computer science, vol. 378).

Kalkbrener, M. (1991): Three contributions to elimination theory. Ph. D. dissertation, Johannes Kepler University Linz, Linz, Austria.

Kalkbrener, M., Sturmfels, B. (1995): Initial complexes of prime ideals. Adv. Math. 116: 365–376.

Kaltofen, E. (1982): On the complexity of factoring polynomials with integer coefficients. Ph. D. thesis, Rensselaer Polytechnic Institute, Troy, New York.

Kaltofen, E. (1983): Factorization of polynomials. In: Buchberger, B., Collins, G. E., Loos, R. (eds.): Computer algebra, symbolic and algebraic computation, 2nd edn. Springer, Wien New York, pp. 95–113.

Kaltofen, E. (1985a): Polynomial-time reductions from multivariate to bi- and univariate integral polynomial factorization. SIAM J. Comput. 14: 469–489.

Kaltofen, E. (1985b): Sparse Hensel lifting. Techn. Rep. 85-12, Rensselaer Polytechnic Institute, Depertment of Computer Science, Troy, New York.

Kaltofen, E. (1985c): Fast parallel irreducibility testing. J. Symb. Comput. 1: 57–67.

Kaltofen, E. (1986): Polynomial factorization 1982–1986. In: Chudnovsky, D. V., Jenks, R. D. (eds.): Computers in mathematics. Marcel Dekker, New York, pp. 285–309 (Lecture notes in pure and applied mathematics, vol. 125).

Kaltofen, E. (1987): Deterministic irreducibility testing of polynomials over large finite fields. J. Symb. Comput. 4: 77–82.

Kaltofen, E. (1988): Greatest common divisors of polynomials given by straight-line programs. J. ACM 35: 231–264.

Kaltofen, E. (ed.) (1990): Computational algebraic complexity. Academic Press, London.

Kaltofen, E. (1992): Polynomial factorization 1987–1991. In: Simon, I. (ed.): LATIN '92. Springer, Berlin Heidelberg New York Tokyo, pp. 294–313 (Lecture notes in computer science, vol. 583).

Kaltofen, E., Watt, S. M. (eds.) (1989): Computers and mathematics. Springer, Berlin Heidelberg New York Tokyo.

Kaminski, M. (1987): A linear time algorithm for residue computation and a fast algorithm for division with a sparse divisor. J. ACM 34: 968–984.

Kaminski, M., Bshouty, N. H. (1989): Multiplicative complexity of polynomial multiplication over finite fields. J. ACM 36: 150–170.

Kandri-Rody, A., Kapur, D. (1984): Algorithms for computing Gröbner bases of polynomial ideals over various Euclidean rings. In: Fitch, J. (ed.): EUROSAM 84. Springer,

Berlin Heidelberg New York Tokyo, pp. 195–206 (Lecture notes in computer science, vol. 174).
Kapur, D. (1986): Using Gröbner bases to reason about geometry problems. J. Symb. Comput. 2: 399–408.
Karatsuba, A., Ofman, Yu. (1962): Multiplication of multidigit numbers on automata. Soviet Phys. Dokl. 7: 595–596. [Dokl. Akad. Nauk USSR 145: 293–294 (1962), in Russian].
Karian, Z. A. (ed.) (1992): Symbolic computation in undergraduate mathematics education. Math. Assoc. Amer. Notes 24.
Karr, M. (1985): Theory of summation in finite terms. J. Symb. Comput. 1: 303–315.
Knuth, D. E. (1973): The art of computer programming, vol. 1, fundamental algorithms, 2nd edn. Addison-Wesley, Reading, MA.
Knuth, D. E. (1981): The art of computer programming, vol. 2, seminumerical algorithms, 2nd edn. Addison-Wesley, Reading, MA.
Knuth, D. E., Bendix, P. B. (1967): Simple word problems in universal algebras. In: Leech, J. (ed.): Computational problems in abstract algebra. Pergamon, Oxford, pp. 263–297.
Koblitz, N. (1977): $p$-adic numbers, $p$-adic analysis, and zeta-functions. Springer, New York Berlin Heidelberg.
Koepf, W. (1992): Power series in computer algebra. J. Symb. Comput. 13: 581–603.
Kovacic, J. J. (1986): An algorithm for solving second order linear homogeneous differential equations. J. Symb. Comput. 2: 3–43.
Kozen, D., Landau, S. (1989): Polynomial decomposition algorithms. J. Symb. Comput. 7: 445–456.
Kredel, H., Weispfenning, V. (1988): Computing dimension and independent sets for polynomial ideals. J. Symb. Comput. 6: 231–248.
Kronecker, L. (1882): Grundzüge einer arithmetischen Theorie der algebraischen Größen. J. Reine Angew. Math. 92: 1–122.
Kutzler, B., Stifter, S. (1986a): Automated geometry theorem proving using Buchberger's algorithm. In: Char, B. W. (ed.): Proceedings of the 1986 Symposium on Symbolic and Algebraic Computation. Association for Computing Machinery, New York, pp. 209–214.
Kutzler, B., Stifter, S. (1986b): On the application of Buchberger's algorithm to automated geometry theorem proving. J. Symb. Comput. 2: 389–397.
Kutzler, B., Wall, B., Winkler, F. (1992): Mathematische Expertensysteme – praktisches Arbeiten mit den Computer-Algebra-Systemen MACSYMA, Mathematica und DERIVE. Expert Verlag, Ehningen bei Böblingen.
Lafon, J. C. (1983): Summation in finite terms. In: Buchberger, B., Collins, G. E., Loos, R. (eds.): Computer algebra, symbolic and algebraic computation, 2nd edn. Springer, Wien New York, pp. 71–77.
Lakshman, Y. N. (1991): A single exponential bound on the complexity of computing Gröbner bases of zero dimensional ideals. In: Mora, T., Traverso, C. (eds.): Effective methods in algebraic geometry. Birkhäuser, Boston, pp. 227–234 (Progress in mathematics, vol. 94).
Lakshman, Y. N., Lazard, D. (1991): On the complexity of zero-dimensional algebraic systems. In: Mora, T., Traverso, C. (eds.): Effective methods in algebraic geometry. Birkhäuser, Boston, pp. 217–225 (Progress in mathematics, vol. 94).

Landau, E. (1905): Sur quelques théorèmes de M. Petrovitch relatifs aux zéros des fonctions analytiques. Bull. Soc. Math. France 33: 251–261.
Landau, S. (1985): Factoring polynomials over algebraic number fields. SIAM J. Comput. 14: 184–195.
Lang, S. (1984): Algebra, 2nd edn. Addison-Wesley, Reading, MA.
Langemyr, L., McCallum, S. (1989): The computation of polynomial greatest common divisors over an algebraic number field. J. Symb. Comput. 8: 429–448.
Lauer, M. (1983): Computing by homomorphic images. In: Buchberger, B., Collins, G. E., Loos, R. (eds.): Computer algebra, symbolic and algebraic computation, 2nd edn. Springer, Wien New York, pp. 139–168.
Lazard, D. (1985): Ideal bases and primary decomposition: case of two variables. J. Symb. Comput. 1: 261–270.
Lazard, D. (1992): A note on upper bounds for ideal-theoretic problems. J. Symb. Comput. 13: 231–233.
Lazard, D., Rioboo, R. (1990): Integration of rational functions: rational computation of the logarithmic part. J. Symb. Comput. 9: 113–115.
Le Chenadec, P. (1986): Canonical forms in finitely presented algebras. Pitman, London.
Leech, J. (ed.) (1970): Computational Problems in Abstract Algebra, Proceedings of the conference held at Oxford in 1967. Pergamon, Oxford.
Lehmer, D. H. (1938): Euclid's algorithm for large numbers. Amer. Math. M. 45: 227–233.
Lenstra, A. K., Lenstra, H. W. Jr., Lovász, L. (1982): Factoring polynomials with rational coefficients. Math. Ann. 261: 515–534.
Levelt, A. H. M. (ed.) (1995): Proceedings of the International Symposium on Symbolic and Algebraic Computation (ISSAC '95). Association for Computing Machinery, New York.
Lewis, V. E. (ed.) (1979): MACSYMA Users' Conference, Proceedings of the conference held at Washington, D.C. MIT, Cambridge, MA.
Lidl, R. (1985): On decomposable and commuting polynomials. In: Caviness, B. F. (ed.): EUROCAL '85. Springer, Berlin Heidelberg New York Tokyo, pp. 148–149 (Lecture notes in computer science, vol. 204).
Lidl, R., Niederreiter, H. (1983): Finite fields. Addison-Wesley, Reading, MA.
Lidl, R., Pilz, G. (1984): Applied abstract algebra. Springer, Berlin Heidelberg New York Tokyo.
Lipson, J. D. (1981): Elements of algebra and algebraic computing. Addison-Wesley, Reading, MA.
Llovet, J., Sendra, J. R. (1989): Hankel matrices and polynomials. In: Huguet, L., Poli, A. (eds.): Applied algebra, algebraic algorithms, and error-correcting codes. Springer, Berlin Heidelberg New York Tokyo, pp. 321–333 (Lecture notes in computer science, vol. 356).
Llovet, J., Sendra, J. R., Martinez, R. (1992): An algorithm for computing the number of real roots using Hankel forms. Linear Algebra Appl. 179: 228–234.
Loos, R. (1983): Generalized polynomial remainder sequences. In: Buchberger, B., Collins, G. E., Loos, R. (eds.): Computer algebra, symbolic and algebraic computation, 2nd edn. Springer, Wien New York, pp. 115–137.
MacCallum, M., Wright, F. (eds.) (1991): Algebraic computing with Reduce. Clarendon, Oxford.

Maeder, R. (1991): Programming in Mathematica, 2nd edn. Addison-Wesley, Redwood City, CA.

MacLane, S., Birkhoff, G. (1979): Algebra, 2nd edn. Macmillan, New York.

Mahler, K. (1973): Introduction to $p$-adic numbers and their functions. Cambridge University Press, Cambridge.

Mattson, H. F., Mora, T. (eds.) (1991): Applied algebra, algebraic algorithms, and error-correcting codes. Discr. Appl. Math. 3 no. 1–3.

Mattson, H. F., Mora, T., Rao, T. R. N. (eds.) (1991): Applied algebra, algebraic algorithms and error-correcting codes. Springer, Berlin Heidelberg New York Tokyo (Lecture notes in computer science, vol. 539).

Mayr, E. W., Meyer, A. R. (1982): The complexity of the word problem for commutative semigroups and polynomial ideals. Adv. Math. 46: 305–329.

McCallum, S. (1988): An improved projection operation for cylindrical algebraic decomposition of three-dimensional space. J. Symb. Comput. 5: 141–161.

McCallum, S. (1996): An improved projection operation for cylindrical algebraic decomposition. In: Caviness, B. F., Johnson, J. R. (eds.): Quantifier elimination and cylindrical algebraic decomposition. Springer, Wien New York (forthcoming).

McCarthy, J., et al. (1962): LISP 1.5 programmer's manual. MIT Press, Cambridge, MA.

Mignotte, M. (1974): An inequality about factors of polynomials. Math. Comput. 28: 1153–1157.

Mignotte, M. (1983): Some useful bounds. In: Buchberger, B., Collins, G. E., Loos, R. (eds.): Computer algebra, symbolic and algebraic computation, 2nd edn. Springer, Wien New York, pp. 259–263.

Mignotte, M. (1992): Mathematics for computer algebra. Springer, New York Berlin Heidelberg Tokyo.

Miola, A. (ed.) (1993): Design and implementation of symbolic computation systems. Springer, New York Berlin Heidelberg Tokyo (Lecture notes in computer science, vol. 722).

Mishra, B. (1993): Algorithmic algebra. Springer, New York Berlin Heidelberg Tokyo.

Möller, H. M. (1988): On the construction of Gröbner bases using syzygies. J. Symb. Comput. 6: 345–359.

Möller, H. M., Mora, F. (1984): Upper and lower bounds for the degree of Groebner bases. In: Fitch, J. P. (ed.): EUROSAM '84. Springer, Berlin Heidelberg New York Tokyo, pp. 172–183 (Lecture notes in computer science, vol. 174).

Mora, T. (ed.) (1989): Applied algebra, algebraic algorithms and error-correcting codes. Springer, Berlin Heidelberg New York Tokyo (Lecture notes in computer science, vol. 357).

Mora, F., Möller, H. M. (1986): New constructive methods in classical ideal theory. J. Algebra 100: 138–178.

Mora, T., Robbiano, L. (1988): The Gröbner fan of an ideal. J. Symb. Comput. 6: 183–208.

Mora, T., Traverso, C. (eds.) (1991): Effective methods in algebraic geometry. Birkhäuser, Boston (Progress in mathematics, vol. 94).

Moses, J. (1971a): Algebraic simplification: a guide for the perplexed. Commun. ACM 14: 527–537.

Moses, J. (1971b): Symbolic integration: the stormy decade. Commun. ACM 14: 548–560.

Moses, J., Yun, D. Y. Y. (1973): The EZ GCD algorithm. In: Proceedings Association for Computing Machinery National Conference, August 1973. Association for Computing Machinery, New York, pp. 159–166.

Musser, D. R. (1976): Multivariate polynomial factorization. J. ACM 22: 291–308.

Newell, A., Shaw, J. C., Simon, H. A. (1957): Empirical explorations of the logic theory machine. In: Proceedings 1957 Western Joint Computer Conference, pp. 218–230.

Ng, W. (ed.) (1979): Symbolic and algebraic computation. Springer, Berlin Heidelberg New York (Lecture notes in computer science, vol. 72).

Niederreiter, H., Göttfert, R. (1993): Factorization of polynomials over finite fields and characteristic sequences. J. Symb. Comput. 16: 401–412.

Noether, M. (1884): Rationale Ausführung der Operationen in der Theorie der algebraischen Functionen. Math. Ann. 23: 311–358.

Pauer, F. (1992): On lucky ideals for Gröbner basis computations. J. Symb. Comput. 14: 471–482.

Pauer, F., Pfeifhofer, M. (1988): The theory of Gröbner bases. Enseign. Math. 34: 215–232.

Paule, P. (1993): Greatest factorial factorization and symbolic summation I. Techn. Rep. RISC Linz 93-02.

Paule, P. (1994): Short and easy computer proofs of the Rogers–Ramanujan identities and of identities of similar type. Electronic J. Combin. 1: R10.

Paule, P., Strehl, V. (1995): Symbolic symmation – some recent developments. In: Fleischer, J., Grabmeier, J., Hehl, F., Kuechlin, W. (eds.): Computer algebra in science and engineering, algorithms, systems, applications. World Scientific, Singapore, pp. 138–162.

Pavelle, R. (ed.) (1985): Applications of computer algebra. Kluwer, Boston.

Pavelle, R., Wang, P. S. (1985): MACSYMA from F to G. J. Symb. Comput. 1: 69–100.

Pavelle, R., Rothstein, M., Fitch, J. (1981): Computer algebra. Sci. Am. 245/6: 102–113.

Petrick, S. R. (ed.) (1971): Proceedings of the ACM Symposium on Symbolic and Algebraic Manipulation (SYMSAM '71). Association for Computing Machinery, New York.

Pohst, M. (1987): A modification of the LLL reduction algorithm. J. Symb. Comput. 4: 123–127.

Pohst, M., Zassenhaus, H. (1989): Algorithmic algebraic number theory. Cambridge University Press, Cambridge.

Pope, D. A., Stein, M. L. (1960): Multiple precision arithmetic. Commun. ACM 3: 652–654.

Rand, R. H. (1984): Computer algebra in applied mathematics: an introduction to MACSYMA. Pitman, London.

Rand, R. H., Armbruster, D. (1987): Perturbation methods, bifurcation theory and computer algebra. Springer, Berlin Heidelberg New York Tokyo.

Rayna, G. (1987): REDUCE: software for algebraic compuation. Springer, Berlin Heidelberg New York Tokyo.

Renegar, J. (1992a): On the computational complexity and geometry of the first-order theory of the reals. Part I: introduction. Preliminaries. The geometry of semi-algebraic sets. The decision problem for the existential theory of the reals. J. Symb. Comput. 13: 255–299.

Renegar, J. (1992b): On the computational complexity and geometry of the first-order

theory of the reals. Part II: the general decision problem. Preliminaries for quantifier elimination. J. Symb. Comput. 13: 301–327.
Renegar, J. (1992c): On the computational complexity and geometry of the first-order theory of the reals. Part III: quantifier elimination. J. Symb. Comput. 13: 329–352.
Rich, A., Rich, J., Stoutemyer, D. (1988): Derive, a mathematical assistant. Soft Warehouse, Honolulu, HI.
Ritt, J. F. (1922): Prime and composite polynomials. Trans. Amer. Math. Soc. 23: 51–66.
Robbiano, L. (1985): Term orderings on the polynomial ring. In: Caviness, B. F. (ed.): EUROCAL '85. Springer, Berlin Heidelberg New York Tokyo, pp. 513–517 (Lecture notes in computer science, vol. 204).
Robbiano, L. (1986): On the theory of graded structures. J. Symb. Comput. 2: 139–170.
Rolletschek, H. (1991): Computer algebra. Techn. Rep. RISC Linz 91-07.
Rothstein, M. (1976): Aspects of symbolic integration and simplification of exponential and primitive functions. Ph. D. thesis, University of Wisconsin, Madison, WI.
Runge, C., König, H. (1924): Vorlesungen über numerisches Rechnen. Springer, Berlin [Courant, R., et al. (eds.): Die Grundlehren der mathematischen Wissenschaften, vol. 11].
Sakata, S. (ed.) (1991): Applied algebra, algebraic algorithms and error-correcting codes. Springer, Berlin Heidelberg New York Tokyo (Lecture notes in computer science, vol. 508).
Schicho, J. (1992): On the choice of pencils in the parametrization of curves. J. Symb. Comput. 14: 557–576.
Schönhage, A., Strassen, V. (1971): Schnelle Multiplikation großer Zahlen. Computing 7: 281–292.
Schwarz, J. T., Sharir, M. (1983): On the "piano movers" problem. II. General techniques for computing topological properties of real algebraic manifolds. Adv. Appl. Math. 4: 298–351.
Sederberg, T. W. (1986): Improperly parametrized rational curves. Comput. Aided Geom. Des. 3: 67–75.
Seidenberg, A. (1954): A new decision method for elementary algebra and geometry. Ann. Math. 60: 365–374.
Sendra, J. R. (1990a): Algoritmos simbólicos de Hankel en algebra computacional. Ph. D. thesis, Universidad de Alcalá de Henares, Madrid, Spain.
Sendra, J. R. (1990b): Hankel matrices and computer algebra. ACM SIGSAM Bull. 24/3: 17–26.
Sendra, J. R., Llovet, J. (1992a): An extended polynomial GCD algorithm using Hankel matrices. J. Symb. Comput. 13: 25–39.
Sendra, J. R., Llovet, J. (1992b): Rank of a Hankel matrix over $Z[x_1, \ldots, x_r]$. J. Appl. Algebra Appl. Algorithms Error Correct. Codes 3: 245–256.
Sendra, J. R., Winkler, F. (1991): Symbolic parametrization of curves. J. Symb. Comput. 12: 607–631.
Sendra, J. R., Winkler, F. (1994): Optimal parametrization of algebraic curves. Techn. Rep. RISC Linz 94-65.
Sharpe, D. (1987): Rings and factorization. Cambridge University Press, Cambridge.
Shirkov, D. V., Rostovtsev, V. A., Gerdt, V. P. (1991): Computer algebra in physical research. World Scientific, Singapore.
Shoup, V. (1991): A fast deterministic algorithm for factoring polynomials over finite

fields of small characteristic. In: Watt, S. M. (ed.): Proceedings of the 1991 International Symposium on Symbolic and Algebraic Computation. Association for Computing Machinery, New York, pp. 14–21.
Simon, B. (1990): Four computer mathematical environments. Notices Amer. Math. Soc. 37: 861–868.
Singer, M. F. (1981): Liouvillian solutions of $n$-th order homogeneous linear differential equations. Am. J. Math. 103: 661–682.
Smedley, T. J. (1989): A new modular algorithm for computation of algebraic number polynomials Gcds. In: Gonnet, G. H. (ed.): Proceedings of the ACM-SIGSAM 1989 International Symposium on Symbolic and Algebraic Computation. Association for Computing Machinery, New York, pp. 91–94.
Stifter, S. (1987): A generalization of reduction rings. J. Symb. Comput. 4: 351–364.
Stifter, S. (1991): The reduction ring property is hereditary. J. Algebra 140: 399–414.
Stifter, S. (1993): Gröbner bases of modules over reduction rings. J. Algebra 159: 54–63.
Strassen, V. (1972): Berechnung und Programm. I. Acta Inf. 1: 320–335.
Tarski, A. (1951): A decision method for elementary algebra and geometry. University of California Press, Berkeley, CA.
Tobey, R. G. (1967): Algorithms for anti-differentiation of rational functions. Ph. D. thesis, Harvard University, Cambridge, MA.
Trager, B. M. (1976): Algebraic factoring and rational function integration. In: Jenks, R. D. (ed.): Proceedings of the ACM Symposium on Symbolic and Algebraic Computation. Association for Computing Machinery, New York, pp. 219–226.
Trager, B. M. (1984): Integration of algebraic functions. Ph. D. thesis, MIT, Cambridge, MA.
Trinks, W. (1978): Über B. Buchbergers Verfahren, Systeme algebraischer Gleichungen zu lösen. J. Number Theory 10: 475–488.
van der Waerden, B. L. (1970): Algebra I, II. Ungar, New York.
van Hoeij, M. (1994): Computing parametrizations of rational algebraic curves. In: von zur Gathen, J., Giesbrecht, M. (eds.): Proceedings of the International Symposium on Symbolic and Algebraic Computation. Association for Computing Machinery, New York, pp. 187–190.
van Hulzen, J. A. (ed.) (1983): Computer algebra. Springer, Berlin Heidelberg New York Tokyo (Lecture notes in computer science, vol. 162).
von zur Gathen, J. (1987): Feasible arithmetic computations: Valiant's hypothesis. J. Symb. Comput. 4: 137–172.
von zur Gathen, J. (1990a): Functional decomposition of polynomials: the tame case. J. Symb. Comput. 9: 281–299.
von zur Gathen, J. (1990b): Functional decomposition of polynomials: the wild case. J. Symb. Comput. 10: 437–452.
von zur Gathen, J., Giesbrecht, M. (eds.) (1994): Proceedings of the International Symposium on Symbolic and Algebraic Computation (ISSAC '94). Association for Computing Machinery, New York.
von zur Gathen, J., Kozen, D., Landau, S. (1987): Functional decomposition of polynomials. In: Proceedings of the 28th Annual IEEE Symposium Foundations of Computer Science, Los Angeles, CA, pp. 127–131.
Walker, R. J. (1950): Algebraic curves. Princeton University Press, Princeton.

Wall, B. (1993): Symbolic computation with algebraic sets. Ph. D. dissertation, Johannes Kepler University Linz, Linz, Austria.

Wang, P. S. (1976): Factoring multivariate polynomials over algebraic number fields. Math. Comput. 30: 324–336.

Wang, P. S. (1978): An improved multivariate polynomial factoring algorithm. Math. Comput. 32: 1215–1231.

Wang, P. S. (1979): Parallel $p$-adic constructions in the univariate polynomial factoring algorithm. In: Lewis, V. E. (ed.): 1979 Macsyma Users' Conference. MIT Laboratory for Computer Science, Cambridge, MA.

Wang, P. S. (ed.) (1981): Proceedings of the ACM Symposium on Symbolic and Algebraic Computation (SYMSAC '81). Association for Computing Machinery, New York.

Wang, P. S. (ed.) (1992): Proceedings of the International Symposium on Symbolic and Algebraic Computation (ISSAC '92). Association for Computing Machinery, New York.

Wang, P. S., Rothschild, L. P. (1975): Factoring multivariate polynomials over the integers. Math. Comput. 29: 935–950.

Watanabe, S., Nagata, M. (eds.) (1990): Proceedings of the International Symposium on Symbolic and Algebraic Computation (ISSAC '90). Association for Computing Machinery, New York.

Watt, S. M. (ed.) (1991): Proceedings of the 1991 International Symposium on Symbolic and Algebraic Computation (ISSAC '91). Association for Computing Machinery, New York.

Weispfenning, V. (1989): Constructing universal Gröbner bases. In: Huguet, L., Poli, A. (eds.): Applied algebra, algebraic algorithms, and error-correcting codes. Springer, Berlin Heidelberg New York Tokyo, pp. 408–417 (Lecture notes in computer science, vol. 356).

Weiß, J. (1992): Homogeneous decomposition of polynomials. In: Wang, P. S. (ed.): Proceedings of the International Symposium on Symbolic and Algebraic Computation. Association for Computing Machinery, New York, pp. 146–151.

Wiedemann, D. (1986): Solving sparse linear equations over finite fields. IEEE Trans. Inf. Theory 32: 54–62.

Wilf, H. S., Zeilberger, D. (1992): An algorithmic proof theory for hypergeometric (ordinary and "q") multisum/integral identities. Invent. Math. 108: 575–633.

Winkler, F. (1984a): The Church–Rosser property in computer algebra and special theorem proving: an investigation of critical-pair/completion algorithms. Ph. D. dissertation, Johannes Kepler University Linz, Linz, Austria.

Winkler, F. (1984b): On the complexity of the Gröbner-bases algorithm over $K[x, y, z]$. In: Fitch, J. (ed.): EUROSAM 84. Springer, Berlin Heidelberg New York Tokyo, pp. 184–194 (Lecture notes in computer science, vol. 174).

Winkler, F. (1987): Computer algebra. In: Meyers, R. A. (ed.): Encyclopedia of physical science and technology, vol. 3. Academic Press, London, pp. 330–356.

Winkler, F. (1988a): A $p$-adic approach to the computation of Gröbner bases. J. Symb. Comput. 6: 287–304.

Winkler, F. (1988b): Computer Algebra I (Algebraische Grundalgorithmen). Techn. Rep. RISC 88-88.

Winkler, F. (1989): A geometrical decision algorithm based on the Gröbner bases algorithm. In: Gianni, P. (ed.): Symbolic and algebraic computation. Springer, Berlin Heidelberg New York Tokyo, pp. 356–363 (Lecture notes in computer science, vol. 358).

Winkler, F. (1990a): Solution of equations I: polynomial ideals and Gröbner bases. In: Chudnovsky, D. V., Jenks, R. D. (eds.): Computers in mathematics. Marcel Dekker, New York, pp. 383–407 (Lecture notes in pure and applied mathematics, vol. 125).

Winkler, F. (1990b): Gröbner bases in geometry theorem proving and simplest degeneracy conditions. Math. Pannonica 1: 15–32.

Winkler, F. (1992): Automated theorem proving in nonlinear geometry. In: Hoffmann, C. (ed.): Issues in robotics and nonlinear geometry. Jai Press, Greenwich, pp. 183–197 (Advances in computing research, vol. 6).

Winkler, F. (1993): Computer algebra. In: Ralston, A., Reilly, E. D. (eds.): Encyclopedia of computer science, 3rd edn. Van Nostrand Reinhold, New York, pp. 227–231.

Winkler, F., Buchberger, B. (1983): A criterion for eliminating unnecessary reductions in the Knuth–Bendix algorithm. In: Demetrovics, J., Katona, G., Salomaa, A. (eds.): Colloquia Mathematica Societatis János Bolyai 42. Algebra, Combinatorics and Logic in Computer Science, vol. 2. North-Holland, Amsterdam, pp. 849–869.

Winkler, F., Buchberger, B., Lichtenberger, F., Rolletschek, H. (1985): Algorithm 628 – an algorithm for constructing canonical bases of polynomial ideals. ACM Trans. Math. Software 11: 66–78.

Wolfram, S. (1991): Mathematica – a system for doing mathematics by computer, 2nd edn. Addison-Wesley, Reading, MA.

Yap, C. K. (1991): A new lower bound construction for commutative Thue systems with applications. J. Symb. Comput. 12: 1–27.

Yokoyama, K., Noro, M., Takeshima, T. (1994): Multi-modular approach to polynomial-time factorization of bivariate integral polynomials. J. Symb. Comput. 17: 545–563.

Yun, D. Y. Y. (ed.) (1980): Short Course: Computer Algebra – Symbolic Mathematical Computation. Workshop held in Ann Arbor, Michigan. AMS Notices, June 1980.

Zacharias, G. (1978): Generalized Gröbner bases in commutative polynomial rings. Bachelor's thesis, MIT, Cambridge, Massachusetts.

Zassenhaus, H. (1969): On Hensel factorization, I. J. Number Theory 1: 291–311.

Zassenhaus, H. (1975): On Hensel factorization II. Symp. Math. 15: 499–513.

Zassenhaus, H. (1978): A remark on the Hensel factorization method. Math. Comput. 32: 287–292.

Zassenhaus, H. (1985): On polynomial factorization. Rocky Mountains J. Math. 15: 657–665.

Zariski, O., Samuel, P. (1958): Commutative algebra, vol. 1. Springer, Berlin Göttingen Heidelberg.

Zeilberger, D. (1990): A holonomic systems approach to special function identities. J. Comput. Appl. Math. 32: 321–368.

Zeilberger, D. (1991): The method of creative telescoping. J. Symb. Comput. 11: 195–204.

Zippel, R. (1991): Rational function decomposition. In: Watt, S. M. (ed.): Proceedings of the 1991 International Symposium on Symbolic and Algebraic Computation. Association for Computing Machinery, New York, pp. 1–6.

Zippel, R. (1993): Effective polynomial computation. Kluwer, Boston.

# Subject index

# Texts and Monographs in Symbolic Computation

Wen-tsün Wu

## Mechanical Theorem Proving in Geometries

Basic Principles

Translated from the Chinese by Xiaofan Jin and Dongming Wang

1994. 120 figures. XIV, 288 pages.
Soft cover DM 98,–, öS 686,–. ISBN 3-211-82506-1

This book is a translation of Professor Wu's seminal Chinese book of 1984 on Automated Geometric Theorem Proving. The translation was done by his former student Dongming Wang jointly with Xiaofan Jin so that authenticity is guaranteed. Meanwhile, automated geometric theorem proving based on Wu's method of characteristic sets has become one of the fundamental, practically successful, methods in this area that has drastically enhanced the scope of what is computationally tractable in automated theorem proving. This book is a source book for students and researchers who want to study both the intuitive first ideas behind the method and the formal details together with many examples.

Bernd Sturmfels

## Algorithms in Invariant Theory

1993. 5 figures. VII, 197 pages.
Soft cover DM 59,–, öS 415,–. ISBN 3-211-82445-6

J. Kung and G.-C. Rota, in their 1984 paper, write: "Like the Arabian phoenix rising out of its ashes, the theory of invariants, pronounced dead at the turn of the century, is once again at the forefront of mathematics."
The book of Sturmfels is both an easy-to-read textbook for invariant theory and a challenging research monograph that introduces a new approach to the algorithmic side of invariant theory. The Groebner bases method is the main tool by which the central problems in invariant theory become amenable to algorithmic solutions. Students will find the book an easy introduction to this "classical and new" area of mathematics. Researchers in mathematics, symbolic computation, and computer science will get access to a wealth of research ideas, hints for applications, outlines and details of algorithms, worked out examples, and research problems.

SpringerWienNewYork

P.O.Box 89, A-1201 Wien • New York, NY 10010, 175 Fifth Avenue
Heidelberger Platz 3, D-14197 Berlin • Tokyo 113, 3-13, Hongo 3-chome, Bunkyo-ku

# Texts and Monographs in Symbolic Computation

Jochen Pfalzgraf, Dongming Wang (eds.)

## Automated Practical Reasoning

Algebraic Approaches

With a Foreword by Jim Cunningham

1995. 23 figures. XI, 223 pages.
Soft cover DM 98,–, öS 686,–. ISBN 3-211-82600-9

This book presents a collection of articles on the general framework of mechanizing deduction in the logics of practical reasoning. Topics treated are novel approaches in the field of constructive algebraic methods (theory and algorithms) to handle geometric reasoning problems, especially in robotics and automated geometry theorem proving; constructive algebraic geometry of curves and surfaces showing some new interesting aspects; implementational issues concerning the use of computer algebra systems to deal with such algebraic methods.
Besides work on nonmonotonic logic and a proposed approach for a unified treatment of critical pair completion procedures, a new semantical modeling approach based on the concept of fibered structures is discussed; an application to cooperating robots is demonstrated.

In preparation:

Alfonso Miola, Marco Temperini (eds.)

## Advances in the Design of Symbolic Computation

Approx. 250 pages. ISBN 3-211-82844-3

Norbert Kajler (ed.)

## Human Interaction in Symbolic Computation

Approx. 250 pages. ISBN 3-211-82843-5

SpringerWienNewYork

P.O.Box 89, A-1201 Wien • New York, NY 10010, 175 Fifth Avenue
Heidelberger Platz 3, D-14197 Berlin • Tokyo 113, 3-13, Hongo 3-chome, Bunkyo-ku

SpringerMathematics

# Collegium Logicum

Annals of the Kurt Gödel Society

## Volume 2

1996. 3 figures. VII, 137 pages.
Soft cover DM 64,–, öS 450,–
ISBN 3-211-82796-X

Contents: H. de Nivelle: Resolution Games and Non-Liftable Resolution Orderings. - M. Kerber, M. Kohlhase: A Tableau Calculus for Partial Functions. - G. Salzer: MUltlog: an Expert System for Multiple-valued Logics. - J. Krajíček: A Fundamental Problem of Mathematical Logic. - P. Pudlák: On the Lengths of Proofs of Consistency. - A. Carbone: The Craig Interpolation Theorem for Schematic Systems. - I.A. Stewart: The Role of Monotonicity in Descriptive Complexity Theory. - R. Freund, L. Staiger: Numbers Defined by Turing Machines.

## Volume 1

1995. 2 figures. VII, 122 pages.
Soft cover DM 64,–, öS 450,–
ISBN 3-211-82646-7

Contents: P. Vihan: The Last Months of Gerhard Gentzen in Prague. - F.A. Rodríguez-Consuegra: Some Issues on Gödel's Unpublished Philosophical Manuscripts. - D.D. Spalt: Vollständigkeit als Ziel historischer Explikation. Eine Fallstudie. - E. Engeler: Existenz und Negation in Mathematik und Logik. - W.J. Gutjahr: Paradoxien der Prognose und der Evaluation: Eine fixpunkttheoretische Analyse. - R. Hähnle: Automated Deduction and Integer Programming. - M. Baaz, A. Leitsch: Methods of Functional Extension.

SpringerWienNewYork

P.O.Box 89, A-1201 Wien • New York, NY 10010, 175 Fifth Avenue
Heidelberger Platz 3, D-14197 Berlin • Tokyo 113, 3-13, Hongo 3-chome, Bunkyo-ku

## SpringerComputerkultur

**Bernhard Dotzler (Hrsg.)**

**Babbages Rechen-Automate**

Ausgewählte Schriften

1996. 22 Abbildungen, 1 Frontispiz. VIII, 502 Seiten.
Broschiert DM 89,–, öS 625,–. ISBN 3-211-82640-8
Computerkultur, Band 6

Charles Babbage, 1791 geboren in Walworth, gestorben in London 1871. Die Liste seiner Beiträge zur Vermehrung des menschlichen Wissens scheint endlos: von der Magenpumpe bis zur Statistik biblischer Wunder. Doch nicht erst die folgende oder unsere Zeit vergaß seinen Wunder(un)glauben wie seine vielen Miszellen zu Mensch und Welt, um ihm stattdessen die eine und einzige Rolle als Pionier der Computerpioniere auf den Grabstein zu schreiben.
„Babbages Rechen-Automate" vereinigt jene Schriften, die erlauben, sich ein detailliertes und vollständiges Bild seiner Rechenmaschinenerfindungen zu machen. Zwar blieben die (von rudimentären Versuchskonstruktionen abgesehen) Entwürfe auf dem Papier, aber sie enthielten doch bereits alle wesentlichen Elemente eines Computers. Nur informiert durch die hier erstmals in so umfassender Form in deutscher Sprache vorliegenden Schriften kann man einen ihrer vorgeblichen Meilensteine in seinen tatsächlichen Konturen erkennen.

SpringerWienNewYork

P.O.Box 89, A-1201 Wien • New York, NY 10010, 175 Fifth Avenue
Heidelberger Platz 3, D-14197 Berlin • Tokyo 113, 3-13, Hongo 3-chome, Bunkyo-ku

*Springer-Verlag*
*and the Environment*

WE AT SPRINGER-VERLAG FIRMLY BELIEVE THAT AN international science publisher has a special obligation to the environment, and our corporate policies consistently reflect this conviction.

WE ALSO EXPECT OUR BUSINESS PARTNERS – PRINTERS, paper mills, packaging manufacturers, etc. – to commit themselves to using environmentally friendly materials and production processes.

THE PAPER IN THIS BOOK IS MADE FROM NO-CHLORINE pulp and is acid free, in conformance with international standards for paper permanency.

LaVergne, TN USA
01 April 2011
222632LV00005B/19/A